Sustainable Development Goals Series

The **Sustainable Development Goals Series** is Springer Nature's inaugural cross-imprint book series that addresses and supports the United Nations' seventeen Sustainable Development Goals. The series fosters comprehensive research focused on these global targets and endeavours to address some of society's greatest grand challenges. The SDGs are inherently multidisciplinary, and they bring people working across different fields together and working towards a common goal. In this spirit, the Sustainable Development Goals series is the first at Springer Nature to publish books under both the Springer and Palgrave Macmillan imprints, bringing the strengths of our imprints together.

The Sustainable Development Goals Series is organized into eighteen subseries: one subseries based around each of the seventeen respective Sustainable Development Goals, and an eighteenth subseries, "Connecting the Goals", which serves as a home for volumes addressing multiple goals or studying the SDGs as a whole. Each subseries is guided by an expert Subseries Advisor with years or decades of experience studying and addressing core components of their respective Goal.

The SDG Series has a remit as broad as the SDGs themselves, and contributions are welcome from scientists, academics, policymakers, and researchers working in fields related to any of the seventeen goals. If you are interested in contributing a monograph or curated volume to the series, please contact the Publishers: Zachary Romano [Springer; zachary.romano@springer.com] and Rachael Ballard [Palgrave Macmillan; rachael.ballard@palgrave.com].

Vania Ceccato · Catherine Sundling · Gabriel Gliori

Transit Safety for a Sustainable City

People, Paths and Places in Railway Stations

Springer

Vania Ceccato
UCS-Urban and community safety research group, Department of urban planning and environment
KTH Royal Institute of Technology
Stockholm, Sweden

Catherine Sundling
Department of Psychology
Södertörn University
Huddinge, Sweden

Gabriel Gliori
UCS-Urban and community safety research group, Department of urban planning and environment
KTH Royal Institute of Technology
Stockholm, Sweden

ISSN 2523-3084 ISSN 2523-3092 (electronic)
Sustainable Development Goals Series
ISBN 978-3-031-97426-7 ISBN 978-3-031-97427-4 (eBook)
https://doi.org/10.1007/978-3-031-97427-4

This work was supported by KTH Royal Institute of Technology.

This Springer imprint is published by the registered company Springer Nature Switzerland AG
The registered company address is: Gewerbestrasse 11, 6330 Cham, Switzerland

This book is dedicated to everyone committed to enhancing safety in public spaces and transit systems, ensuring the safety needs of all are met from an equitable perspective.

Preface

Our collaboration began in the late 2010s in Stockholm, bringing together complementary perspectives from different fields. Catherine had completed her PhD on accessibility for older adults in public transport, while I was studying safety and the environment of the underground system. With her background as a psychologist and police educator and mine as a geographer focused on criminology in the school of architecture and urban planning, our expertise naturally complemented each other. Despite much of our early work being unfunded, we published chapters on transit safety in Stockholm in one of the most comprehensive books on young people's transit experiences worldwide (Transit Crime and Sexual Violence in Cities: International Evidence and Prevention, by Routledge, 2020). In 2020, we secured funding from the Swedish Transport Administration (Trafikverket) to develop the first-ever study about transit safety in railway systems in the Nordic countries. With Gabriel joining as a research assistant in 2022, we started analysing transit safety in southern Sweden's railway stations. This book builds on our collaborative research.

We reframe stations as interconnected systems embedded within cities and towns—linked like pearls across 28 municipalities, composed of towns and villages in southern Sweden. A key feature of our study is the creation of a unique geodatabase, integrating x and y coordinates and DeSO areas—the smallest units of statistical analysis in Sweden—across the entire study region. This unprecedented resource allows us to link survey responses with spatial data, providing a robust foundation for analysis and evidence-based recommendations to enhance safety within the railway transit system and beyond. Adopting a systems thinking perspective, we examine both users' experiences and the challenges faced by station managers. Our research highlights four key findings: (1) station environments and their surroundings significantly shape safety perceptions, even when controlling for other factors; (2) transit safety is intersectional—it varies not only according to individual characteristics of users (gender, age, etc), but also across time and location within city and regional contexts; (3) the deregulation of railway systems across Europe and in Sweden has reshaped station management, indirectly affecting safety conditions of train users; and finally, (4) effective safety planning requires a comprehensive, unified incident reporting system—something currently absent in the railway system in Sweden. This final point drives our ongoing efforts to investigate the conditions for developing a nationwide incident reporting system in the country.

We connect our research to the 2030 Sustainable Development Goals (SDGs) by emphasising that transit safety is essential for creating safer, more sustainable cities. The SDGs provide a roadmap for addressing interconnected challenges, and our exploration of transit safety is linked to several of these goals—from ensuring equity, justice, and strong institutions to fostering sustainable communities beyond big cities and across areas on the rural-urban continuum. Recognising that safety is experienced differently by different groups, we move away from gender-neutral recommendations and instead give voice to the distinct needs of not only women, LGBTQI+ people, those living in more remote places in Sweden and individuals with physical or neurological disabilities, but also people living with conditions such as asthma, allergies, depression or anxiety. This volume highlights the teamwork and collaboration that inspired us to surpass geographical limits and cross disciplinary divides. We hope that this book will serve as inspiration for new research questions and foster dialogue between practitioners and researchers. By exploring transit spaces—from remote, desolate stations in rural areas to bustling transport hubs in major cities—we aim to contribute to the global conversation on urban mobility and safety. We hope that this knowledge can be helpful to ensure individuals' mobility and create safer, more sustainable cities here in Sweden and elsewhere.

Stockholm, Sweden — Vania Ceccato
Huddinge, Sweden — Catherine Sundling
Stockholm, Sweden — Gabriel Gliori

Acknowledgements This book would not have been possible without the support, contributions, and encouragement of numerous individuals and organisations. We would like to take this opportunity to express our deepest gratitude to all those who contributed to making this project a reality.

First and foremost, we extend our sincere thanks to all the individuals who took the time to participate in our survey. Your insights and experiences were invaluable in shaping the foundation of this work and providing a comprehensive understanding of the issues at hand. We are equally grateful to the experts who generously shared their time and expertise by participating in our interviews. Your perspectives enriched our understanding and added depth to our research, offering critical insights that were essential to the development of this book.

This project was made possible through the generous funding provided by the Swedish Transport Administration, Trafikverket. Their support was instrumental in enabling us to carry out this research, bring our ideas to fruition, and cover the Open Access fee, ensuring this book is accessible to all readers. A special thank you goes to our esteemed colleague, Charlotta Palmlund, whose unwavering support has been a cornerstone of this research. Her guidance and encouragement were fundamental in aligning our research objectives with the practical needs of the field, ensuring the relevance and impact of our work.

We would also like to acknowledge our colleagues at KTH, the Royal Institute of Technology, and Södertörn University for their steadfast support.

Special gratitude is extended to Springer Nature for their invaluable assistance in publishing this book, with particular thanks to Evelien Bakker and Belsiya Catherin for their guidance, patience, and dedication throughout the publication process. Their expertise and commitment were instrumental in bringing this project to completion.

Additionally, thanks go to Russell James for his meticulous proofreading. His attention to detail and dedication significantly enhanced the quality of this book. Any mistakes that remain are entirely ours.

Finally, we thank our families and friends for their patience, understanding, and encouragement during the countless hours we dedicated to this project. Your support has meant the world to us.

To all who contributed in big and small ways—thank you.

Vania Ceccato
Catherine Sundling
Gabriel Gliori

Competing Interests The authors have no competing interests to declare that are relevant to the content of this manuscript.

Ethics Approval The Swedish Ethical Review Authority approved the study (including the survey conducted in 2022), Dnr 2021-06393-01, 2021-12-20.

Praise for *Transit Safety for a Sustainable City*

"A good public transit system is indispensable for a sustainable city. A good transit system is also a safe transit system. And a system is safe if its most vulnerable riders are feeling safe. Focusing on railway stations, this book provides a valuable interdisciplinary look at their safety and how—through a systems approach, design, and policy—they can be made safe for even the most vulnerable travellers. An important read for policy makers, transportation planners, and criminologists, and everyone caring about the safety of our urban environments."

—Anastasia Loukaitou-Sideris, *Distinguished Professor of Urban Planning, UCLA*

"This book provides a comprehensive examination of transit safety in Sweden, considering the functions and purpose of transit stations, the context within which they are situated, and most importantly, the users and providers of the transit system. It draws from systems thinking to offer a broader perspective from which to understand transit safety and to support more sustainable efforts to enhance the safety of the transit network. The authors draw from detailed survey work to provide strong empirical evidence to support their findings and recommendations. This is a superb offering for those interested in sustainable cities and safe transit and is applicable to global audiences outside of Sweden."

—Andy Newton, *Professor of Criminology and Policing, Nottingham Trent University, the United Kingdom*

"*Transit Safety for a Sustainable City* provides a comprehensive and multidimensional analysis of public transport safety, focusing on railway stations. The authors combine empirical research, theoretical perspectives, and international case studies to explore challenges and opportunities to enhance passenger safety. It is an essential contribution for professionals, researchers, and policymakers committed to creating safer, more inclusive, and sustainable transportation systems."

—Antonio Iudici, *Department of Philosophy, Sociology, Education and Applied Psychology, University of Padua, Italy*

"Provocative recipe: mix authors with different trajectories, critics of gender-neutral technicality, with a pinch of rigour and adventure with data and an intersectional approach, will result in this book thought-provoking: can different women, LGBTQI people, residents of remote areas, who have asthma and allergies, with physical and neurological disabilities, experience spaces, and the feeling of safety differently? Please, read!"

—Paula Santoro, *School of Architecture and Urbanism, São Paulo University (FAUUSP), Brazil*

Contents

List of Figures

List of Tables

Part I
Stations as Central Nodes

1 Stations in Cities: A Systems-Thinking Approach

Abstract

This chapter introduces the book's exploration of the vital role stations play in urban life, focusing on transit users' victimisation and their perceptions of safety during their journeys. It emphasises the importance of transit safety in its own right while framing stations and cities as interconnected systems. The chapter also examines the application of systems thinking to the governance of stations within a broader planning context. We link the challenges posed by conditions at the stations and beyond to broader sustainability goals, highlighting the need to tackle inequalities in victimisation and the vulnerabilities of certain groups. Finally, we outline the structure of the book.

Keywords

Transit safety · Rail-bound · Crime attractors · Systems perspective · Sustainable development goals—SDGs

1.1 Introduction

Cities are centres of social and economic life, serving as hubs of innovation, commerce, politics, and cultural exchange (UN-Habitat 2025a). Cities host 56% of the global population, which is forecasted to rise to 68% by 2050 (UN-Habitat 2025b). In this dynamic environment, a well-functioning and safe public transport system is crucial in connecting people and places and reducing social inequalities. Cities are also pivotal for the operationalisation of the Paris Agreement as they are responsible for 70 per cent of global primary energy consumption (UN-Habitat 2025b). The United Nations has launched several platforms for sustainability, all of which propose many interrelated Sustainable Development Goals (SDGs) that emphasise the need for action (UN 2015). SDG 11 especially advocates universal access to safe, inclusive public spaces, with a particular focus on ensuring accessibility for women, older people, and individuals with disabilities (UN 2017).

Thus, the primary aim of this book is to demonstrate the importance of safe public transport environments in promoting sustainable cities. The book investigates the impact of environments on transit safety and offers strategies to improve it, addressing the diverse safety needs of city users. Transit safety encompasses both the experience of crime victimisation and safety perceptions among individuals using public transportation, from their homes to transport nodes, particularly stations.

V. Ceccato et al., *Transit Safety for a Sustainable City*, Sustainable Development Goals Series,
https://doi.org/10.1007/978-3-031-97427-4_1

The book distinguishes itself through its unique focus on railway stations, which are viewed as complex systems with interwoven physical, social, economic, technological, and organisational dimensions. Our novel approach offers researchers and practitioners an in-depth perspective on the diverse factors influencing security and safety in public spaces and on the way to and from public transportation. We use the terms "security" and "safety" interchangeably while acknowledging that "security" is more commonly associated with the absence of crime, whereas "safety" is typically linked to a lack of fear or concern about crime.

Stations play an important role when planning safe environments because they have an absolute location in space; they are linked to human activities regulated by a rhythmic schedule of trains, buses and people. A railway station is often more than "a place with one or more buildings and platforms where trains stop for people to get on or off" (Cambridge University Press 2014); it often reflects the dynamicity of the city as a whole. Stations are often called crime generators and attractors (Brantingham and Brantingham 1993). Like any other transport nodes, stations concentrate large flows of people, making it easier for offenders to commit crimes. Some physical and social characteristics found in stations might draw the attention of people with high levels of criminal motivation. They can potentially pull motivated offenders towards them.

In this book, we focus on transit crimes, offences committed against individuals while moving towards a destination. They target people and environments within public transportation systems, those on their way to these systems, and those who use or work within them. They include violence, property crimes, vandalism, and social disorder, ranging from stalking and harassment to theft, robbery, and drug-related activities. These crimes, committed by both passengers and non-passengers, tend to concentrate on specific environments, such as stations, platforms, and parking areas, and on the way to and from the stations, making some locations more criminogenic or at least perceived as less safe than others (Ceccato 2025). For a more detailed discussion of the concepts used in this book, see the glossary.

The gender dimension of transit safety is particularly important. It is, in fact, crucial. Regardless of their location, women consistently feel less safe in public spaces than men (UN Women 2017). At the turn of the millennium, Hamilton and Jenkins (2000) argued that neglecting women's transportation needs not only heightens exclusion but also undermines commercial success, as women are the primary users of public transport worldwide. It is concerning this that the 2023 UN Commission on the Status of Women, despite its focus on empowering rural women and girls, fails to address critical issues such as "mobility" and "safety" (UN 2023). In this book, we adopt an intersectional approach to safety (Crenshaw 1989), considering individuals' multiple identities and experiences as crime victims or in fear of crime. A distinctive feature of our work is the exploration of how the station environment impacts crime and passenger safety, not just locally but throughout the city, focusing particularly on groups such as women, LGBTQI+ people, and not least, people with disabilities such as serious allergies. Thus, we follow research on transitions and sustainability (e.g., Mapar et al. 2017) that adopts the term "safety" (e.g. women's safety, travellers' safety) as a synonym for "security" due to the interdisciplinary nature of the term "safety", recognising that these terms often overlap.

This study stands out by presenting evidence from the Nordic countries, often cited as models of sustainable development, in contrast to most of the literature in this field which focuses on other countries of Western Europe and North America. The breadth of its investigation, covering 47 stations, including a quarter located in smaller municipalities, and spanning diverse regional contexts in Sweden, further highlights its unique contribution to transit safety research and sustainability studies.

This book calls for a paradigm shift in crime prevention in transit environments by embracing systems thinking to be applied in public spaces.

Systems thinking highlights the interconnectedness of various factors relevant for the understanding of transit safety, offering new insights for environmental criminology, urban planning, and related disciplines. We hope that with this book, we can encourage researchers, policymakers, and practitioners to adopt a broader perspective to address safety in urban and regional planning as a key element of sustainability by focusing on problem detection, and ways of solving problems following an agenda of shared vision among stakeholders and continuous assessment.

1.1.1 Aims

This book has three aims. The first is to assess the impact of transit environments on safety and provide strategies to enhance it, catering to the diverse safety needs of city users utilising a case study in Sweden, the first in the context of the Nordic countries. We focus on safety in mobility across different pieces of the puzzle—nodes, peoples, paths, edges, and districts in urban and rural environments—while maintaining a whole journey perspective that integrates these interconnected elements. A second aim is to advocate systems thinking to foster a longer-term approach to understanding; rather than relying on conventional crime prevention approaches, which often fail to deliver lasting solutions and are limited to rigid system boundaries, we advocate a more integrated perspective. Another key goal is to draw lessons from the Swedish case and offer recommendations for both research and practice, particularly regarding current and future transport nodes. This introductory chapter is relevant for other cities worldwide because it draws attention to the universal challenges and opportunities faced by cities of different sizes regarding transportation infrastructure and safety. Not all cities have railways or metro systems, but similar dynamics can be observed in other means of transportation, such as bus stops, tram lines, or even informal transit hubs. The quality of the transit environment plays an important role in facilitating public transportation, thereby indirectly contributing to an increase in ridership and promoting sustainability by encouraging a shift towards greener mobility options.

1.1.2 Why is Transit Safety Important?

Transit safety is essential for ensuring equitable access to public transportation, fostering social and gender inclusion, and enhancing urban mobility. While it is important, transit safety means different things for different people. Tackling the risk of accidents, crime, and terror attacks can be challenging because there are various types of risks and many ways to measure them (Litman 2014). Transit operators and safety experts must address diverse safety concerns while considering the needs of both passengers and employees.

Focusing solely on crime and the fear of becoming a victim while in transit, evidence shows that a significant number of riders avoid public transportation due to safety concerns, often because of internal safety conditions (29 per cent of former New York subway riders stopped using public transit, while riders feared crime more than traffic hazards, often leading to risky pedestrian behaviours to avoid high-crime areas) (TransitCenter 2021). This "exaggerated fear" is arguably associated with the fact that transit travel often means sharing crowded, uncomfortable spaces with strangers. Although most passengers are considerate, a small minority may be rude, anti-social, or dirty, which can create feelings of discomfort and powerlessness (Litman 2014). Others may avoid public transportation due to the routes and locations they must navigate to reach stations or bus stops (Ceccato et al. 2020; Currie et al. 2013). People starting their trips in high-crime neighbourhoods tend to drive to transit stops and are discouraged from walking or cycling to them (Ferrell et al. 2008). When safety concerns go unaddressed, they can disproportionately impact vulnerable groups, such as women, the elderly, and marginalised communities, restricting their mobility and access to essential services and maybe affecting their life chances

(Natarajan et al. 2017; Sánchez-de Madariaga and Zucchini 2020).

Despite these challenges, investing in transit safety is worthwhile, as a safe transit system affects ridership and can promote sustainable transport choices. Trends show that many individuals, particularly millennials, prefer public transportation to automobiles as it allows them to use electronic devices, stay connected, and make productive use of their travel time (Litman 2014; Malokin et al. 2021). Investing in public transit encouragement can help make transit a more sustainable travel choice by enhancing stops and stations, reducing fares, and improving safety and walkability towards these transit nodes. This means that knowing where and when problems happen is fundamental to ensuring people's safety. Certain times, such as late at night often see more crime incidents due to there being fewer people and less guardianship. Some stations are more criminogenic than others, influenced by factors such as size, ridership, and design. Poorly designed areas (e.g., entrances, lobbies, platforms, and surrounding environments) can increase crime, heighten fear, and affect people's decisions to use public transportation. We revisit this topic in Chap. 16, where we explore various strategies to attract new travellers.

1.1.3 Cities, Systems Thinking and Governance

1.1.3.1 The City as a System

A city can be viewed as a system composed of interconnected elements such as people, places, paths, and traffic flows, where changes in one part can affect the whole (Batty 2013; Gustafsson et al. 1982). These elements form distinct structures and subsystems that work together to maintain the city's overall organisation (Batty 2013). Systems consist of elements such as stations, neighbourhoods, and residents, which are connected through channels such as transportation networks, paths and information flows. Donella Meadows, a pioneer in systems thinking, defined a system as "an interconnected set of elements that is coherently organised in a way that achieves something" (Meadows 2008, p. 18), and it is composed of parts:

- "Elements" are the individual parts of a system and are usually the easiest to identify, such as a station in a rail-bound system across several municipalities or neighbourhoods in a city. Intangible elements, including trust, social cohesion, or fear of crime, can also be elements of a system. Elements may include law enforcement agencies, planners, community organisations, and residents. The elements work together and are linked by interconnections. A station, for example, is part of a broader system—the city—which is itself embedded within a physical and digital regional system.
- "Interconnections" refer to communication channels, collaboration efforts, and information sharing. Some of them, like information flows, can be difficult to detect. They represent "the relationships that hold the elements together (Meadows 2008, p. 13). "Feedback loops" are recurring interactions where outputs influence inputs. For example, an increased police presence in an area may reduce crime, reinforcing safety conditions. Yet, this intervention should not be carried out in isolation as crime itself is an event, often a symptom of systemic issues rather than an isolated occurrence.
- A system's "function" or "purpose" can be to achieve a safer rail transport system for all users by reducing crime or decreasing incidents that cause fear and reluctance to use trains. Meadows (2008) emphasised that understanding complex, interconnected problems requires recognising how their structural components relate to their behaviours (Meadows 2008), which are aligned to their purpose.

Donella Meadows recommends that instead of tackling the problem through the whole system, we should identify "leverage points", which are strategic intervention areas in a system where small changes can create significant impact (Meadows 2008). She argued that systems thinking helps uncover the causes of issues by identifying their elements, interconnections, and purposes. Understanding how these components interact—what she termed "the beat of the system"—is essential for identifying patterns and leverage points for change. If the purpose of the system is to facilitate safe, efficient mobility, but if the resources or communication within the system are misaligned, the system may fail to meet its objectives. Observing the system's functioning reveals its true purpose, often exposed by discrepancies between stated goals and actual outcomes. We will return to this issue in Part III of this book.

In crime prevention in a station area, modifying urban design or management of the place could curb crime and shift safety dynamics, at least for some of them, for some time. Feedback loops shape system dynamics, reinforcing or balancing behaviours, while beneficiaries experience the system's positive or negative outcomes depending on their role within it. Beneficiaries are those who gain from a system's operation, such as commuters benefiting from efficient transport. Depending on the system's scale and context, beneficiaries involve other actors, ranging from direct users (e.g., restaurant patrons) to stakeholders such as the police, managers, and users (Ceccato and Newton 2024).

Contemporary views of cities as systems see them as organic, emphasising actions, interactions, and transactions rather than machine-like structures. While traditional approaches focus on fixed locations and physical attributes, analysing cities through a mechanistic, top-down process, the complex systems perspective highlights the dynamic, interconnected nature of urban environments. Cities are seen as constantly evolving, with networks, flows, and organic processes driving change, making them difficult to understand or predict fully (Batty 2013). Stations and public transportation are integral parts of the towns, acting as key "nodes" and "places" (Bertolini 1996) within the complex web of communication, interaction, and exchange of the systems. They serve as locations where various processes converge, enabling physical and digital exchanges, including illicit activities. As part of the urban network, stations facilitate the movement of people, goods, and information, connecting different districts and influencing social and economic activities.

1.1.3.2 The City's Elements and Crime

Kevin Lynch (1976), in his seminal work "The Image of the City", identified five key elements that shape how people perceive and navigate urban environments: paths, edges, districts, nodes, and landmarks. While most activities in these spaces are routine, some are illicit. Nodes, such as stations and transport hubs, concentrate activity, making them potential crime attractors or generators. Edges, such as railway cuttings or development perimeters, create boundaries that can influence crime by restricting surveillance or access. Paths, including streets and underpasses, guide movement and shape awareness spaces, affecting both offenders and potential victims (Brantingham and Brantingham 1993).

Human behaviour typically follows predictable spatial patterns, allowing researchers to better estimate spatial crime processes by considering these trends. Torsten Hägerstrand's (1970) time geography highlights those human activities, including crime, which are constrained by space and time. Individuals navigate these limitations, making crime risks dependent on movement patterns and the opportunities they present, reinforcing the connection between urban form and crime opportunities. For a crime to happen there must be convergence in space and time of *motivated individuals* to consider crime as an alternative, *a suitable target or victim*, and the *absence of a capable guardian*, whether a person or an object, to deter that particular criminal activity (Cohen and Felson 1979). However, individuals do not randomly search an entire city for targets but focus on places within their familiar surroundings or awareness space (Brantingham and Brantingham 1995, p. 365). Their patterns of offending are shaped by their daily routines, even when they actively and intentionally search for a target after deciding to commit a crime. It is important also to recognise that even individuals who engage in crime may function as guardians in their own immediate neighbourhoods (Pattillo 1998), which indicates an element's adaptability in a system.

Crime attractors are locations known for criminal activity, drawing offenders who intentionally seek out these areas, while crime generators are busy places where large crowds create unplanned opportunities for crime. Some stations may become crime attractors if they develop a reputation for criminal activity (Brantingham and Brantingham 1993). Offenders may specifically target these stations because they know that the location is associated with crime, for instance, open drug markets outside or at the station. While attractors tend to have fewer targets but higher crime rates, generators have more potential victims but lower crime rates. Stations may also be crime generators, as their high extent of foot traffic creates opportunities for crime (Brantingham and Brantingham 1995) or what Bowers (2014) calls crime radiators and absorbers within their neighbourhood contexts. Yet, crime is often disproportionately concentrated in small geographical places.

Indeed, this led Weisburd (2015) to declare the "law of crime concentration", which he links to the criminology of place. The author suggests that within a city, 50% of crime is concentrated within approximately two to 6% of a micro-unit of geography, street segments, for example. Stations behave like risky facilities (Eck et al. 2007) because a few stations attract a disproportionately large number of incidents compared to the rest of the facilities of the same type. For instance, 10% of stations encompass 43% of violence directed towards personnel (Ceccato and Ali 2025). When a crime occurs at one station, nearby stations may experience a spike in similar criminal activity shortly afterwards. This is due to the concentration of potential offenders in a relatively small area and the fact that offenders may target similar environments where crime is more likely to go undetected. Similarly, individuals who are victimised at a station may be more likely to experience subsequent victimisation at the exact same location, especially if the station has particular vulnerabilities, such as poor surveillance or high levels of anonymity, making it an ideal target for repeated crimes (Townsley et al. 2003). Various approaches address these issues, including security guards, police presence, and unarmed customer service personnel. However, these methods can be burdensome for vulnerable riders and fail to ensure consistent safety. A more effective approach includes improving station environments with better lighting, awareness campaigns, and station managers following national guidelines to create a safer, more supportive atmosphere. In Chap. 3, we discuss this issue in more detail.

1.1.4 The Novelty of the Book

This book offers a mixture of theoretically informed empirical research and suggestions for practice, focusing on the safety of public spaces with railway stations at the centre, making it relevant for professionals in urban planning, architecture and engineering, transportation, geography, criminology, environmental psychology, gender studies, and sustainability studies.

One special feature is the focus on crime and/or the perceived safety of passengers. Crime and perceived safety are important indicators of transit safety because safety concerns, whether 'real' or 'perceived', can deter individuals from using public transit, reducing ridership.

Second, stations are taken as units of analysis, reflecting the dynamicity of the city as a whole. By centring its analysis on railway stations as the pivotal hubs of urban life, the book examines their role and impact on enhancing safety and sustainability, positioning these transit points as critical to the advancement of safer, more sustainable cities, sometimes located in rural surroundings.

Third, by employing a diverse array of methodologies, including fieldwork, data analysis, and stakeholder engagement, the book unveils the complex dynamics between physical and social factors that influence safety, paving the way for innovative safety and inclusive strategies in public transport.

Fourth, the book not only deepens the understanding of urban safety through rigorous research and literature review but also provides concrete, practical recommendations for improving transit environments from the perspective of users (such as women, LGBTQI+ people, and people with disabilities or allergies) in cities and rural contexts. The book considers "dispositional" (individuals' propensities to experience) and "situational" (a transitory state) dimensions (Gabriel and Greve 2003) of victimisation and fear. In this book, we also adopt an intersectional perspective on safety (Crenshaw 1989) that, among other things, considers individuals' multiple identities and experiences as victims of crime (and/or experiences of fear of crime or other threats). Of relevance is the fact that the book also considers the temporal dimension of transit safety, examining it from an hourly to a weekly and seasonal perspective. It provides a comprehensive understanding of how time influences users' experiences in transit environments.

Additionally, an important feature of the book is its multidisciplinary perspective. The analysis is primarily based on architecture, urban planning, and engineering, as well as transportation, human geography, and environmental psychology, which together form the core of its analytical framework. It further integrates insights from sociology, criminology, and crime prevention, alongside perspectives from gender studies, to explore safety, security, and urban dynamics.

Finally, the book places stations at the center of discussions on governance and sustainability studies highlighting the broader policy and planning implications, ensuring a comprehensive approach to understanding urban environments. We conclude this chapter by emphasising the need to integrate security within the broader social sustainability framework, which is in line with the United Nations 2030 Agenda for Sustainable Development.

1.1.5 What This Book Does not Address

While this book covers a wide range of potential threats in transit environments, including violence, theft, and health-related risks for individuals with allergies and disabilities, it does not address terrorism in railway systems as a distinct threat. Additionally, it does not cover cybercrime or cyberattacks targeting transit systems and passengers. This book does not explore emerging technologies like AI-driven surveillance, biometric ticketing, and smart sensors for crime prevention. Although important, the financial and policy dimensions of transit safety, such as funding for infrastructure improvements, fare-free public transport (e.g., see Gillard, 2025), insurance implications, and cost–benefit analyses of safety measures, are not the focus of the book. Moreover, while the systems-thinking approach provides a valuable framework for describing and analysing the problem, the book does not include practical, long-term examples of interventions based on this approach. Finally, the book does not explore the interactions between different transport modes. Examining interactions between transport modes is vital for transit safety, as poorly designed transfers can create risks like overcrowding, poor surveillance, and accessibility gaps. Coordination between providers also impacts crime prevention and emergency response. While important, this topic falls outside the book's scope.

1.1.6 The Structure of the Book

The book explores the role of stations in urban environments, examining their function as central nodes within transportation systems. The first section establishes the book's scope and foundation, presenting (train) stations both as physical spaces and as integral components (nodes) of city infrastructure. This opening chapter introduces key concepts, including how we define "a system" and what transit safety means from a systems-thinking perspective, while additional definitions can be found in the book's glossary. It also outlines the book's scope and limitations, as well as the research design and study area, which are further explained in detail in later chapters.

The second part focuses on the design, safety, and socioeconomic factors influencing train stations. It highlights the importance of station design and how the surrounding environment can either foster or hinder safety and accessibility. Regression models are used to identify the most important aspects of the station and surrounding areas, following the theoretical model introduced in Chap. 3. The book also explores crime and safety perceptions around train stations, with specific attention to rural stations, which face different challenges compared to their urban counterparts. This section also examines the temporal variations of safety perceptions at train stations and on the way to them. The book includes comparative case studies from Sweden, Denmark, and the Netherlands, highlighting how each country manages and organises safety within their railway systems, with a focus on station's management. It also explores common and perhaps unique challenges these countries face in delivering transportation services.

The third part shifts its focus to the diverse user groups that interact with train stations. With an intersectional perspective, the book emphasises how different groups of individuals—such as women, individuals with disabilities, and LGBTQI+ individuals—experience safety and accessibility in these spaces. The experiences of women, in particular, are examined in terms of safety concerns, with an exploration of how public spaces such as train stations (and the way to them) can be made safer for them. Additionally, the book looks at the health implications for passengers with asthma or allergies and discusses the importance of inclusivity in public transport for individuals with disabilities. This part also has a chapter that explores the characteristics, motivations, and barriers faced by non-train travellers in Sweden through an explorative analysis. The chapter highlights potential improvements that could encourage public transportation, shedding light on the key aspects that need to be addressed to make train travel a more viable option for a wider audience.

The final section addresses the governance and planning of train stations, offering recommendations for making stations more inclusive and safer. It covers the planning process for new stations, examining how local, regional, and national factors shape their development. Through practical examples and case studies, the book intends to provide a roadmap for professionals involved in the design and management of train stations while also suggesting avenues for future research. In doing that, the book offers a comprehensive examination of train stations in urban life by combining theoretical insights, real-world examples, and practical guidance.

1.1.7 Conclusions

In this chapter, we have introduced the theme of this book and discussed its novelty. This study explores the safety of urban public spaces, focusing on railway stations as the centre of analysis. In this chapter, we have defined key terms used in the book, providing a foundation for readers to navigate the diverse topics presented in the 17 chapters. Using diverse methodologies, it highlights the interplay between physical and social factors while offering practical recommendations for inclusive transit environments. Systems thinking is utilised as a reference for fostering a longer-term understanding of transit safety and promoting a more integrated perspective on the governance of transport nodes. With a multidisciplinary approach, the book attempts to provide a selective but still comprehensive perspective on transit safety, governance, and sustainability issues in urban and rural contexts.

References

Batty M (2013) The new science of cities. MIT press

Bertolini L (1996) Nodes and places: complexities of railway station redevelopment. Eur Plan Stud 4(3):331–345

Bowers K (2014) Risky facilities: crime radiators or crime absorbers? A comparison of internal and external levels of theft. J Quant Criminol 30:389–414

Brantingham PL, Brantingham PJ (1993) Nodes, paths and edges: considerations on the complexity of crime and the physical environment. J Environ Psychol 13(1):3–28. https://doi.org/10.1016/S0272-4944(05)80212-9

Brantingham P, Brantingham P (1995) Criminality of place: Crime generators and crime attractors. Eur J Crim Policy Res 3:5–26

Cambridge University Press (Ed.) (2014) Railway station. Cambridge learner's dictionary (4th ed.) Cambridge

University Press. https://dictionary.cambridge.org/dictionary/english/railway-station

Ceccato V (2025) Transit crime. In: Nelson J, Mulley C, Ison S (eds) Encyclopedia of Transport and Society

Ceccato V, Ali T (2025) Violence against transit workers in Stockholm's metro. Case Studies on Transport Policy.

Ceccato V, Newton A (2024) Systems thinking for sustainable crime prevention in risky places. Routledge

Ceccato V, Näsman P, Langefors L (2020) Sexual violence on the move: an assessment of youth's victimization in public transportation [Article]. Women and Criminal Justice. https://www.scopus.com/inward/record.uri?eid=2-s2.0-85083363335&doi=10.1080%2f08974454.2020.1733732&partnerID=40&md5=dacd634042c106569323ebcf6ff11050

Cohen LE, Felson M (1979) Social change and crime rate trends: a routine activity approach. Am Sociol Rev 588–608

Crenshaw KW (1989) Demarginalizing the intersection of race and sex: a black feminist critique of antidiscrimination doctrine. University of Chicago legal forum, 1(8):139–168. http://chicagounbound.uchicago.edu/uclf/vol1989/iss1/8

Currie G, Delbosc A, Mahmoud S (2013) Factors influencing young peoples' perceptions of personal safety on public transport [Article]. J Public Transp 16(1):1–19. https://www.scopus.com/inward/record.uri?eid=2-s2.0-84875747951&doi=10.5038%2f2375-0901.16.1.1&partnerID=40&md5=9ea06b6352ed48b8c2d04a64a254fec1

Debrincat M, Dupart S, Moggio C (2017) Harcèlement sexiste dans les transports collectifs routiers et les pôles d'échanges multimodaux: L'analyse de la FNAUT, Paris, France. FNAUT.

Eck JE, Clarke RV, Guerette RT (2007) Risky facilities: crime concentration in homogeneous sets of establishments and facilities. Crime prev stud 21:225–264.

Ferrell CE, Mathur S, Mendoza E (2008) Neighborhood crime and travel behavior: an investigation of the influence of neighborhood crime rates on mode choice

Gabriel U, Greve W (2003) The psychology of fear of crime. Conceptual and methodological perspectives. Br J Criminol 43(3):600–614

Gillard M (2025) Producing public transport: the roles of fares in mobility governance, labour, and policing (Doctoral dissertation). Vrije Universiteit Brussel.

Gustafsson L, Lanshammar H, Sandblad B (1982) System och modell: en introduktion till systemanalysen. Studentlitteratur, Lund, Sweden

Hägerstrand T (1970) What about people in regional science. Transport Sociology: Social aspects of transport planning, pp 143–158

Hamilton K, Jenkins L (2000) A gender audit for public transport: a new policy tool in the tackling of social exclusion. Urban Stud 37(10):1793–1800. https://doi.org/10.1080/00420980020080411

Litman T (2014) A new transit safety narrative. J Public Transp 17(4):114–135. https://doi.org/10.5038/2375-0901.17.4.7

Lynch K (1976) What time is this place? MIT Press

Malokin A, Circella G, Mokhtarian PL (2021) Do millennials value travel time differently because of productive multitasking? A revealed-preference study of Northern California commuters. Transportation 48(5):2787–2823. https://doi.org/10.1007/s11116-020-10148-2

Mapar M, Jafari MJ, Mansouri N, Arjmandi R, Azizinejad R, Ramos TB (2017) Sustainability indicators for municipalities of megacities: integrating health, safety and environmental performance. Ecol Indic 83:271–91. https://doi.org/10.1016/j.ecolind.2017.08.012

Meadows DH (2008) Thinking in systems: a primer. Chelsea Green Publishing

Natarajan M, Schmuhl M, Sudula S, Mandala M (2017) Sexual victimization of college students in public transport environments: A whole journey approach. Crime Prevent Community Saf 19(3):168–182

Pattillo ME (1998) Sweet mothers and gangbangers: managing crime in a black middle-class neighborhood. Soc Forces 76(3):747–774

Sánchez-de Madariaga I, Zucchini E (2020) "Movilidad del cuidado" en Madrid: nuevos criterios para las políticas de transporte. Ciudad y Territorio Estudios Territoriales 52(203):89–102

Townsley M, Homel R, Chaseling J (2003) Infectious burglaries. A test of the near repeat hypothesis. Br J Criminol 43(3):615–633

TransitCenter (2021) Safety for all. https://transitcenter.org/wp-content/uploads/2021/07/SafetyForAll.pdf

UN-Habitat (2025a) Healthier cities and communities through public spaces: A guidance paper. https://unhabitat.org/sites/default/files/2025/01/final_public_space_and_urban_health.pdf?trk=public_post_comment-text

UN-Habitat (2025b) Urban content of NDCs - local climate action explored through in-depth country analyses, 2024 Report (UN Habitat, Issue. https://unhabitat.org/sites/default/files/2024/06/ndc_global_report_2023_v4_watermark.pdf

UN, United Nations (2015) Transforming our world: the 2030 Agenda for Sustainable Development. https://sds.un.org/2030agenda

UN, United Nations (2017) New urban agenda. Habitat III. https://habitat3.ort/the-new-urban-agenda/

UN, United Nations (2023) Innovation and technological change, and education in the digital age for achieving gender equality and the empowerment of all women and girls. Agreed Conclusions. https://www.un.org/ga/search/view_doc.asp?symbol=E/2023/27

UN Women (2017) Safe cities and safe public spaces: Global results report. New York

Weisburd D (2015) The law of crime concentration and the criminology of place. Criminology 53(2):133–157

2 Safety in Stations in the City and Beyond: Evidence from the Literature

Abstract

Stations are more than just transportation hubs; they are places of convergence, settings that create both opportunities and challenges for safety. As cities grow, the design, management, and social dynamics of stations assume central importance when it comes to ensuring the well-being of those who use them. However, not all areas experience such growth, and some transport nodes may become isolated spots with low passenger volumes, which can impact safety conditions, both in terms of victimisation and safety perceptions. This chapter reviews the recent literature on transit safety in varied contexts, focusing on stations as key transit nodes. Topics of interest include victimisation and its determinants and impacts, situationally risky conditions for crime, examples of interventions, and the possible role of technology in ensuring safety in transit environments. Finally, the chapter also examines how victimisation and safety perceptions fluctuate over time and how these variations differ across various traveller groups.

Keywords

Review of studies · Transit crime · Disorder · Safety perceptions · Fear · Risk factors

2.1 Introduction

The environment of trains and buses, along with the physical and social settings of transport nodes, play a crucial role in shaping the extent and nature of crime in transit environments. Research highlights that station design, surveillance measures, and social cohesion in adjacent areas influence safety in urban and peripheral transit hubs (Ceccato et al. 2022; Sundling and Ceccato 2022). In particular, crime varies across the city and outlying stations, with urban hubs often experiencing higher levels of opportunistic offences. This chapter reviews the international literature on transit safety, both rail-bound and on buses, and focuses on stations as key transit hubs. It reviews the literature according to types of transit users and pays particular attention to temporal variations of the flow of passengers and safety. It covers crime victimisation and its determinants, situational crime risks, interventions, and the role of technology in enhancing safety. Understanding what measures are effective—and which are not—is important for developing strategies that enhance safety across transit networks.

The chapter begins with an introduction that provides a brief overview of key environmental criminological theories. Next, it examines the factors influencing victimisation and safety in

V. Ceccato et al., *Transit Safety for a Sustainable City*, Sustainable Development Goals Series,
https://doi.org/10.1007/978-3-031-97427-4_2

transit environments, addressing issues such as the determinants and impacts of victimisation, situational risk factors for crime, intervention strategies, and the potential role of technology in enhancing transit safety.

2.2 Environmental Criminology and Transit Crime

Environmental criminology provides a framework for understanding crime victimisation in transit environments by examining both individual and situational factors. Traditional criminological theories have focused on individual motivations, but more recent approaches highlight how environmental conditions interact with personal characteristics to facilitate crime. Rational choice theory (see Cornish and Clarke, 2016; Ceccato 2022; Farrell 2017) and routine activity approach (Cohen and Felson 1979) suggest that individuals weigh up the risks and rewards of their actions, often taking advantage of crowded or poorly monitored spaces like bus stops and train stations. Crime pattern theory further explains how offenders' movements and awareness spaces intersect with potential victims, shaping the likelihood of crime occurring in certain locations (Brantingham and Brantingham 1993).

Social disorganisation theory links crime to neighbourhood instability, suggesting that areas with weak social control and high turnover are more prone to crime (Shaw and McKay 1942). Situational action theory expands on this by considering how individuals' moral perceptions and immediate surroundings influence their behaviour (Wikström and Treiber 2017). The physical design of transit spaces also plays a crucial role in shaping people's behaviour and safety, as highlighted by Crime Prevention Through Environmental Design (CPTED) and situational crime prevention strategies (Cozens 2002; Jeffery 1977). Practical measures such as improved lighting, increased surveillance, and strategic design interventions can reduce opportunities for crime and enhance perceived safety in transit environments (Ceccato et al. 2022).

Maintenance of transit environments is also important for safety. According to the "broken window theory" (Kelling and Wilson), unrepaired damage can encourage crime through the encouragement of further vandalism and other crimes. It may indicate that no one has control of the area. In line with this perspective, restoration and development of vacant properties can be effective in reducing crime (Branas et al. 2018).

Through the lens of planning and management of railway station environments, this chapter intersects with several Sustainable Development Goals (SDGs) by offering strategies and insights that contribute to broader global goals. First is SDG 11, as the core focus of this chapter on enhancing safety in railway stations directly contributes to making cities inclusive, safe, resilient, and sustainable. By focusing on highlighting research on creating safer railway stations, the chapter contributes to efforts to eliminate gender-based violence and harassment in public spaces, thus advancing gender equality, which is SDG 5, on gender equality. This extends to other vulnerable groups of passengers too, such as those with disabilities. By advocating safety measures that consider the needs of all users, including marginalised and vulnerable groups, the chapter aligns with the aim of SDG 10, which is to reduce inequalities within and among countries.

2.3 Crime and Social Disorder at the Station

The physical and social environments of trains and transport nodes directly influence the occurrence and nature of crime in transit environments. Table 2.1 shows how research has found evidence of the positive and negative impacts of the transit environment on crime and safety. For instance, visibility decreases the possibility of committing crimes unseen and is usually helpful for increasing the users' perceived safety (Cozens et al. 2004b).

Good visibility can also relate to open environments without hidden corners. Sharp corners restrict sight lines. Here, convex mirrors make it possible to see around corners (Felson et al.

Table 2.1 Positive and negative impact of transit environment on crime and safety

Lighting and design	Visibility, obstacles, and hiding places	Passenger flow and maintenance	Technology
Positive impact			
–High domed and/or white ceilings in stations are encouraged	–Lifts with large glass side panels	–Having visual control over the presence of others	–Convex mirrors for corners
–Curved shelter structures (bus stops)	–Columns no broader than necessary	–Safest in places with a "medium level" of crowding	–Access-control systems, smartphone apps, emergency alarms, crime-resistant materials, ITMs
–Open design allowing clear sightlines	–Glass interior walls	–Perceived safety increases when other passengers are present, especially for women	–"Real-time information (RTI)"
–Light intensity increases perceived safety up to a certain threshold	–Ticket sales at a central location	–Crowded places are linked to fewer crimes: violence, vandalism, and total crimes	–Information and Communication Technology (ICT) improves mobility for visually impaired individuals
–Street lighting around metro stations positively and significantly impacts perceived safety	–Bus stops should be located where visibility and natural surveillance are possible	–Good cleaning	–CCTV for certain crimes and safety, particularly for women and the elderly
–Varying effect depending on passenger type	–Repair of uneven surfaces	–High service frequency, especially in the evening	–High-quality CCTV/ context-dependent
		–Presence of hosts and other types of formal social control	–Body-worn cameras reduce assaults
Negative impact			
–Spiral ramps	–Concealed areas, corners, hiding places, and enclosed spaces	–Empty spaces increase insecurity, especially for women and young people	–Mobile phones as crime targets
–Enclosed brick shelters (bus stops)	–Vegetation should not obstruct visibility	–Crowded platforms are perceived as riskier for sexual harassment than other locations	–Mobile phones enable cyber theft via Wi-Fi or cyber harassment
–Tunnels or underground passages		–Crowded places linked to an increase in several crimes: fights, pickpocketing, total crimes	–CCTV has negative effects, such as privacy concerns for some passenger groups
–Poor lighting (whether artificial or sunlight)		–Unrepaired property damage	
–Height and type of lighting, "fishbowl effect"		–Poor management and maintenance	
–Direct glare negatively impacts safety			

Source: Based on a systematic literature review by Ceccato et al. (2023), p.13

1996). Station design should prioritise visibility by minimising obstacles such as pillars and enclosed spaces. Features such as glass sided lifts, narrow columns, and centralised ticketing enhance sightlines and improve safety for public transport users (La Vigne 1997). Transparent shelters on the platform, lifts with glass-sided panels, and glassed internal walls are ways of increasing visibility. Trimming vegetation that blocks sightlines can also help improve visibility (Cozens et al. 2004a; Felson et al. 1996; La Vigne 1996).

Visibility can also be promoted by good lighting that makes it possible to see potential threats and to be seen by others if something happens (Ceccato et al. 2013). Overall, lighting seems to affect crime reduction (Ceccato et al. 2022). However, the lighting must also avoid a "fishbowl effect". According to Ceccato et al. 2022, other transit areas should be included around the station, such as a car park (Libardo and Nocera 2012). Previous studies indicate that underpass platforms are often related to higher rates of disorder and crime and poor safety perceptions (e.g. Ceccato 2013; Loukaitou-Sideris et al. 2002).

Research shows that virtually empty transit environments (e.g. platforms, stations, and bus stops) increase certain types of crimes in Western countries, such as exhibitionism and sexual assaults, (e.g. Beller et al. 1980; Mohamed and Stanek 2020), as well as robberies, (Clarke et al. 1996), social disorder, and violent assaults (Noble 2015).

CCTV has also been found to be associated with a decrease in crime in some studies; however, the evidence is inconsistent (Ceccato et al. 2022). CCTV can reduce the risk of violence in metro stations in São Paulo, Brazil (Moreira and Ceccato 2021) and car crimes, such as breaking into cars in car parks in the United Kingdom (Tilley 1993b; Webb and Laycock 1992).

Crime in transportation nodes such as stations also influences the local and regional contexts in the city. Ceccato et al. (2011) suggested that larger subway stations attract more crime, controlling for station and neighbourhood contexts. More interestingly, there seem to be gender differences in victimisation according to the location of the station. Moreira and Ceccato (2021) found that women are at a higher risk of victimisation than men in centrally located metro stations, while men run a higher risk of violence at end stations—both notably during late-night periods. Some suggest that the surrounding contexts affect crime and/or the perceived safety of riders. While Abenoza et al. (2018) showed that mixed land use tends to increase the perceived safety of users at night, two other studies found that shops selling alcohol, bars, and pawnshops near transit areas increased the frequency of robberies (Loukaitou-Sideris et al. 2001; Yu and Smith 2014). They also found that transport nodes near vacant lots had higher property crime rates.

Empirical studies show the need to consider safety from a "whole journey perspective". Kim et al. (2007) analysed factors that influence the choice of mode of transport for trips between home and light rail stations and found that the crime risk made it more likely for women to be picked up and dropped off at the station, especially when it was dark. Recent research shows that the perceived "walking accessibility" of metro stations was found to be highly influenced by micro-aspects of the landscape, such as streetlights, police patrols, and traffic signs and signals (Bivina et al. 2020).

2.4 The Temporal Dimension of Transit Safety

The temporal dimension of crime has been recognised as a key to understanding crime conditions in transit environments. For example, peak and off-peak hours attract different kinds of crime. In peak hours, when stations are crowded, offences such as pickpocketing are more common (Zahnow and Corcoran 2019), while empty stations at off-peak hours increase the risk of crimes that demand anonymity, such as robbery (Clarke et al. 1996). Robberies appear to be more common late at night than earlier in the evening (Clarke et al. 1996). When weekdays and weekends are compared, more crimes seem to be committed at weekends (Ceccato et al. 2013). There is a considerable variation in crime type and time windows, from morning to evening, night, week, and month, as well as seasonal variations.

In the systematic review by Ceccato et al. (2022), the authors highlight how different crime types impact transit safety in the morning; for example, theft, robbery, sexual crimes, and harassment, is frequent during peak morning hours in transit environments. Poor lighting in the early morning contributes to increased risks of victimisation. There is also evidence that women feel less safe on public transport, particularly during off-peak hours in the morning.

These studies also show the temporal variation of victimisation and safety perceptions in transit environments during the afternoon and early evening. They indicate that thefts, robberies, pickpocketing, and sexual crimes increase across various transit modes, particularly in subways and buses. Crowding contributes to offences, while reduced surveillance between 2 p.m. and

9 p.m. creates further risks. Women and elderly passengers report a lower level of perceived safety, with many avoiding transit during these hours. In the late evening and early night, violent crimes, victimisation, and life-threatening incidents increase, especially after rush hours and between 6 p.m. and midnight. Thefts, robberies, and vandalism peak during this period, with transport nodes such as subway stations being particularly vulnerable. Riders' perceived safety declines significantly in the evening, particularly for women and during off-peak hours.

Finally, Ceccato et al. (2022) summarise the results of studies reporting crime and safety in transit environments at night. Robberies and violent crimes seem to be more prevalent at night across all types of public transport, particularly in subways and areas near train stations, with peaks between 9 p.m. and early morning. Poor lighting, social interactions/gatherings during night hours, and leisure activities contribute to heightened feelings of insecurity, especially for women, who tend to avoid public transport during these hours. Studies consistently show that night travel is perceived to be far less safe than daytime travel, with one in five women feeling unsafe. Violent incidents rise after 6 p.m. with critical periods between 2 a.m. and 4 a.m. in subways and streets near campuses. They also indicate that waiting at night, proximity to nightclubs, and the general atmosphere of nighttime travel exacerbate safety concerns.

2.5 Safety Perceptions at Stations

The individual's safety perceptions are a product of individual and environmental factors. In this book, the concept of "safety perceptions" is used as a general term that indicates "declared perceived safety" by passengers in the railway system. Thus, safety perceptions (or perceived safety) can be at a high level when an individual declares feeling safe or at a low level when an individual declares feeling unsafe, being fearful or feeling at risk that something might happen. Individuals' safety perceptions are considered an overall safety indicator, recognising the long tradition in the methodological work on fear of crime that recognises the cognitive (e.g., risk perception) (Kurtz and Clow 1997), and the emotional (e.g., being afraid) (Warr 2000) components of fear.

Individual factors play an important role in defining safety perceptions. Gender and age are perhaps the strongest factors concerning safety (Box et al. 1988; Koskela 1999; Loukaitou-Sideris 2005). Among the individual factors, prior victimisation (or awareness of others' victimisation) is often considered one determinant of a person's safety perceptions but also disability (Iudici et al. 2017; Killias and Clerici 2000; Yin 1980), sexual orientation, and social-economic disadvantage (Box et al. 1988; Garofalo and Laub 1979; Pain and Smith 2008). These and other individual factors will be discussed in Sect. 2.6, which focuses on the differences in safety between groups of passengers.

Safety perceptions are also a result of the environment, of what one sees and perceives with one's other senses, an individual's emotional reactions to a place at a particular moment in time and/or memories and associations that a particular setting brings to the surface. For example, exposure to certain criminogenic places may lead to lower declared safety perceptions. However, more importantly, the presence of visual cues in the environment that, if unrepaired or missing, may appear to encourage crime and trigger low safety perceptions (Kelling and Wilson 1982). Signs of disorder, such as litter, vandalism and loitering, are thought to trigger a fear of crime (Lewis and Maxfield 1980) and may be indicators of more serious crimes. How does the environment affect safety perceptions in stations in particular?

The research raises the question of visibility inside stations as an important factor, which can include a wide-open design, which the straightening of sightlines provides, and avoiding obscured areas, corners, hiding spots, and enclosed spaces (e.g. Ceccato et al. 2011). In addition, plants should not become obstacles and impede the field of view (Crime-Concern 2004; Diec et al. 2010; Mohamed and Stanek 2020). Visibility demands good illumination. In one seminal study by Richards (1980), 93% of respondents suggested lighting improvements to improve station safety. Stations that are well-lit and well-maintained contribute to the overall feeling of safety among metro users in Washington, DC (LaVigne 1996). Research also indicates that good lighting increases women's perceived safety more than men's, and results indicate that the effect of lighting on safety is dependent on its type and height. Direct glare has a negative impact on perceived safety, as do different brightness levels at exits and entrances. Brightness increases perceived safety until a certain threshold of sideway cornice-mounted lighting design has a greater impact on perceived safety (Burnett and Pang 2004).

Passengers often highly value surveillance, particularly the presence of staff. However, it is important that staff are readily available, either physically or electronically (Sundling and Ceccato 2022). Likewise, guards and police patrols, if they are available, often positively affect perceived safety (Ceccato 2013; Sundling 2020).

Remote surveillance, such as CCTV cameras, often increases perceived safety. CCTVs are depicted as efficient in increasing perceived safety in several studies (Cozens et al. 2003; Debrincat et al. 2017). On the other hand, their impact on crime reduction and/or perceived safety was not evident in transit systems in Sweden or the United States (Priks 2009; Shellow et al. 1974). One reason may be that the isolated impact of CCTV may be challenging to evaluate as it is often installed at the same time as other interventions (Tilley 1993a). Another reason is that the effectiveness of CCTV depends on the situation (Riley and Dean 1985), for instance, on how much potential criminals fear the system and see in it an increased risk of getting caught (Webb and Laycock 1992). Another prerequisite Another prerequisite is that they are monitored by a staff member (Thomas et al. 2006). Immediate help must be available. Otherwise, they are less valued with regard to perceived safety. Help buttons, emergency alarms, and safety apps (Debrincat et al. 2017; Libardo and Nocera 2012; McCarthy et al. 2016) may be other ways of getting in touch with staff when feeling unsafe. Real-time information (RTI) (Abenoza et al. 2018; Newton 2018) also increases safety perceptions in transit environments.

Platforms and stations overlooked by other activities, such as cafés and shops, are often associated with passengers' higher safety perceptions (Libardo and Nocera 2012). On the contrary, isolated, empty stations are perceived to be more unsafe.

One can feel like an easier target when no one is around (Noble 2015). Thus, other people present are a source of safety (Ouali 2012). However, people who behave in an uncivil manner, intoxicated people or "gangs" hanging around (Kennedy 2008) increase feelings of poor safety in others. The impact of management and maintenance on safety perceptions is well documented according to the "broken window theory" (Kelling and Wilson 1982), see Sect. 2.1, which states that a lack of maintenance may attract crime. Therefore, cleanliness contributes to safety and feelings of safety decrease when cleanliness levels fall (Loukaitou-Sideris 2014)—this is especially the case for casual users (Vanier and d'Arbois de Jubainville 2017). Lack of maintenance may indicate other underlying problems (Eck 2019).

Safety perceptions vary over time. Different "time windows" are important both with regard to actual crime and perceived safety. Peak hours can be perceived as unsafe for women (Mitra-Sarkar and Partheeban 2011) and for people with disabilities (Wayland et al. 2020). However, in a study by Yavuz and Welch (2010), peak-hour travelling was safer for men. This was not found to be true for women.

Much of the research compares day and night travel. Nighttime travel usually feels unsafe, especially for women (e.g., Ceccato and Loukaitou-Sideris 2022). Also, at nighttime, waiting times are generally longer, and longer waiting times have been found to increase feelings of poor safety. The same is true for travel time: the longer the travel time, the more people feel unsafe.

2.6 Safety and Needs of User Groups

Women generally feel more unsafe than men in transit environments (see e.g. Ceccato and Loukaitou-Sideris 2020). There is also a gender difference in what is feared. While men tend to fear theft or violence, women often fear sexual attacks (Ceccato and Loukaitou-Sideris 2022). Women are also more often than men victims of sexual crime as well as of recurring sexual harassment that can arise fear of more serious crime (Lorenc et al. 2013). The fear of sexual crime can also induce more fear than fear of other crimes (Mellgren and Ivert 2019). Furthermore, women are more often "transit captives" (Ceccato 2017), which means that they have no alternative to public transport because they do not have access to, or cannot afford to have a driving licence and/ or a car in their household (Sundling et al. 2014).

People who identify themselves LGBTQI + have often been found to feel unsafe in transit environments. For example, lesbian women were more fearful of assault in a study by Valentine (1997). In a more recent study, Lubitow et al. (2020) interviewed transgender people who felt vulnerable to hostile looks and comments while waiting for the train. However, a study with students from Mexico, and the United States did, however, not find that safety was significantly dependent on sexual orientation (Santoro et al. 2020).

The association between age and perceived safety varies. In some studies, young people feel most unsafe (Kennedy 2008), while in others, older people feel more unsafe (Zegras et al. 2015). In some studies, those between 26 and 65 years old feel more unsafe than both younger and older people (Ceccato 2013). One reason might be that interaction factors such as gender and previous victimisation can impact the result.

Travellers with disabilities often feel more unsafe than others (e.g. Yavuz and Welch 2010). Evidence indicates that they feel more unsafe than others after dark. This can have to do with the attitudes of fellow passengers and with harassment (Sundling et al. 2014). Women with disabilities may be especially vulnerable (Iudici et al. 2017). However, not only other passengers but also staff can be a source of poor safety, for example, when not providing the help needed (Weiland et al. 2020).

The risk of victimisation varies across groups by age, gender, (dis)ability, LGBTQI + status, IT literacy, previous victimisation, frequency of travel on public transport, trip length, etc. (Gekoski et al. 2015). Being victimised tends to lower perceived safety (e.g. Skogan and Maxfield 1981), although mixed results exist (Farrall et al. 2007). The association between previous victimisation and a lack of safety should be seen in interaction with other variables and is dependent on, for example, the kind of victimisation (Rountree 1998). For example, women are more affected by previous victimisation (Yavuz and Welch 2010). A reason could be that women, to a greater extent, are victims of sexual crime, and Ceccato and Loukaitou-Sideris (2021) found that only sexual victimisation resulted in lower safety levels. Therefore, the kind of victimisation should be assessed in order to understand more about possible consequences for safety. Even so, due to the complexity, interaction effects need further research.

An individual's ethnic background may also be a key factor in explaining one's experiences of sexual violence (Ceccato and Paz 2017). A previous study by Loukaitou-Sideris in 2005 showed, for instance, that non-white women often experience higher levels of fear than white women and tend to be fearful of white men, while white women tend to fear non-white men, which may be the result of both racial prejudice and persistent racial hierarchies (Ceccato and Paz 2017; Newton 2018). A recent study across six continents on sexual violence shows evidence of how ethnicity contributes to an increased risk of sexual victimisation; such risk varies according to transportation mode, during the trip, and according to the city and country. For example, interaction effects have been suggested, such as ethnic minorities living in areas with higher crime rates (Yavuz and Welch 2010) or that, for example, people of colour tend to be more comfortable in areas where there are other people of colour (Lubitow et al. 2020).

There seems to be no consensus regarding the potential link between travel frequency and perceived safety. For example, safety has been found to increase with travel frequency (Ceccato et al. 2024; Delbosc and Currie 2014), but also to decrease (Vanier and De Jubainville 2017). This is probably dependent on many different factors, such as many of those mentioned above.

Feeling unsafe impacts both short-term and long-term behaviour. Short-term behaviour adaptations during travel can include using the mobile phone to connect with family and friends (Kim 2021), remaining alert (Ceccato and Loukaitou-Sideris 2022), and pretending to be confident (Chowdhury and Van Wee 2020). Long-term adaptations can be, for example, not travelling at all on public transport (Shiwakoti et al. 2019), travelling only at certain times of the day or avoiding specific routes or places, or dressing in a particular way (Ceccato and Loukaitou-Sideris 2022). The following section summarises safety interventions intended to improve transit safety conditions.

2.7 Strategies and Interventions in Transit Safety

Recent research highlights that transit safety relies on the coordinated efforts of multiple stakeholders, including transportation service providers, safety professionals (such as security guards and hosts), and municipal authorities. After reviewing several decades of international literature from 1970 to 2020, Ceccato et al. (2022) identified five key types of safety interventions to improve transit safety across the world:

- Design—Enhancing infrastructure to improve visibility and safety, including minimising barriers, ensuring good lighting, reducing vegetation, and incorporating emergency buttons. Strategies include CPTED principles, well-maintained environments, and better integration of commercial spaces for informal surveillance.
- Policy—Implementing measures such as introducing female police officers, visible transit staff, stop-on-demand services, women-only transit spaces, stricter penalties for misconduct, and increased public transport supply. Emphasis is placed on adapting transit networks to women's needs.
- Campaigns/education—Raising awareness through bystander intervention training, safety promotion campaigns, and publicising safety improvements. Campaigns should focus on changing victim and perpetrator perceptions, addressing sexism in advertisements, and improving trust among passengers.
- Data/knowledge—Strengthening data collection through gendered statistics, surveys, and knowledge exchange between agencies, including police, planners, and municipalities. Encouraging cooperation between public and private security forces.
- Technology—Leveraging advancements such as real-time bus arrival information, CCTV, body-worn cameras, smart bus stops, vandal-resistant materials, and social media for safety communication.

These interventions indicate the need for a comprehensive approach to transit safety, where infrastructure, policy, education, data, and technology work together. By integrating these strategies, cities can address safety problems in transit environments whenever possible, focusing on the best interventions. Still, many of these interventions have not been systematically tested for their effectiveness. Implementing mechanisms to track conditions before and after interventions and long-term follow-ups is important for assessing their impact and prioritising strategies.

2.8 Conclusions and Recommendations

This chapter has examined recent literature on transit safety across diverse contexts, with a particular focus on stations as critical transit hubs. It has explored key topics such as victimisation, its causes and impacts, situational risk factors for crime, intervention strategies, and the potential role of technology in enhancing transit safety. The chapter also highlights the importance of the physical and social environment in shaping transit safety, emphasising how station design, maintenance, and staff presence contribute to perceptions of safety. It also considers how people adapt their daily routines, during peak and off-peak hours, weekdays and weekends, and across different seasons, affecting activity levels, surveillance, and potential risks. These variations influence station conditions and play a crucial role in shaping safety outcomes in transit environments. Safety is a complex phenomenon that often involves interactions between many factors. These complex interactions need to be further studied. Before interventions are planned, it is necessary to map where people feel unsafe, who feels unsafe, and how unsafe behaviour impacts future behaviour. In the next chapter, we propose a conceptual model that will serve as the theoretical reference for the current study.

References

Abenoza RF, Ceccato V, Susilo YO, Cats O (2018) Individual, travel, and bus stop characteristics influencing travelers' safety perceptions. Transp Res Rec 2672(8):19–28

Beller A, Garelik S, Cooper S (1980) Sex crimes in the subway. Criminology 18(1):35–52

Bivina G, Gupta A, Parida M (2020) Walk accessibility to metro stations: an analysis based on meso-or micro-scale built environment factors. Sustain Cities Soc 55:102047

Box S, Hale C, Andrews G (1988) Explaining fear of crime. Br J Criminol 28:340–356

Branas CC, South E, Kondo MC, Hohl BC, Bourgois P, Wiebe DJ, MacDonald JM (2018) Citywide cluster randomized trial to restore blighted vacant land and its effects on violence, crime, and fear. Proc Natl Acad Sci 115(12):2946–2951

Brantingham PL, Brantingham PJ (1993) Nodes, paths and edges: considerations on the complexity of crime and the physical environment. J Environ Psychol 13(1):3–28. https://doi.org/10.1016/S0272-4944(05)80212-9

Burnett J, Pang AYH (2004) Design and performance of pedestrian subway lighting systems. Tunn Undergr Space Technol 19(6):619–628

Ceccato V, Uittenbogaard AC, Bamzar R (2011) Safety in Stockholm's underground stations: the importance of environmental attributes and context. Research report, School of Architecture and the built environment, Stockholm, KTH

Ceccato V (2013) Moving safely: crime and perceived safety in Stockholm's subway stations. Lexington books

Ceccato V (2017) Women's victimisation and safety in transit environments. Crime Prevent Community Saf 19:163–167

Ceccato V, Loukaitou-Sideris A (2021) Sexual harassment in transit environments among college students in the# MeToo era: reporting evidence from six continents. Am J Crim Justice 46(1):107–129

Ceccato V (2022) Rational choice, routine activity and situational crime prevention. The encyclopedia of rural crime. Bristol University Press, pp 46–49

Ceccato V, Gaudelet N, Graf G (2022) Crime and safety in transit environments: a systematic review of the English and the French literature, 1970–2020. Public Transport 14(1):105–153

Ceccato V, Loukaitou-Sideris A (eds) (2020) Transit crime and sexual violence in cities: international evidence and prevention. Routledge

Ceccato V, Loukaitou-Sideris A (2022) Fear of sexual harassment and its impact on safety perceptions in transit environments: a global perspective. Violence against Women 28(1):26–48

Ceccato V, Paz Y (2017) Crime in São Paulo's metro system: sexual crimes against women. Crime Prevent Community Saf 19:211–226

Ceccato V, Sundling C, Gliori G (2023) Stationsmiljön, brott och resenärers trygghet

Ceccato V, Sundling C, Gliori G (2024) What makes a railway station safe and for whom? The impact of transit environments on passengers' victimisation and safety perceptions. Eur Transp Res Rev 16(1):21. https://doi.org/10.1186/s12544-024-00641-5

Ceccato V, Uittenbogaard A, Bamzar R (2013) Security in Stockholm's underground stations: the importance of environmental attributes and context [Article]. Secur J, 26(1):33–59. https://www.scopus.com/inward/record.uri?eid=2-s2.0-84872438643&doi=10.1057%2fsj.2011.32&partnerID=40&md5=80da5e578838da28b6d3e00f9060f24a

Chowdhury S, Van Wee B (2020) Examining women's perception of safety during waiting times at public transport terminals. Transp Policy 94:102–108

Clarke RV, Belanger M, Eastman J (1996) Where angels fear to tread: a test in the New York City subway of the robbery/density hypothesis. Prevent Mass Transit Crime: Crime Prevent Stud 6:217–236

Cohen LE, Felson M (1979) Social change and crime rate trends: a routine activity approach. Am Sociol Rev 588–608

Cozens P, Neale R, Whitaker J, Hillier D (2003) Managing crime and the fear of crime at railway stations—a case study in South Wales (UK). Int J Transp

Manag 1(3):121–132. https://doi.org/10.1016/j.ijtm.2003.10.001
Cozens P, Neale R, Whitaker J, Hillier D (2004a) Investigating 'cartographies of fear' in and around Valley Lines railways stations (UK) [Review]. Public Transp Int, 53(5):22–24. https://www.scopus.com/inward/record.uri?eid=2-s2.0-11844278444&partnerID=40&md5=1c05d9fc03840689e276b8c7ba3b39f3
Cozens P, Neale R, Whitaker J, Hillier D (2004b) Tackling crime and fear of crime whilst waiting at Britain's Railway Stations
Cozens PM (2002) Sustainable urban development and crime prevention through environmental design for the British City. towards an effective urban environmentalism for the 21st century. Cities 19(2):129–137. https://doi.org/10.1016/S0264-2751(02)00008-2
Concern C (2004) People's perceptions of personal security and their about crime on public transport: research findings. Dft, London
Cornish DB, Clarke RV (2016) The rational choice perspective. In: Environmental criminology and crime analysis, Routledge, pp 8–80
Debrincat M, Dupart S, Moggio C (2017) Harcelement sexiste dans les transports collectifs routiers et les pôles d'échanges multimodaux. L'analyse de la Fnaut
Diec J, Coxon S, De Bono A (2010) Designing a public train station shelter to minimise anti-social behaviour and crime in Melbourne's metropolitan rail environment
Delbosc A, Currie G (2014) Modelling the causes and consequences of perceptions of personal safety on public transport ridership
Eck JE (2019) Place managers and crime places. In: Oxford Research Encyclopedia of Criminology and Criminal Justice
Farrall S, Gray E, Jackson J (2007) Theorising the fear of crime: the cultural and social significance of insecurities about crime. Experience and expression in the fear of crime working paper(5)
Farrell G (2017) Situational crime prevention and its discontents: Rational choice and harm reduction versus 'cultural criminology.' Crime opportunity theories. Routledge, Cham, pp 343–369
Felson M, Belanger M, Bichler G, Bruzinski C, Campbell G, Fried C, Grofik K, Mazur I, O'Regan A, Sweeney P (1996) Redesigning hell: preventing crime and disorder at the port authority bus terminal (from preventing mass transit crime. In: Clarke RV (ed) -See NCJ-168781. pp 5–92
Garofalo J, Laub J (1979) Fear of crime—broadening our perspective. Victomology 3:242–253
Gekoski A, Gray JM, Horvath MA, Edwards S, Emirali A, Adler JR (2015) What works' in reducing sexual harassment and sexual offences on public transport nationally and internationally: A rapid evidence assessment. British Transport Police and Department for Transport, London
Iudici A, Bertoli L, Faccio E (2017) The 'invisible' needs of women with disabilities in transportation systems. Crime Prevent Community Saf 19(3):264–275. https://doi.org/10.1057/s41300-017-0031-6
Jeffery CR (1977) Crime prevention through environmental design (Vol. 524) Sage Publications Beverly Hills
Kelling GL, Wilson JQ (1982) Broken windows. Atl Mon 249(3):29–38
Kennedy D (2008) Personal security in public transport travel in New Zealand: Problems, issues and solutions. Land Transport New Zealand Research Report, 344
Killias M, Clerici C (2000) Different measures of vulnerability in their relation to different dimensions of fear of crime. Br J Criminol 40(3):437–450. https://doi.org/10.1093/bjc/40.3.437
Kim H (2021) Service design for public transportation to address the issue of females' fear of crime. Transportation 48(1):167–192. https://doi.org/10.1007/s11116-019-10043-5
Kim S, Ulfarsson GF, Hennessy JT (2007) Analysis of light rail rider travel behavior: impacts of individual, built environment, and crime characteristics on transit access. Transp Res Part a: Policy Pract 41(6):511–522
Koskela H (1999) Fear, control and space: geographies of gender, fear of violence, and video surveillance [PhD, University of Helsinki]. Helsinki
Kurtz DL, Clow KE (1997) Services marketing. John Wiley and Sons
La Vigne NG (1996) Safe transport: security by design on the Washington Metro. Preventing Mass Transit Crime 6:163–197
La Vigne NG (1997) Visibility and vigilance: Metro's situational approach to preventing subway crime. US Department of Justice, Office of Justice Programs, National Institute of…
Lewis DA, Maxfield MG (1980) Fear in the neighborhoods: an investigation of the impact of crime. J Res Crime Delinq 17(2):160–189. https://doi.org/10.1177/002242788001700203
Libardo A, Nocera S (2012) Exploring the perceived security in transit: the Venetian students' perspective. WIT Trans Ecol Environ 155:943–954
Lorenc T, Petticrew M, Whitehead M, Neary D, Clayton S, Wright K, Thomson H, Cummins S, Sowden A, Renton A (2013) Fear of crime and the environment: systematic review of UK qualitative evidence [Article]. BMC Public Health 13(1):496. https://www.scopus.com/inward/record.uri?eid=2-s2.0-84878005591&doi=10.1186%2f1471-2458-13-496&partnerID=40&md5=d6ef0c3d1dc05ff358c78773ba3946f1
Loukaitou-Sideris A (2005) Is it safe to walk here? Research on Women's Issues in Transportation. Conference Proceedings
Loukaitou-Sideris A (2014) Fear and safety in transit environments from the women's perspective. Secur J 27(2):242–256
Loukaitou-Sideris A, Liggett R, Iseki H (2002) The geography of transit crime: Documentation and evaluation of crime incidence on and around the green line stations in Los Angeles [Article]. J Plann Educ Res

22(2):135–151. https://www.scopus.com/inward/record.uri?eid=2-s2.0-1642430431&doi=10.1177%2f0739456X02238443&partnerID=40&md5=ff8e1533a0914e78d4393fc7f24ac488

Loukaitou-Sideris A, Liggett R, Iseki H, Thurlow W (2001) Measuring the effects of built environment on bus stop crime. Environ Plann B Plann des 28(2):255–280

Lubitow A, Abelson MJ, Carpenter E (2020) Transforming mobility justice: gendered harassment and violence on transit. J Transport Geogr 82:102601. https://doi.org/10.1016/j.jtrangeo.2019.102601

McCarthy OT, Caulfield B, O'Mahony M (2016) How transport users perceive personal safety apps. Transp Res Part F Traffic Psychol Behav 43:166–182

Mellgren C, Ivert A-K (2019) Is women's fear of crime fear of sexual assault? A test of the shadow of sexual assault hypothesis in a sample of Swedish university students. Violence against Women 25(5):511–527

Mitra-Sarkar S, Partheeban P (2011) Abandon all hope, ye who enter here: understanding the problem of "eve teasing" in Chennai, India. Transportation Research Board Conference Proceedings

Mohamed AA, Stanek D (2020) The influence of street network configuration on sexual harassment patterns in Cairo. Cities 98:102583

Moreira GC, Ceccato VA (2021) Gendered mobility and violence in the São Paulo metro, Brazil [Article]. Urban Stud 58(1):203–222. https://www.scopus.com/inward/record.uri?eid=2-s2.0-85078046345&doi=10.1177%2f0042098019885552&partnerID=40&md5=9dd5e490b23e89792fca65e03c9f75e7

Newton A (2018) 497Macro-level generators of crime, including parks, stadiums, and transit stations. In: Bruinsma GJN, Johnson SD (eds) The oxford handbook of environmental criminology. Oxford University Press

Noble J (2015) L'insécurité personnelle des jeunes dans les transports en commun franciliens: une étude dispositionnelle du sentiment d'insécurité Université Sorbonne Paris Cité]

Pain R, Smith SJ (2008) Fear: critical geopolitics and everyday life. In: Pain R, Smith SJ (eds) Fear: critical geopolitics and everyday life. Ashgate, pp 1–24

Priks M (2009) The effect of surveillance cameras on crime: evidence from the Stockholm subway

Riley NE, Dean DL (1985) Bus station security: crime at intercity bus stations. Transp Res Rec 1012:56–64

Richards LJI (1980) Passenger value structure model. Final report DOT contract TSC-1314. Dunlap and Associates, Inc., Darien

Rountree PW (1998) A reexamination of the crime-fear linkage. J Res Crime Delinq 35(3):341–372

Santoro P, Harkot M, Loukaitou-Sideris A, Ceccato V, Sundling C, Romero-Torres J, Agrawal AW (2020) Intersectionality of transit safety. Transit Crime and Sexual Violence in Cities: International Evidence and Prevention

Shaw CR, McKay HD (1942) Juvenile delinquency and urban areas

Shellow R, Romualdi JP, Bartel EW (1974) Crime in rapid transit systems: an analysis and a recommended security and surveillance system. Transp Res Rec 487:1–12

Skogan WG, Maxfield MG (1981) Coping with crime: individual and neighborhood reactions, vol 124. Sage Publications, Beverly Hills, pp 74–77

Sundling C (2020) Huddinge, Sweden. In: Transit Crime and Sexual Violence in Cities, Routledge, pp 167–175

Sundling C, Berglund B, Nilsson ME, Emardson R, Pendrill LR (2014) Overall accessibility to traveling by rail for the elderly with and without functional limitations: the whole-trip perspective. Int J Environ Res Public Health 11(12):12938–12968

Sundling C, Ceccato V (2022) The impact of rail-based stations on passengers' safety perceptions. A systematic review of international evidence. Transport Res F: Traffic Psychol Behav 86:99–120

Tilley N (1993a) Understanding car parks, crime, and CCTV: evaluation lessons from safer cities. Vol. 42, Home Office Police Department London

Tilley,N (1993b) Understanding car parks, crime, and CCTV: evaluation lessons from safer cities. Home Office Police Department London.

Valentine G (1997) A safe place to grow up? Parenting, perceptions of children's safety and the rural idyll. J Rural Stud 13(2):137–148. https://doi.org/10.1016/S0743-0167(97)83094-X

Vanier C, De Jubainville HD (2017) Feeling unsafe in public transportation: a profile analysis of female users in the Parisian region [Article]. Crime Prevent Community Saf 19(3–4):251–263. https://www.scopus.com/inward/record.uri?eid=2-s2.0-85028965378&doi=10.1057%2fs41300-017-0030-7&partnerID=40&md5=8b6479dafc2287288e88f043232c6335

Warr M (2000) Fear of crime in the United States: avenues for research and policy. Criminal Justice 4(4):451–489

Wayland S, McKay K, Maple M (2020) How is participating in suicide prevention activities experienced by those with lived and living experiences of suicide in australia? A qualitative study [Article]. Int J Environ Res Public Health 17(13):1–16. https://doi.org/10.3390/ijerph17134635

Webb B, Laycock G (1992) Reducing crime on the London underground. Crime prevention unit paper

Wikström P-OH, Treiber K (2017) Beyond risk factors: an analytical approach to crime prevention. In: Teasdale B, Bradley MS (eds) Preventing crime and violence. Springer International Publishing, pp 73–87

Yavuz N, Welch EW (2010) Addressing fear of crime in public space: gender differences in reaction to safety measures in train transit [Article]. Urban Stud 47(12):2491–2515. https://www.scopus.com/inward/record.uri?eid=2-s2.0-77958571531&doi=10.1177%2f0042098009359033&partnerID=40&md5=1ce9441e7a6ae8050c8d5c6fb5094173

Yin PP (1980) Fear of crime among the elderly: some issues and suggestions. Soc Probl 27(4):492–504. https://doi.org/10.2307/800177

Yu S-SV, Smith MJ (2014) Commuters using public transit in New York City: using area-level data to identify neighbourhoods with vulnerable riders. Secur J 27(2):194–209. https://doi.org/10.1057/sj.2014.6

Zahnow R, Corcoran J (2019) Crime and bus stops: an examination using transit smart card and crime data. Environ Plann b: Urban Anal City Sci 48(4):706–723. https://doi.org/10.1177/2399808319890614

Zegras PC, Butts K, Cadena A, Palencia D (2015) Spatiotemporal dynamics in public transport personal security perceptions: digital evidence from Mexico City's periphery. In ICT for Transport. Edward Elgar Publishing.

3 The Station in Context: A Systemic Model of Analysis

Abstract

Building on the theoretical foundations established in the previous chapter, we propose a conceptual framework for analysing safety at the station and its context in the city. Based on systems thinking, we propose identifying key dimensions that influence station safety: (a) the internal physical and social environment of the station, (b) the condition of surrounding neighbourhoods—including their socio-economic makeup—and (c) the station's strategic role within the city's overall layout as well as the regional, national, and the global context. The station, viewed as a system, may exhibit non-uniform safety conditions across different temporal scales, such as peak and off-peak hours and day versus night traffic. This framework addresses the safety needs of different groups of travellers and recognises stations as a hyperspace. This framework serves as a theoretical foundation and aligns with several Sustainable Development Goals (SDGs) encompassing the physical and social environment, digital technologies, place management, and stakeholder engagement to foster safer station settings for and with various stakeholders.

Keywords

Stations · City · Node · Place · Crime · Safety · Systems thinking

3.1 Introduction

This book considers the "station" as a "node" and "place" rather than merely a building. Bertolini (1996) suggests that stations function both as nodes of transport and as places embedded in the city. By being a node, a station has a transport function as an exchange point where networks of different scales and modalities are connected, and flows are exchanged between them. By being a place, the station, especially the larger ones, becomes multifunctional and might accommodate a coffee shop and a restaurant, which might work as a meeting point for residents. A station can become a complex system, extending beyond its location and encompassing the surrounding area.

A station can be described as a system (Ackoff 1999; Meadows 2008) by understanding its various interconnected elements—platform, transition area, ticket hall, and exit—as components that work together to facilitate the smooth movement of travellers and services. Each element plays a specific role but also depends on the others to maintain the overall function of the station. Viewing a station in a city through the lens of systems thinking allows us to understand a station's dynamic role as an interconnected node within a broader urban ecosystem.

A station is an active component of a network of physical, social, and economic flows (Trip 2007). Cities themselves are complex systems

V. Ceccato et al., *Transit Safety for a Sustainable City*, Sustainable Development Goals Series,
https://doi.org/10.1007/978-3-031-97427-4_3

characterised by interdependent layers—transportation, housing, commerce, and governance—all of which interact with one another (Batty 2013). Stations facilitate movement, connecting cities and neighbourhoods to the broader city and beyond and acting as conduits for human interaction, commerce, and cultural exchange. From this perspective, a station is a place of convergence (travellers, goods) and redistribution of people, urban activities, and services. It both influences and is influenced by the city system around it. It reflects the economic vitality of an area but also interacts with its surrounding context, regardless of the size of the flow it accommodates. Some stations are large, others relatively small. The relationship between the station and its environment shapes the station's function and the dynamics of movement within it. Whether in areas of rapid growth or more stagnant regions, the context influences the activities within these nodes.

Most people spend some time commuting to and from the station. On their way to the station or when at the station, they are exposed to various environments, each influenced by the time of day. During the morning rush hour, many are headed to the station, and platforms can become crowded as people wait for the train to arrive. Although the wait may only be a few minutes, it is often enough time for people to make phone calls, meet colleagues, or grab a coffee. However, these moments can also expose them to risks, including being a victim of crime or being involved in an accident. While such events are rare and often unexpected—sometimes difficult to prevent—they occur within transit systems, and their impact can vary significantly, affecting specific individuals more than others. Identifying areas where risks are greater or where people feel fearful, regardless of the actual risk, is crucial for encouraging and maintaining public transportation usage. Thus, one of our contributions to this book is to provide information on how the physical and social environments of stations affect people's safety, considering the quality of these transit environments (and on the way to them) in relation to people's safety needs. We now turn to the station as a system within multiple complex systems that is the city. This extends beyond the station's physical quality to encompass its management and governance, considering multiple local and national layers. Building on the theoretical foundations established in the previous chapter, we propose a conceptual framework for analysing safety at the station and its context in the city.

3.2 The Conceptual Model of the Station

Based on previous literature on transit safety, Ceccato (2014a) identified three dimensions of influence that shape station safety (e.g. Ceccato et al. 2022a; Loukaitou-Sideris 1999; Loukaitou-Sideris et al. 2001; Newton 2004) namely the station's environment, the neighbourhood, and the city. In this book, we extend this conceptual model to encompass the system's other "leverage points": the people (users, employees) and the digital and physical realities as presented below:

The Internal Physical and Social Environment of the Station—This encompasses elements like lighting, accessibility, design, maintenance, security measures, and the presence of staff or monitoring systems. The social environment, for example, the presence of other passengers and the overall atmosphere, can also significantly affect transit safety (Ceccato 2014b).

The Condition of Surrounding Neighbourhoods of the Station—This includes the conditions around the stations. Inner city areas, as do end station hubs, offer more safety challenges, with multiple public transportation interchanges (Liggett et al. 2003). Regardless of location, neighbourhoods marked by social fragmentation or economic disadvantage can present more safety challenges (Ferrell and Mathur 2012). The local and regional context matters because some stations attract or absorb crime (Moreira and Ceccato 2021), while others radiate crime beyond their surroundings.

The Station's Strategic Role in the City and Other Contexts—Transit stations often evolve into commercial and cultural hubs, bringing economic opportunities and security challenges associated with the location of new transportation nodes (Ceccato 2014b). What happens at these nodes is a function of these embedded "global contexts" (Robertson 1992) when globalisation in terms of the "compression of the world" is visible to the human eye at or around the station. For examples, drugs from South America being distributed across the railway system as "county lines" (Wroe 2021), sold at the stations or delivered by post after being ordered by internet.

The Station and Peoples—The experience of those most affected by crime and safety issues will vary across the system and over time. For example, women may face higher levels of harassment during the morning rush hours (Ceccato and Paz 2017), while they might be more vulnerable to sexual assault, such as rape, in the evenings on the way to the platform if the pedestrian underpass is poorly lit (Ceccato 2014a; Ceccato et al. 2019). The needs of individuals with disabilities, diverse sexual orientations, and other vulnerabilities must be considered to address inequalities in mobility and safety (Iudici et al. 2017; Sundling et al. 2024).

The Stations as Hyperspaces—Advancements in technology allow us to stay connected to the internet while travelling, and the entire transit system becomes "hyperspaces" (Jameson 2016). In hyperspaces, digital and physical realities intersect and are a backdrop for human interactions. We can connect to chat rooms, cyber markets, social media, and Wi-Fi. Some of these interactions are illicit activities (Bichler 2019). We remain linked to corresponding digital environments as we move through physical spaces. This opens the door to risks such as cyber pickpocketing while expanding our digital footprint throughout our journeys. The security of our mobile smart devices can fluctuate, especially as they interact with unfamiliar devices along the way (Ceccato and Newton 2024). Nevertheless, apps on mobile phones are used to inform travellers about particular risks in the system, while CCTVs and body-worn cameras might prevent crime from happening (Ariel et al. 2019). In random acts of sousveillance (Mann et al. 2003), the presence of travellers filming a particular scene can also be crime preventive or help to solve a crime.

Current developments impose new challenges to track illicit activities and prevent them from happening in critical infrastructures (Lehto 2022; Marassi and Marrone 2023). Transit systems can also become a target for hacking and attacks (Huq et al. 2017). Consequently, our understanding of crime prevention in these places needs to evolve to encompass the physical and digital landscapes we navigate.

If we now return to the context of station safety within a city, identifying and focusing on leverage points can help transport operators and safety experts to prioritise interventions that will have the most significant impact and lasting change with fewer resources. The following section begins by examining the station as a system, highlighting the significance of its physical and social environment for the safety of those who travel or work within it.

3.3 The Station as a System

A station's size, type, location, and design style directly influence the safety conditions experienced by users (Ceccato 2013; Loukaitou-Sideris 2008; Newton and Ceccato 2015). These physical characteristics shape the station's functionality and the perceived and actual safety of those who use it. The surroundings of a station play a crucial role in shaping its crime risks. Stations in high-crime areas may experience spill-over effects, particularly in inner-city locations where mixed land use—such as bars, restaurants, and nightlife—can contribute to heightened vulnerability. As a result, an individual's safety at a station is influenced not only by the station's own physical and social environment but also by the characteristics of its immediate surroundings, its broader neighbourhood, and its position within the city. These risk factors are not uniform across all stations; safety conditions can vary significantly between different sections of the same station and fluctuate over time.

3.3.1 The Station's Physical and Social Environment

The station's vulnerability to crime depends on its physical environment and the type of social interactions that take place at this transportation node. The physical environment refers to the hardware of the station. Traditionally, railway stations tend to be larger than those of commuter or metro stations. From their entrances/exits (e.g., tunnels, stairs, lifts, shops, and restaurants), lobbies (e.g. ticket booths, automatic controls, and commercial shops), transit areas (e.g. stairs and lifts), platforms (e.g. single and multiple), to the immediate surroundings (e.g. streets and car parks).

A railway station can encompass various settings: the platform, the transition area, the lobby, the entrances, and the immediate surroundings. The station is constituted by the platform where the trains arrive, and passengers exit or wait for the train. The transit area commonly includes stairs and lifts from the platform to the lobby, where control gates/ticket booths are located. The lobby may be an open area that ends at the entrances/exits. Commercial shops may be found in lobbies and entrance/exit areas. The entrances/exits are limited to entering the lobby directly from the street or via doorways, stairs, lifts, or tunnels. The immediate surroundings are what individuals see from the station, from a field of view of about 25 to 50 m from the entrances (Ceccato 2013), where factors such as visibility, lighting, and levels of social interaction contribute to the perception of safety and potential crime risks (Fig. 3.1).

For a safe station, previous research shows the importance of good lighting, a design that allows visibility, clear sightlines in the platforms and transition areas, such as corridors and stairs (Cozens et al. 2003), accessible entrances/exits (Deníz 2019), and well-placed security technology. CCTV (closed-circuit television), a surveillance system that uses video cameras to monitor and record activity in a specific area, is associated with a decrease in crime in some studies (Moreira and Ceccato 2021), depending on the crime type. Such conditions may vary across the stations; platforms may be safer than their exits because of the constant flow of passengers or security guards, or it is possible that poor guardianship during the off-peak hours makes them more susceptible to particular crimes than during rush hours. The presence of place managers or other professionals who can keep an eye on what happens on the platform is also associated with the safety perceptions of users (Ceccato et al. 2022b; Sundling and Ceccato 2022). Smith and Cornish (2012) point out that overcrowding and lack of supervision are important environmental features that contribute to increased crime risk. Overcrowding, particularly in peak hours, facilitates theft and other property crimes. During off-peak hours, poor social control contributes to vandalism and graffiti, robbery of staff and passengers, assaults on staff and passengers, or fare evasion.

Visible ticket booths in the lobbies and overpass walkways with good surveyability and separation of passenger flows have a positive effect on station safety (Gaylord and Galliher 1991; La Vigne 1997; Myhre and Rosso 1996). Although the platform constitutes an open space with good visibility, it can still be the most feared place in the station (Cozens et al. 2003). The presence of station management or personnel impacts victimisation and safety. The presence of other travellers who might be able to intervene if anything happens is also important for safety (Ceccato et al. 2024). Homelessness and panhandling at stations can influence individuals' perceptions of safety and comfort, sometimes creating a sense of unease or discouraging transit use, especially in areas where panhandling is persistent or aggressive (Ceccato et al. 2024; Ding et al. 2022).

3.3.2 The Station in the Local and Neighbourhood Context

Evidence from the USA suggests that areas surrounding metro stations host twice as many businesses and three times as many jobs as areas without station access, highlighting the economic benefits of transit-oriented development (WMATA 2024). Involving residents in the planning process is important for ensuring that the station aligns with people's needs and fosters a

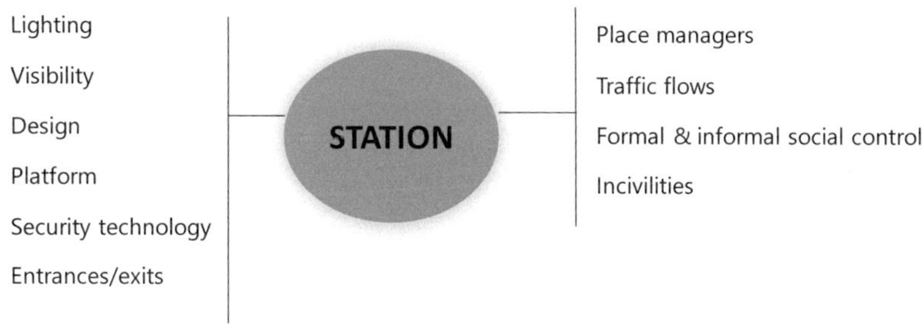

Fig. 3.1 The station's physical and social environment. Source: Ceccato (2013), p. 38

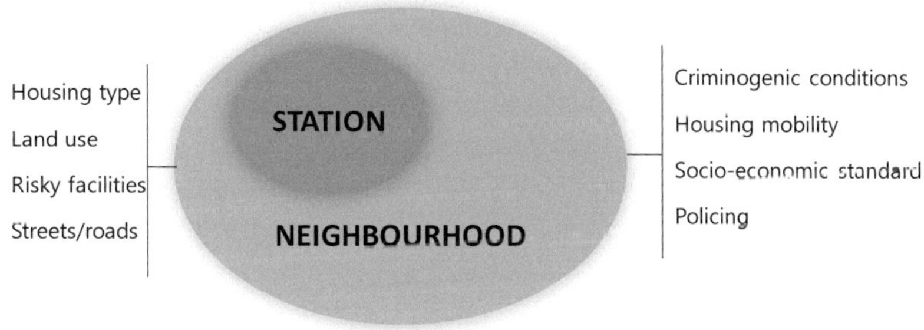

Fig. 3.2 The station in the neighbourhood. (*Source* Ceccato (2013), p. 38)

sense of ownership and social cohesion among those most affected.

Research has shown that the type of land use influences safety conditions at the station in the immediate surrounding area, the social activities it may attract, and the land use and the demographic and socioeconomic characteristics of the population residing or working in the neighbourhood. Loukaitou-Sideris (1999) suggests that the relationship between surrounding land uses and crime incidents at stations tends to be statistically significant as certain environmental features either attract offenders or influence criminal activities by concentrating potential offenders and encouraging anti-social behaviour (Fig. 3.2).

Research has so far shown that high-crime concentrations are often found in commercial, entertainment, and multifunctional areas (Kinney et al. 2008). Studies in California and the U.K. highlight how transit crime correlates with neighbourhood crime levels and socio-environmental factors such as truancy and the presence of young people (Newton and Bowers 2007; Pearlstein and Wachs 1982). There is a need to encourage development around station areas that combine residential, commercial, and recreational spaces, which can increase pedestrian traffic and reduce the sense of isolation, making the area more vibrant and secure. Prioritising the needs of those who use the station, including accessibility and consideration of the safety needs of older adults, women, LGBTQI + people, and others such as individuals with disabilities, is also relevant in the context of management of existing stations and the planning of new ones.

3.3.3 The Station in the City Context

Inner city stations are expected to be more criminogenic than any other city stations. There is plenty of evidence of how inner-city areas are more criminogenic than other parts of the city (Smith and Clarke 2000; Wikström 1991). However, end stations, often located in the most peripheral parts of the city, can be more criminogenic than those found along the railway system and inner city areas." Inte ändrat. Låter fortfarande som att "end stations" inte är en del av systemet. Som jag skrev förra gången borde det vara mer korrekt med "in inner city areas.

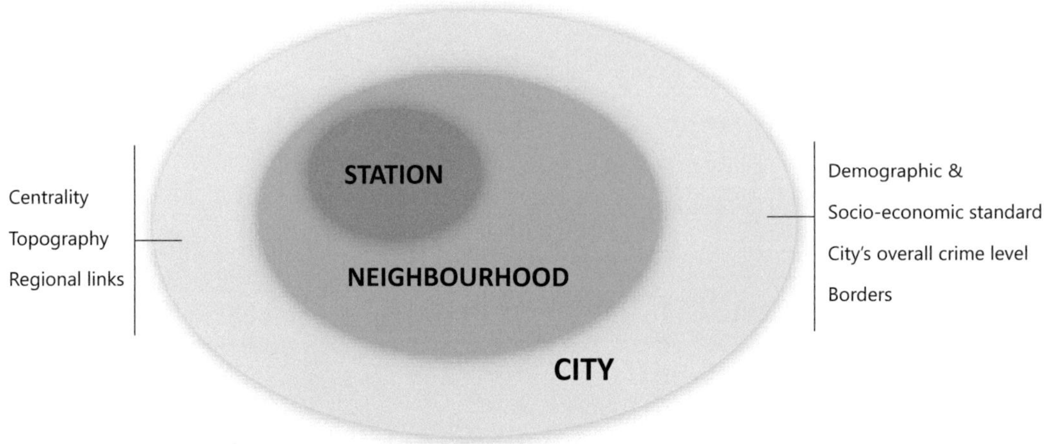

Fig. 3.3 The station and the neighbourhood in the city context. (*Source* Ceccato (2013), p. 38)

These end stations are often linked to other transport hubs and modes, with large flows of passengers, and may adjoin parking lots and commercial areas that may also attract crime (Fig. 3.3). Moreover, the city's geography and the presence of different geographical barriers, such as lakes, mountains, rivers, or parks, are also influential in defining regional patterns of offences, which may indirectly affect transit safety. These physical features might provide hidden places and escape opportunities for motivated offenders at the stations.

The city's population composition, income disparities, employment opportunities, and social cohesion affect crime patterns and safety in public spaces, including transit hubs. Areas with higher socioeconomic inequalities may experience greater safety challenges. The general crime rate within the city sets a baseline for what happens in smaller units such as neighbourhoods and stations. If the city has a high prevalence of crime, stations within it are more likely to experience similar trends, "absorbing" from the surrounding areas (Bowers 2014). Subway stations can act as crime attractors, with robberies clustering near them (Block and Davis 1996). The presence of bus stops and commercial hubs is also linked to elevated crime levels (Vu 2009). However, there are situations when stations become "oases" of safety in neighbourhood criminogenic contexts (Ceccato 2014b).

3.3.4 The Station in National and Other Contexts

Large cities host unique activities and services, such as sports arenas, conference centres, and universities, which attract large crowds and influence crime levels. Studies show that sporting events, in particular, lead to spikes in assaults, vandalism, and alcohol-related offences (Rees and Schnepel 2009). Public transportation plays a crucial role in connecting these locations to the rest of the city, but it also becomes a means for offenders to travel and a target for criminal activity during such events.

These events are only examples of people's movement across the city. Monthly and seasonal variations in crime have long been documented in the international literature and are often associated with changes in people's routine activities (Fig. 3.4). Differences are often found in the criminogenic conditions between peak and off-peak hours and daily and weekly or seasonal variations (Ceccato et al. 2022a). Violent crime and domestic violence increase during the summer months (Farrell and Pease 1994), while commercial robberies can increase during the winter (Van

Koppen and Jansen 1999). These temporal shifts also affect station safety. During off-peak hours or seasons, fewer people use the station due to fear of victimisation, making it more desolate and vulnerable to crimes such as vandalism. The absence of law-abiding citizens creates a reinforcing loop—crime increases due to fewer witnesses, further discouraging use.

With globalisation, the interconnectedness makes stations attractive points for illicit trade, as criminals can easily transit through them, facilitated by anonymity, and easy access to broader networks. An example is when stations and their surrounding areas themselves can become a marketplace for drugs coming from other parts of the world.

As previously pointed out, human activity increasingly shifts between physical and digital realms, and hyperspace becomes more vital to urban systems. Stations can also be considered a hyperspace (Fig. 3.4). Hyperspaces in architecture have been explored in relation to the idea of complex, layered spaces that transcend traditional spatial boundaries. Jameson (2016) makes use of the term hyperspace when physical spaces are entangled with technological and media networks. Crime may be one of the activities that cross these layered spaces, when information flows more seamlessly between the system and passengers, as well as among riders themselves. These systems can also become targets for cyberattacks (Lehto 2022; Marassi and Marrone 2023) or allow more mundane events to happen, such as pickpocketing, as thieves can exploit real-time information for their own purposes (Bichler 2019).

3.4 Urban Planning and Crime Prevention

In a complex world where various social, economic, and technological forces intersect, the question arises: How can stations and surrounding areas be planned to remain attractive and safe ? In Part III of this book, we will review examples from Sweden and other locations, focusing on alterations in the station's environment and initiatives involving local neighbourhoods and municipalities. As previously mentioned, we adopt a systems-thinking approach to problem-solving that emphasises viewing a station as a whole, understanding the interdependencies, and recognising the relationships between parts (Ackoff 1994; Ceccato and Newton 2024; OECD and Analysis 2020). This approach considers the physical environment and social and economic factors interconnectedly. It goes beyond individual components, focusing on how elements influence each other (Meadows 2008; Stroh 2015). For example, rather than solely improving lighting in a high-crime area, systems thinking encourages broader strategies to reduce crime, such as managing the station and surrounding areas and community engagement, focusing particularly on those with specific safety needs, in other words, the governance of public places. We will turn to this topic in Sect. 3.5.

The conceptual model of the station presented in this chapter highlights key leverage points within the station system and beyond, identifying areas across the transit system, city, region, country, and even global connections that require attention. Safety interventions can focus on critical points, times, and those most affected. To achieve this, a unified incident-reporting system must cover all aspects of the station and its surrounding areas. Data should be integrated, organised, and managed by experts who can identify appropriate actions, which are then implemented to resolve the issue.

This approach is not new; since the 1960s, such frameworks have evolved into strategic tools, not merely data repositories, to support informed decision-making. Known as Decision Support Systems (DSS), these tools were designed to assist decision-makers at all levels within an organisation (Power 2002). In the next sections, we discuss the evolution of DSS in urban planning, with a particular focus on safety intervention models.

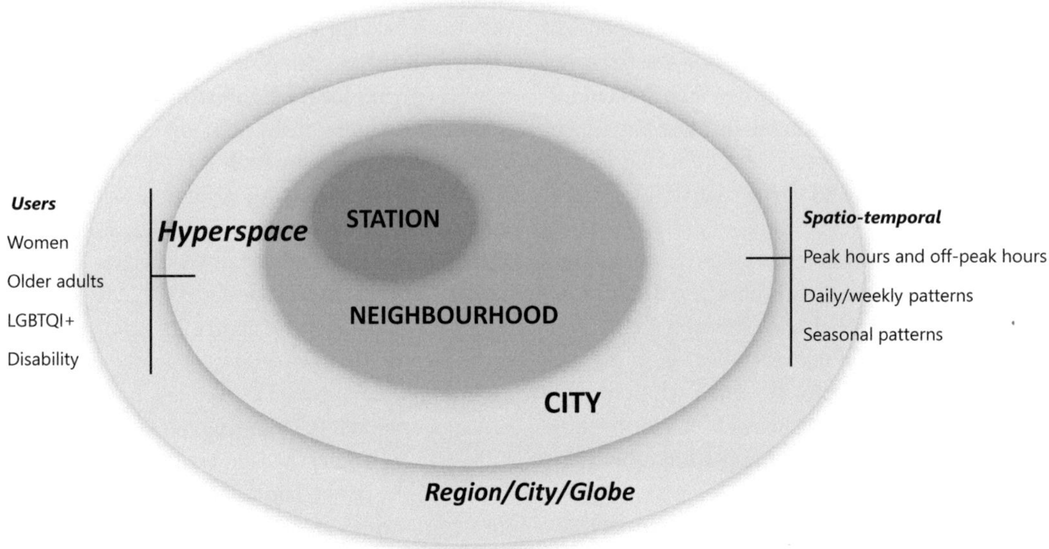

Fig. 3.4 The station and the neighbourhood in the city context

3.4.1 Decision Support Systems (DSS)

Decision Support Systems (DSS) integrates several components that can facilitate decision-making. The concept of DSS began in the 1960s and 1970s and was used to assist in land-use planning and transportation modelling, gradually expanding to include more complex urban challenges. In the 1980s and 1990s, Decision Support Systems, DSS, had become more integrated into planning practices, also involving Geographical Information Systems, GIS (Goodchild 2000), a technology for capturing, storing, analysing, and visualising spatial or geographic data. In the 2000s, computer-based DSS started to support decision-making with more sophisticated data models and tools. In the last few decades, DSS models in urban planning have included various stakeholders from participating schemes (Ceccato and Snickars 2000; Huang et al. 2022). Currently, DSS includes data management for gathering and organising accurate data, model management for storing predictive tools such as statistical and machine-learning models, an intuitive user interface for easy interaction with the system, and what-if modelling for scenario simulation. Real-time processing and automated alerts improve responsiveness (Banik and Huq 2020). In criminal justice and criminology, there have been examples of DSS devoted to automated crime report analysis and classification (Hariguna and Ruangkanjanases 2023), detecting serial crimes (Chi et al. 2017) and identification of predictors of traditional and/or cyber offences (Guo and Wang 2024).

3.4.2 Situational Crime Prevention, SARA Model, and Systems-Thinking Approach

Situational crime prevention (SCP) is a framework aimed at reducing criminal opportunities by modifying the immediate environment where crimes occur (Clarke 1995). This approach seeks to make committing offences less attractive, to make it more difficult for them to happen, or riskier for potential offenders by means of surveillance cameras, improved lighting, and access controls. SCP is closely related to crime prevention through environmental design (CPTED), which focuses on design-based solutions in architecture and planning (Clarke 1989).

Unlike CPTED, SCP is broader, encompassing management and policing strategies. Additionally, situational action theory (SAT) emphasises the interaction between individuals and their environments, suggesting that effective crime prevention should address both environmental conditions and how individuals interpret them (Wikström and Treiber 2017). For a review, see Ceccato and Newton (2024).

Several problem-solving tools support SCP interventions, including the SARA model (Eck and Spelman 1987), the CRAVED model for theft-prone goods (Clarke and Webb 1999), the 5Is framework (Ekblom et al. 2013), the VOLTAGE framework for crime analysis (Ratcliffe 2018), and the EMMIE framework for evaluating crime prevention interventions (Johnson et al. 2015). Most of these models are designed to ensure that crime prevention strategies are targeted and effective by focusing on specific crimes, times, locations, and contextual factors.

The SARA model is one of the most popular SCP models and is a widely used problem solving framework in situational crime prevention and community policing (Fig. 3.5). The SARA model (Scanning, Analysis, Response, and Assessment) arose out of problem-oriented policing, POP (Goldstein 1979). Analysis is integral to both approaches, providing an understanding of the events and conditions that precede and accompany the problem. Each stage of the SARA model follows a structured approach: "Scanning" involves identifying recurring crime problems and assessing their impact; "Analysis" examines the underlying causes, affected individuals, and contributing factors; "Response" focuses on implementing targeted interventions to address the identified issues; and "Assessment" evaluates the effectiveness of these interventions, making necessary adjustments based on the outcomes. The model is intended to include continuous evaluation and adjustment by regularly assessing outcomes and refining strategies as needed. If interventions do not work, adjustments are made by revisiting earlier SARA stages to adapt to

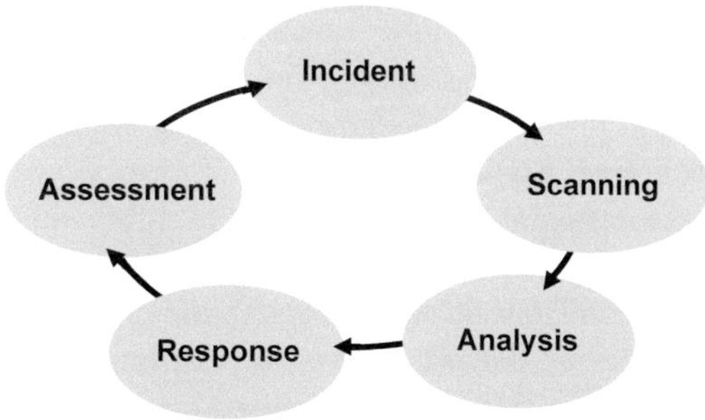

Fig. 3.5 The SARA model

changing crime patterns. Diagrams play a key role in SCP, such as the crime triangle, which illustrates the convergence of key factors necessary for crime. Another influential model is the conjunction of criminal opportunity (CCO) framework (Ekblom 2013), which extends opportunity theories to include crime promoters. The 25 techniques of situational prevention, developed by Clarke and refined by Eck and Clarke (2019), categorise intervention strategies into a structured format. Despite its practical applications, SCP has been criticised for oversimplifying social issues by focusing on immediate symptoms rather than structural causes (Clarke and Bowers 2017).

Concerns about SCP also include crime displacement and unequal impacts on different communities (Garland 2000). Another criticism against SCP is that crime prevention's "chant of specificity" focuses on targeted interventions (crime specific, space-specific, etc.) but risks fragmentation. However, proponents argue that the strength of SCP lies in its place-based focus, aiming to understand why crime (or fear of crime) occurs at specific locations rather than explaining criminal motivation (Laycock and Tilley 1995). Yet, conventional approaches to situational crime prevention may be too specific, and the boundaries may be too rigid. This becomes an issue when the "parts" do not provide enough information about the system. Situational crime prevention, while effective in the short term in specific situations, does not reflect the broader system of risky places. The parts, specific interventions, are treated independently, which may miss essential relationships, spill-over effects, or unintended consequences. Systems thinking promotes "reassembly", ensuring interventions fit into a broader strategy (Ceccato and Newton 2024), which is a fundamental aspect when we study transit safety.

Moreover, a quick fix in terms of safety interventions in a station might lead to unintended long-term consequences. Imagine the crime rate in a city as the water level in a bathtub, with contributing factors like organised crime around stations as the tap adding the water. The police and crime prevention measures act as the drain, removing water by acting upon gangs around the stations from the local area. If the inflow of water (structural causes) exceeds the drain's capacity (crime prevention efforts), the water level (overall crime rate) will not come down. This highlights the need to address both immediate symptoms and long-term causes of crime. This means that all stakeholders should aim having the same set of goals and a shared vision. In many interventions, individuals may have different goals and are unaware of each other's priorities, and intended outcomes are not realised, creating a typical situation for "unintentional adversaries". Authorities may set police targets to reduce crime, increase arrests, or rank performance across forces. We will turn to these barriers in safety interventions in Chapter 15 of this book.

Systems thinking also offers a thorough analysis of the interconnected factors contributing to crime and fear of crime in transit environments, which is a basis for the long-term impact of safety interventions. By understanding relationships between the parts, we can develop crime prevention strategies that tackle the causes in a long-term, systematic manner. Identifying key leverage points allows us to target areas within the system where we can have the most significant impact rather than attempting to address the entire complexity. This approach leads to more inclusive, nuanced solutions for reducing crime and alleviating fear in stations and on the way to them while considering the broader context in which these issues arise. Systems thinking can help address the safety needs of diverse users by enabling targeted interventions that consider the specific vulnerabilities of groups such as women, older adults, LGBTQI + people, people with disabilities, and other vulnerable populations. In rural settings, for instance, it can help address issues such as limited emergency services, inadequate lighting, and the challenges of large open spaces (Ceccato and Newton 2024).

Finally, systems thinking can be used by transport operators, municipalities, and the private sector to improve crime prevention frameworks. These key actors can collaborate to address safety needs through targeted interventions. For instance, practitioners and safety experts can apply systems thinking, while researchers develop evidence-based models to inform strategies. Those affected by crime can offer valuable insights to shape solutions and reassess interventions. Policymakers can, for example, foster collaboration by allocating resources and encouraging partnerships (Ceccato and Newton 2024). All stakeholders can work towards shared goals, which demands collaboration and coordination, as we will discuss in more detail in the next section.

3.5 The Governance of the Station in the City: Policing and Safety Planning

Linking sustainability with crime and safety influences how social, political, and economic relationships are governed (UNODC 2020). Urban planning attempts to align community needs with broader policy goals, particularly the UN's 2030 Agenda, which stresses that reducing crime, violence, and discrimination—while promoting inclusion, good governance, and the rule of law—is key to sustainable development and well-being (UN 2015). While contemporary problems demand more integrated forms of governance, classical reductionist urban planning approaches still oversimplify the complexities of cities as systems. We adopt here Kjaer's (2023) definition of governance to refer to the processes, structures, and mechanisms through which individuals and institutions exercise authority, make decisions, implement policies, and manage resources within an organisation, community, or society. International literature indicates principles of effective governance, including transparency, accountability, and responsiveness to the needs and interests of constituents, which can take various forms when applied to cities.

Thus, a city can be understood as a system of interconnected subsystems, each with its distinct elements, relationships, and objectives. By viewing cities as problems of organised complexity and highlighting the intricate interconnections within urban environments, Jacobs (1961) once wrote that "cities happen to be problems in organised complexity….. Cities, again like the life sciences, do not exhibit one problem in organised complexity, which, if understood, explains all. They can be analysed into many such problems or segments, which, as in the case of the life sciences, are also related with one another. The variables are many, but they are not helter-skelter; they are 'interrelated into an organic whole (Jacobs 1961, p. 432)".

This complexity is further explored in Batty's (2013) work five decades later. The author highlights how the classical systems approach treats cities as static entities with fixed attributes, relying on a mechanistic, top-down analysis that assumes equilibrium and clearly defined boundaries. In contrast, the complex systems approach sees cities as dynamic and interconnected, emphasising networks, flows, and unpredictable interactions that keep them in constant flux. Ceccato and Newton (2025) summarise these contrasting qualities in Table 3.1, which compares classic and contemporary approaches to viewing cities as systems.

Table 3.1 Comparison of classical and contemporary complex systems approaches

Classical system approach	Complex systems approach
• Location, places	• Network flows
• Physical morphologies	• Communications and exchange
• Sets of spaces, places and locations	• Sets of actions, interactions and transactions
• Mechanic, linear processes	• Organic, stochastic processes
• Cartesian structure	• Fractal structure
• Top-down, centralised	• Bottom-up, hierarchical organisations
• Defined as if they were closed off	• Defined as if they were open, "glocal links"
• Tendency to equilibrium	• Far from equilibrium status
• Uni/bi-disciplinary	• Inter/multidisciplinary

Source Ceccato and Newton (2025), p. 30

A system's organisation comprises aggregated elements at different scales, forming distinct structures and subsystems that can be arranged in a hierarchy to hold these parts together (Batty 2013). This perspective acknowledges the ever-changing nature of urban environments (Batty 2013). Open to global and local influences, the complex systems approach recognises cities as being far from a state of equilibrium. Decision-making is decentralised, recognising emergent urban dynamics and constant change. The complex systems approach views cities as open, ever-evolving entities shaped by global and local influences; it promotes interdisciplinary analysis for a broader understanding of urban challenges (Ceccato and Newton 2024).

An example of a subsystem in the city can be the rail system, in which the station is a system itself, a node and a place (Bertolini 1996), where different subsystems intersect, such as people, transport infrastructure, and law enforcement, to name but a few. Activities must be coordinated when many actors from different arenas are involved. These arenas are often managed by a dedicated project organ that is semi-independent yet primarily public. The actors and their relationships vary, and external trends and shifts in institutions can influence the decision-making process and interactions between public and private arenas (Trip 2007). To better understand the complexity of the governance of railway stations, we draw on Trip's (2007) model and propose four key decision-making arenas (Fig. 3.6):

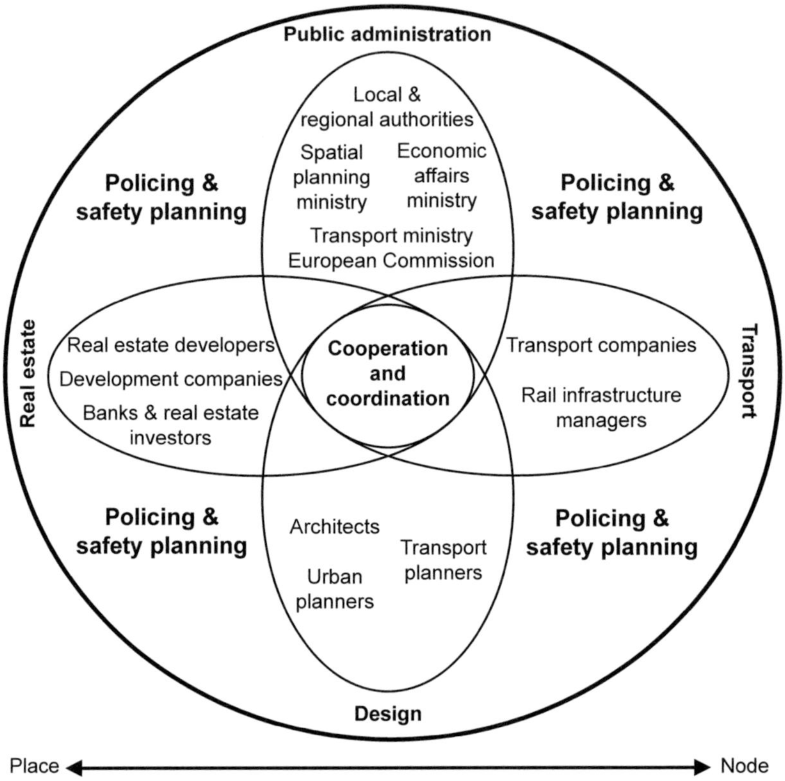

Fig. 3.6 Arenas of decision-making involved in railway station area development. (*Source* Adapted from Trip (2007), p. 53)

1. The public administration arena: defines policy objectives through various public bodies.
2. The real estate development arena: focuses on the physical development of the area, involving both private and public actors.
3. The transport arena: centres on the station as a transport node and address conflicts between transport functions and the quality of place.
4. The design arena: engages planners and architects in the detailed planning and design of the station area.

Each arena includes a blend of public, semi-public, and private actors with varying levels of involvement based on their objectives. The design arena plays a crucial role in shaping the quality of the place, as it is responsible for the physical realisation of the development project.

Building on Trip's model, we propose a fifth arena, "the policing and safety planning arena", which includes crime prevention strategies, surveillance measures, and coordination between law enforcement agencies, urban planners and architects, community stakeholders and civil society. According to Trip (2007), a distinct design arena emerges, comprising urban planners, transport planners, architects, and other professionals and includes public and private actors, as planners and architects may work for public institutions—such as the spatial planning department of a local authority—and private architectural and engineering firms. Together with the policing and safety arena, the design arena, and transport providers, the state and municipalities will play an important role in shaping the quality of the place where a station and its surrounding areas will be located, focusing on the detailed planning and design of the physical project.

Several barriers prevent effective collaboration among stakeholders and professionals. First, architects, urban planners, and designers often overlook crime-related considerations in their decision-making despite influencing criminogenic conditions through their city designs. By shaping urban spaces, they impact governance, socio-economic conditions, and environmental factors that affect crime and safety. Second, criminologists, the police, and safety experts tend to underestimate the role of urban planning professionals due to the ongoing siloed work practices in the field (for an in-depth discussion of these barriers and implications for practice, see Ceccato and Brantingham (2024)).

Architects, urban planners, designers, and engineers—the "built environment professionals" have many roles, including the power to affect how the urban environment is shaped, indirectly affecting crime. Their responsibilities are diverse and interdisciplinary, strategically using land and resources to meet social, economic, environmental, and aesthetic goals. In general, they have two main areas of impact. First, they can incorporate safe features during planning and construction by thinking situationally and designing spaces to minimise crime, although their ability to impact these plans often depends on their political power and ability to balance conflicting interests. Second, once a (station) area is built, they are tasked with modifying problem areas to reduce crime, often through situational crime prevention strategies (Ceccato and Brantingham 2024). While architects and urban planners are responsible for designing buildings and urban spaces, Sampson et al. (2010) call for the recognition of other actors, the "supercontrollers", which are individuals, institutions, and organisations that indirectly influence crime prevention. While not directly controlling crime, they shape incentives for those who do, such as handlers, guardians, and managers. Other actors include financial and government organisations that regulate property ownership, and managers, like property owners who oversee properties or municipal officials, who adjust settings to deter crime. Their policies, norms, and environmental designs are crucial in crime prevention (Eck and Madensen-Herold 2018; Rothstein 2017; Sampson et al. 2010). We return to this topic in Chapter 16, where we discuss their role in partnerships for future stations and new railway projects.

The construction of transport nodes is important, but their long-term management should not be overlooked. In recent decades, new public management reforms of the last decades have reshaped the public sector, emphasising competition, performance measurement, and private-sector management practices (Sørensen and Longva 2011). EU and North American authorities have increasingly relied on public–private cooperation in urban and transport planning, sharing risks and costs through private co-financing. As a result, there is unanimous recognition that private and public–private control over transport and urban spaces has expanded while the public sector's traditional role has declined. These changes have brought not only opportunities, such as improved efficiency and investment in modern infrastructure (Boyne 1998), but also challenges, including fragmented responsibilities, reduced public oversight, and inequalities in service access (Hodge et al. 2010; Klijn and Teisman 2000). Chapter 8 examines this shift by briefly showing the historical development of European railway systems, focusing on Sweden, the Netherlands, and Denmark. Within the transport sector, there have been calls for reforms, including better coordination (Sørensen and Longva 2011) between different partners to tackle problems of fragmented responsibilities, reduced public oversight of what happens (in the railway system) and inequalities in service access, particularly for vulnerable transit users.

In the following chapters, we indicate how these challenges manifest themselves in Sweden, where responsibility for a single station can be divided among five or more stakeholders. If decision support systems (DSS) exist, they are confined to individual stakeholders rather than operating at a system-wide level. This fragmentation results in weak safety oversight at the station level, a lack of a unified incident-reporting system—where each stakeholder collects information independently—and little to no comprehensive overview of safety conditions across the entire railway system. Consequently, this creates a weak foundation for evidence-based interventions that other mechanisms of the planning system must compensate for. The following section examines the Swedish planning context and crime prevention work.

3.5.1 The Swedish Planning Context and Crime Prevention Work

Historically, crime prevention integration into urban planning has taken a unique approach in Sweden. Rather than being a certification system as was adopted in other countries such as the Netherlands and the United Kingdom (Ceccato and Newton 2024), the Swedish model has taken advantage of the strength of the historical legacy of local democracy, focusing on skills development and information sharing. This model has emphasised municipal-level collaboration and is aligned with Nordic planning tradition. For instance, The National Council for Crime Prevention (Brå) has implemented many initiatives over several decades to strengthen crime prevention efforts in Sweden.

"Everyone's Responsibility" ("Allas vårt ansvar") was a national crime prevention programme launched in Sweden in the 1990s. The idea was that public and private initiatives, along with community engagement, should work together to combat crime. This programme created local crime prevention councils in many municipalities, but not all municipalities had the same level of engagement. These programmes were inspired by the North American Problem-Oriented Policing (POP) model and facilitated and supported the work of local councils. More than thirty years later, the "Together Against Crime" ("Tillsammans mot brott") programme prioritised cooperation and knowledge-based approaches, with Brå supporting municipalities through initiatives such as the National Crime Victim and Safety Survey. Since then, Brå and the County Administrative Boards have been given expanded responsibilities in crime prevention work.

On the part of the Police Authority, the BoTryggt initiative, which was started by the police in 2000, has promoted situational crime prevention through environmental design. It has evolved into BoTryggt2030 and now addresses more general societal challenges. Meanwhile, Sweden's police force has grown by 20% since 2000, though this increase has mainly benefited larger municipalities, with 25% of municipalities lacking a permanent police presence. Additionally, the Swedish police force has been restructured into a national authority in recent years, which may also affect the overall local crime prevention work.

In 2023, a new law mandates municipalities to take responsibility for crime prevention, requiring a designated safety coordinator. This has impacted planning at all levels, especially locally, where crime prevention efforts need to be more focused. The new law mandates municipalities to prevent crime through systematic, evidence-based approaches, meaning they must map crime trends, develop action plans, and coordinate efforts with the police and other actors. The law aims to enhance safety by ensuring municipalities are the leaders when it comes to crime prevention with national support from Brå. In this new framework, a proposed council could help monitor and address crime prevention continuously, aiming to improve knowledge about high-crime areas and their connections within the municipality. For instance, stations and other transportation hubs could become target areas for safety interventions.

The Swedish National Board of Housing, Building, and Planning (Boverket) has also been important in guiding municipalities when it comes to integrating crime prevention into urban planning. In 2023, new rules in the Planning and Building Act required municipalities to update their comprehensive master plans. These plans are a key political document outlining the use, development, and preservation of the physical environment. They are also important for long-term sustainable development and housing strategies aligned with the Agenda 2030 goals. In practice, architects, urban planners, designers, and engineers—"the shapers"—often consider issues of crime and safety to be outside their area of action, but in reality, they inadvertently influence criminogenic conditions when shaping the city (Ceccato and Brantingham 2024).

According to Ceccato and Brantingham (2024), the shapers play key roles in shaping the urban environment, indirectly influencing crime. Their responsibilities are interdisciplinary, aiming to meet social, economic, environmental, and aesthetic goals. From a criminological perspective, they can influence safety in two ways: by incorporating crime-reducing features during the planning or construction of a station or residential area and/or by modifying existing areas using strategies such as situational crime prevention. While collaboration with safety experts and the police is becoming more common in many Swedish municipalities, communities are still struggling to adopt this approach and reducing social disparities in safety.

3.6 Conclusions

In this chapter, we have proposed a conceptual framework for analysing station safety within its urban context. Based on systems thinking, the framework identifies key dimensions affecting safety: the station's internal physical and social environment, the surrounding neighbourhood's socioeconomic characteristics and social cohesion, and the station's role within the city's layout, as well as the stations' regional, national, and global contexts. We conceptualise a station not merely as a physical location but as a dynamic hub where various flows intersect. This means that its safety is shaped by its immediate surroundings and broader social, economic, and technological forces, sometimes far beyond the locality. In this chapter, we have also explored several key topics related to urban planning and crime prevention. We addressed the role of urban planning in crime prevention, including the need to integrate data and knowledge into a unified evidence-based Decision Support System (DSS).

This chapter has explored several key topics related to urban planning and crime prevention. We introduced and applied the Situational Crime Prevention approach, incorporating the SARA model and systems thinking. While the SARA model is important for addressing immediate, localised issues, it is necessary to adopt an integrated and long-term perspective to address the complexities of urban safety fully. This is where systems thinking becomes important, as it allows for a broader, more holistic understanding of how various factors within urban environments interact and affect crime over time. These frameworks collectively highlight how planning decisions can influence safety outcomes.

We then shifted focus to the governance of stations within the urban context, examining the roles of multiple stakeholders involved in managing these spaces. Lastly, we discussed the Swedish planning context, specifically how crime prevention has been established and implemented within the country. As is illustrated throughout the book, efforts from these collaborations demonstrate ongoing attempts to integrate public transportation into cities and regions sustainably, with a strong emphasis on prioritising safety as a key dimension.

References

Ackoff RL (1994) Systems thinking and thinking systems. Syst Dyn Rev 10(2–3):175–188

Ackoff RL (1999) Ackoff's best: His classic writings on management. John Wiley & Sons

Ariel B, Newton M, McEwan L, Ashbridge GA, Weinborn C, Brants HS (2019) Reducing assaults against staff using body-worn cameras (BWCs) in railway stations. Crim Justice Rev 44(1):76–93

Banik S, Huq MR (2020) Geolocation based decision making for appropriate land use using machine learning model. Int J Sci Technol Res 9(2):3084–3090

Batty M (2013) The new science of cities. MIT press

Bertolini L (1996) Nodes and places: complexities of railway station redevelopment. Eur Plan Stud 4(3):331–345

Bichler G (2019) Understanding criminal networks: a research guide. University of California Press

Block R, Davis S (1996) The environs of rapid transit stations: a focus for street crime or just another risky place. Crime Prevent Stud 6:237–257

Bowers K (2014) Risky facilities: crime radiators or crime absorbers? A comparison of internal and external levels of theft. J Quant Criminol 30:389–414

Boyne GA (1998) Competitive tendering in local government: a review of theory and evidence. Public Admin 76(4):695–712

Ceccato V (2013) Moving safely: crime and perceived safety in Stockholm's subway stations. Lexington books

Ceccato V (2014a) The nature of rape places. J Environ Psychol 40:97–107. https://doi.org/10.1016/j.jenvp.2014.05.006

Ceccato V (2014b) Safety on the move: Crime and perceived safety in transit environments. Springer

Ceccato V, Brantingham P (2024) What is the role of architects and urban planners in crime prevention? Secur

J 37(3):489–514. https://doi.org/10.1057/s41284-024-00442-4
Ceccato V, Gaudelet N, Graf G (2022a) Crime and safety in transit environments: a systematic review of the English and the French literature, 1970–2020. Public Transport 14(1):105–153
Ceccato V, Gaudelet N, Graf G (2022b) Crime and safety in transit environments: a systematic review of the English and the French literature, 1970–2020. Public Transport 14(1):105–153. https://doi.org/10.1007/s12469-021-00265-1
Ceccato V, Li G, Haining R (2019) The ecology of outdoor rape: the case of Stockholm, Sweden. Eur J Criminol 16(2):210–236
Ceccato V, Newton A (2024) Systems thinking for sustainable crime prevention in risky places. Routledge
Ceccato V, Newton A (2025) Systems thinking for sustainable crime prevention: Planning for risky places. Taylor & Francis
Ceccato V, Paz Y (2017) Crime in São Paulo's metro system: sexual crimes against women. Crime Prevent Community Saf 19(3):211–226
Ceccato V, Sundling C, Gliori G (2024) What makes a railway station safe and for whom? The impact of transit environments on passengers' victimisation and safety perceptions. Eur Transp Res Rev 16(1):21
Ceccato VA, Snickars F (2000) Adapting GIS technology to the needs of local planning. Environ Plann B Plann des 27(6):923–937
Chi H, Lin Z, Jin H, Xu B, Qi M (2017) A decision support system for detecting serial crimes. Knowl-Based Syst 123:88–101
Clarke RV (1989) Theoreticall background for crime prevention through environmental design and situational prevention designing out crime: crime prevention through environmental design. Brisbane, Australia
Clarke RV (1995) Situational crime prevention. Crime Justice 19:91–150
Clarke RV, Bowers K (2017) Seven misconceptions of situational crime prevention. In: Handbook of crime prevention and community safety, pp 109–142
Clarke RVG, Webb B (1999) Hot products: understanding, anticipating and reducing demand for stolen goods, vol 112. Citeseer
Cozens P, Neale R, Whitaker J, Hillier D (2003) Managing crime and the fear of crime at railway stations—a case study in South Wales (UK). Int J Transport Manag 1(3):121–132. https://doi.org/10.1016/j.ijtm.2003.10.001
Deníz D (2019) Improving perceived safety in public transportation through design. WIT Trans Built Environ 182:199–208
Ding H, Loukaitou-Sideris A, Wasserman JL (2022) Homelessness on public transit: a review of problems and responses. Transp Rev 42(2):134–156
Eck JE, Clarke RV (2019) Situational crime prevention: theory, practice and evidence. Handbook on crime and deviance. Springer International Publishing, Cham, pp 355–376
Eck J, Spelman W (1987) Problem-oriented policing. US Department of Justice, National Institute of Justice
Eck JE, Madensen-Herold TD (2018) Place management, guardianship, and the establishment of order. Deterrence, choice, and crime, vol 23. Routledge, pp 269–307
Ekblom P, Armitage R, Monchuk L, Castell B (2013) Crime prevention through environmental design in the United Arab Emirates: a suitable case for reorientation? Built Environ 39(1):92–113
Farrell G, Pease P (1994) CRIM SEASONALITY: domestic disputes and residential burglary in Merseyside 1988–90. Br J Criminol 34(4):487–498
Ferrell C, Mathur S (2012) Neighborhood crime and travel behavior: an investigation of the influence of neighborhood crime rates on mode choice–Phase ii
Garland D (2000) Ideas, institutions and situational crime prevention. Ethical and social perspectives on situational crime prevention, pp 1–16
Gaylord MS, Galliher JF (1991) Riding the underground dragon: crime control and public order on Hong Kong's mass transit railway. Br J Criminol 31(1):15–26
Goldstein H (1979) Improving policing: a problem-oriented approach. Crime Delinq 25(2):236–258. https://doi.org/10.1177/001112877902500207
Goodchild MF (2000) Part 1 Spatial analysts and GIS practitioners: the current status of GIS and spatial analysis. J Geogr Syst 2(1):5–10
Guo S, Wang Y (2024) Investigating predictors of juvenile traditional and/or cyber offense using machine learning by constructing a decision support system. Comput Hum Behav 152:108079
Hariguna T, Ruangkanjanases A (2023) Adaptive decision-support system model for automated analysis and classification of crime reports for E-Government. J Appl Data Sci 4(3):303–316
Hodge GA, Greve C, Boardman AE (2010) International handbook on public-private partnership. Edward Elgar Publishing
Huang D, Ceccato V, Kyttä M (2022) Safety perceptions in university campuses: the role of environment. Crime Prevent Community Saf 24(3):266–285
Huq N, Vosseler R, Swimmer M (2017) Cyberattacks against intelligent transportation systems. TrendLabs Research Paper
Iudici A, Bertoli L, Faccio E (2017) The 'invisible' needs of women with disabilities in transportation systems. Crime Prevent Community Saf 19:264–275
Jacobs J (1961) The death and life of great American cities. Vintage books
Jameson, F (2016) Postmodernism, or the cultural logic of late capitalism. In Postmodernism (pp. 62–92). Routledge.
Johnson SD, Tilley N, Bowers KJ (2015) Introducing EMMIE: an evidence rating scale to encourage mixed-method crime prevention synthesis reviews. J Exp Criminol 11(3):459–473. https://doi.org/10.1007/s11292-015-9238-7
Kinney JB, Brantingham PL, Wuschke K, Kirk MG, Brantingham PJ (2008) Crime attractors, generators and

detractors: Land use and urban crime opportunities. Built Environ 34(1):62–74
Kjaer AM (2023) Governance. John Wiley & Sons
Klijn E-H, Teisman GR (2000) Governing public-private partnerships: analysing and managing the processes and institutional characteristics of public-private partnerships. In: Public-private partnerships. Routledge, pp 102–120
La Vigne NG (1997) Visibility and vigilance: Metro's situational approach to preventing subway crime. US Department of Justice, Office of Justice Programs, National Institute of Justice
Lehto M (2022) Cyber-attacks against critical infrastructure. In: Cyber security: Critical infrastructure protection. Springer, pp 3–42
Laycock G, Tilley N (1995) Policing and neighbourhood watch. Strategic issues. London: Home Office Police Research Group.
Liggett R, Loukaitou-Sideris A, Iseki H (2003) Journeys to crime: assessing the effects of a light rail line on crime in the neighborhoods
Loukaitou-Sideris A (1999) Hot spots of bus stop crime: the importance of environmental attributes. J Am Plann Assoc 65(4):395–411
Loukaitou-Sideris A, Liggett R, Iseki H, Thurlow W (2001) Measuring the effects of built environment on bus stop crime. Environ Plann B Plann des 28(2):255–280
Loukaitou-Sideris AL (2008) How safe is the ride? Evaluation of design and policy responses to women's fear of victimization and crime
Mann S, Nolan J, Wellman B (2003) Sousveillance: inventing and using wearable computing devices for data collection in surveillance environments. Surveill Soc 1(3):331–355
Marassi L, Marrone S (2023) What would happen if hackers attacked the railways? Consideration of the need for ethical codes in the railway transport systems. In: Applications of Artificial Intelligence and Neural Systems to Data Science. Springer, pp 289–296
Meadows DH (2008) Thinking in systems: a primer. Chelsea Green Publishing
Moreira GC, Ceccato VA (2021) Gendered mobility and violence in the São Paulo metro Brazil. Urban Stud 58(1):203–222
Myhre M, Rosso F (1996) Designing for security in meteor: a projected new metro line in Paris. Prevent Mass Transit Crime 6:199–216
Newton A, Ceccato V (2015) Theoretical perspectives of safety and security in transit environments. In: Safety and security in transit environments. Springer, pp 23–36
Newton AD (2004) Crime on public transport: 'Static' and 'non-static (moving) crime events [Article]. West Criminol Rev 5(3):25–42. https://www.scopus.com/inward/record.uri?eid=2-s2.0-27544458202&partnerID=40&md5=d7757906390b64f246e17542c39581c1
Newton AD, and Bowers KJ (2007) The geography of bus shelter damage: the influence of crime, neighbourhood characteristics and land-use. Internet J Criminol
OECD, and Analysis IIfAS (2020) Systemic thinking for policy making. https://doi.org/10.1787/879c4f7a-en
Pearlstein A, Wachs M (1982) Crime in public transit systems: an environmental design perspective. Transportation 11(3):277–297
Power DJ (2002) Decision support systems: concepts and resources for managers, vol 13. Quorum Books Westport
Ratcliffe J (2018) Reducing crime: a companion for police leaders. Routledge
Rees DI, Schnepel KT (2009) College football games and crime. J Sports Econ 10(1):68–87
Robertson R (1992) Globalization: social theory and global culture
Rothstein R (2017) The color of law: a forgotten history of how our government segregated America. Liveright Publishing
Sampson R, Eck JE, Dunham J (2010) Super controllers and crime prevention: a routine activity explanation of crime prevention success and failure. Secur J 23(1):37–51. https://doi.org/10.1057/sj.2009.17
Smith MJ, Clarke RV (2000) Crime and public transport. Crime Justice 27:169–233
Smith MJ, Cornish DB (2012) Secure and tranquil travel: preventing crime and disorder on public transport [Book]. https://www.scopus.com/inward/record.uri?eid=2-s2.0-84920442791&doi=10.4324%2f9780203728406&partnerID=40&md5=5bddf043a8bcf09b3c160fffe4855f12
Stroh DP (2015) Systems thinking for social change: a practical guide to solving complex problems, avoiding unintended consequences, and achieving lasting results. Chelsea Green Publishing
Sundling C, Ceccato V (2022) The impact of rail-based stations on passengers' safety perceptions. A systematic review of international evidence. Transp Res Part F: Traffic Psychol Behav 86:99–120. https://doi.org/10.1016/j.trf.2022.02.011
Sundling C, Ceccato V, Gliori G (2024) Disability, victimisation, and safety in train travel. Transp Res Interdiscip Perspect 25:101131
Sørensen CH, Longva F (2011) Increased coordination in public transport—which mechanisms are available? Transp Policy 18(1):117–125
Trip JJ (2007) What makes a city? Planning for'quality of place: the case of high-speed train station area development, vol 12. IOS press
UN United Nations (2015) Sustainable development goals: 17 goals to transform our world. United Nations, [Online]. Available: https://www.un.org/sustainabledevelopment/energy/. [Accessed 04 June 2018]
UNODC United Nations Office on Drugs and Crime (2020) UNODCITY Urban safety governance initiative. https://www.unodc.org/documents/Urban-security/Safety_Governance_Assessment_Guidance_final.pdf
Van Koppen PJ, Jansen RW (1999) The time to rob: variations in time of number of commercial robberies. J Res Crime Delinq 36(1):7–29

Vu Y (2009) Bus stops and crime: do bus stops increase crime opportunities in local neighbourhoods. Unpublished PhD Thesis, 14
Wikström P-OH (1991) Urban crime, criminals, and victims: the Swedish experience in an Anglo-American comparative perspective. Springer Science & Business Media
Wikström P-OH, Treiber K (2017) Beyond risk factors: an analytical approach to crime prevention. In: Teasdale B, Bradley MS (eds) Preventing crime and violence. Springer International Publishing, pp 73–87
WMATA Washington Metropolitan Area Transit authority (2024) When we invest in transit, our community thrives. https://wmata.com/about/news/upload/BenefitsOfTransit_Report_240304_Web.pdf
Wroe LE (2021) Young people and "county lines": a contextual and social account. J Children's Serv 16(1):39–55

The Research Design and Study Area

4

Abstract

This chapter details the step-by-step creation of a database and methodology used to analyse safety conditions in the rail system of southern Sweden. This experience is important not only for understanding the empirical analysis and recommendations made in this book but also for offering valuable insights into the opportunities and challenges of developing a national incident-reporting system. This chapter begins by outlining the research design of the study, covering the study area, methods, key terms, data collection (surveys, interviews, observations, official statistics), and analytical tools such as Geographical Information Systems (GIS). We then compare the effectiveness of tools such as paper versus online surveys and include a glossary and analysis of limitations. Finally, we reflect on whether the lessons learnt from this case study in Sweden, particularly when creating this unique database, would be a valuable step in implementing an incident-reporting system for decision support of safety evidence-based interventions.

Keywords

Study area · Research design · Survey · Police record · GIS · Evidence-based interventions

4.1 Introduction

Sweden lacks a unified reporting system for security and safety issues, such as crime, impacting passengers and staff across the national public transportation network. As a result, public transportation users often receive little or no assistance when experiencing crime or incidents of incivility, especially when stations lack staff. Similarly, acts of physical damage to stations as public property go unaddressed unless reported to the police. While traffic safety incidents must be reported by law, there is no equivalent requirement for recording crimes or public disorder that happens in transit environments, apart from calling the police. Conditions vary widely between large and small stations and across urban and rural areas, leaving rail safety planning largely unsupported by evidence. This issue is not unique to Sweden, as it is a common challenge in many other rail systems worldwide.

Mapping crime and safety related incidents is essential for evidence-based interventions, helping planners and safety experts to identify unsafe areas and underlying causes for targeted responses (Lab 2014; Clarke 1995). A long-term, data-driven strategy based on situational crime prevention requires comprehensive incident records to create shared awareness and provide

V. Ceccato et al., *Transit Safety for a Sustainable City*, Sustainable Development Goals Series,
https://doi.org/10.1007/978-3-031-97427-4_4

information for necessary interventions. While individual transport operators and related companies have their own reporting systems, these operate in isolation, preventing a comprehensive understanding of safety concerns throughout the rail system. When an intervention is carried out, the effectiveness of safety measures is compromised because rarely are they assessed systematically or cover the whole system. A fragmented approach hinders long-term planning and sustainable, inclusive systems.

A systems-thinking approach is key to improving rail safety by means of a unified reporting system. If passengers feel unsafe and lack support, they may turn to cars, increasing congestion, emissions, and social exclusion. Addressing these challenges requires an integrated strategy incorporating crime prevention in national infrastructure planning, with a shared vision among stakeholders. Safety concerns are interconnected with urban mobility, social equity, and environmental sustainability, which should be a priority in building a safer, more integrated, and sustainable rail system. Later in Chap. 6 of this book, we show the interlinkages between accessibility and safety experiences along the trip, from home to station and from station to station.

This chapter details the step-by-step creation of a database and methodology used to analyse safety conditions in the rail system of southern Sweden. This experience is crucial not only for understanding the empirical analysis and recommendations in the book but also for offering valuable insights into the opportunities and challenges of developing a national incident-reporting system.

We start by outlining the study area selection and research design underpinning the study, clarifying key terms and methodological choices. The chapter provides a comprehensive account of data collection methods—including surveys, interviews, observational studies, and official statistics—alongside analytical techniques such as Geographical Information Systems (GIS). It also compares the effectiveness of different tools, such as paper-based versus online surveys. A glossary of terms and a critical analysis of the study's limitations are included. The result is a unique dataset containing thousands of numerical (crime, safety, land use, and socioeconomic) variables, visual materials, and extensive documentation designed to register, detect, and create a foundational base for safety interventions. Indirectly, this chapter also offers insights into how a nationwide Decision Support System (DSS) could be if implemented in Sweden or in other countries that still today lack an incident-reporting system. Finally, in this chapter, we reflect on whether the lessons from this case study in Sweden, particularly the creation of this unique database, would be a valuable step when implementing an incident-reporting system for decision support for evidence-based safety interventions.

Studying a topic such as transit safety empirically with multiple data sources presents great opportunities for analysis but also some challenges. This chapter has three main parts: we first present the study area, then report the data used in the study and methods applied in the empirical study conducted in the context of three railway lines in Sweden with 47 different stations and then discuss the challenges we encountered during the execution of the study.

4.2 The Research Design

The execution of the project covered about three years, starting in January 2021 and finishing in December 2023, including a series of varying activities as displayed in Fig. 4.1 and detailed in the following sections.

A project reference group (functioning as an advisory group) was established to provide comments on the development of the project and support the research work, as well as help us interpret the results of the study. The reference group included most of the various Swedish national and regional transit stakeholders, municipal and regional actors, and related organisations, such as stakeholder organisations for travellers and individuals with disabilities. The research established a direct communication between the research team and these stakeholders, ensuring that their insights and experiences provided necessary information for the methodological steps and outcomes.

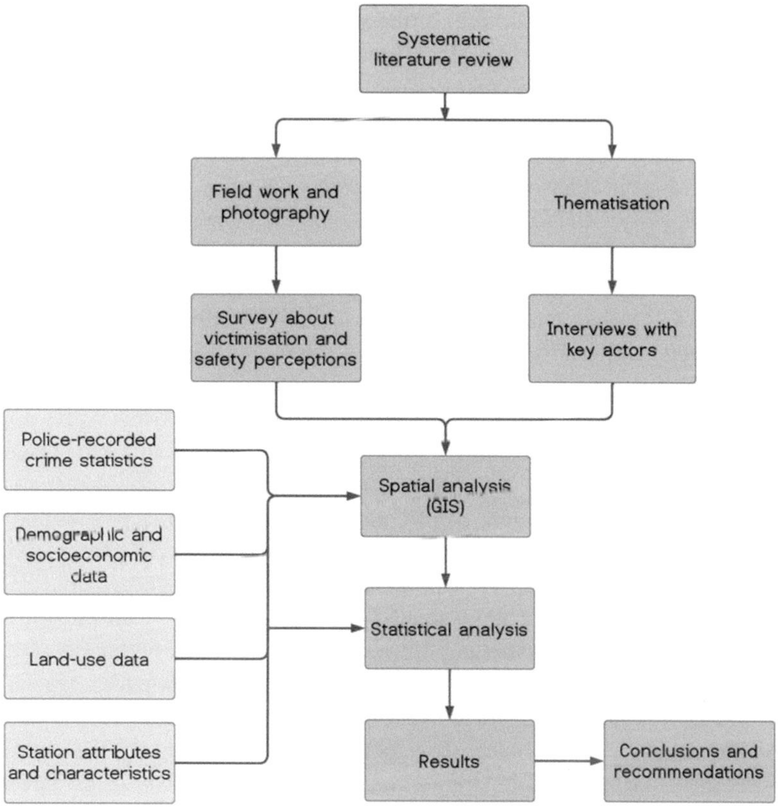

Fig. 4.1 Flowchart of activities of the study. (*Source* Authors)

The objective of the literature review was to collect existing knowledge and identify knowledge gaps through two systematic literature reviews. We wanted to know what knowledge already exists about factors that affect crime and safety perceptions in stations. Particular focus was given to the role of the physical and social environments of the stations. The systematic reviews aimed at identifying the most important characteristics of the stations and/or of the public transport system that affect crime and safety perceptions, behavioural consequences of poor safety, and how different groups of travellers experience safety conditions at stations.

The systematic literature reviews were carried out during 2021—one with a broader focus on crime and security in the transport system as a whole and the other, more specifically, focused on perceived safety in railway station environments. Both collected international literature via electronic databases (Scopus, ScienceDirect, Google Scholar). A protocol (PRISMA-P 2015; Moher et al. 2015) for systematic literature searches was used in both cases. The analysis was later split into two parts: a bibliometric analysis and an in-depth analysis of materials. The first review covered studies in English and French from 1970 to 2020, while the second review focused on literature in English from 1980 to 2020. We used VOSviewer (https://www.vosviewer.com) to organise and visualise this vast amount of bibliographic material. For more details about the methodology, see Ceccato et al. (2022a), Sundling and Ceccato (2022) and, in Swedish, Ceccato et al. (2022b).

4.2.1 The Study Area

The study area is composed of 47 stations distributed over three railway lines stretching through central and southern Sweden, serving 28 municipalities, which together have a total population of 1.78 million (Statistics Sweden

2020). Figure 4.2 shows the three different railway lines and the number of travellers per day for each station. In red, the first railway line has nine stations, running between Nykvarn and Kumla, consisting of two different parts, the Svealandsbanan which runs between Nykvarn and Kungsör, where the Mälarbanan then connects and continues to Kumla. The other is the Western Main Line (Västra stambanan), which includes a total of 29 stations between Södertälje and Gothenburg, which is thus the longest of the three. The third railway line in green is the Southern Main Line (Södra stambanan) between Nyköping and Mjölby, with nine stations. The three lines are served primarily by five different regional transport operators, namely Sörmlandstrafiken, Storstockholms Lokaltrafik (SL), Länstrafiken Örebro, Västtrafiken and Östgötatrafiken, as well as the state-owned Statens Järnvägar (SJ), which is the largest operator of passenger traffic in Sweden (Government Offices, n.d).

We aspired to study the topic in many contexts on the urban-rural continuum. Together with the Swedish Transport Administration, we selected the three railway lines to serve as case studies to further identify which characteristics of the environment of stations most affect passenger safety. Size, location, and context were used as criteria for the selection of the stations. Altogether, the three lines serve approximately 64,000 passengers a day, which is, according to the Swedish Transport Administration, a rough estimate of the annual average day per station (Swedish Transport Administration 2009). A categorisation of station size with respect to passenger flow shows that seven stations have more than 2,000 passengers per day, and then there are 18 stations with fewer than 500 passengers per day, while 22 stations serve 500–2,000 passengers per day, covering a variety of rural contexts.

Gothenburg Central Station on Sweden's west coast, with around 20,000 passengers per day, is by far the largest station in terms of passenger flow, followed by Linköping Central Station, with about 4,600, and Norrköping Central Station, with about 4,000. On an average day in a year, approximately 64,000 passengers travel on all three lines combined (Swedish Transport Administration 2009). A categorisation of station size for passenger flow shows that seven stations have more than 2,000 passengers per day; then there are 18 stations with fewer than 500 passengers per day, while 22 stations serve 500–2,000 passengers per day, covering a variety of rural contexts.

Gothenburg is the only large municipality. There are a number of medium-sized municipalities along the three railway lines, such as Södertälje, Eskilstuna, Örebro, Norrköping, and Linköping, but most of the stations are located in small towns with a population of less than 15,000 inhabitants, but close to at least one larger municipality, and where most of the daily commuting takes place. Stations vary in terms of land use and socio-demographic contexts where they are embedded. Many stations are located in areas with several other facilities nearby, such as restaurants, bars or nightclubs, but also schools, parks, forests and car parks. However, there are differences in the distances to these facilities depending on the station size. For instance, alcohol-selling establishments (e.g., licenced bars and restaurants), but also schools and parks usually have a greater distance to smaller stations than to larger ones. However, the distance to the nearest car park is generally short for all stations and is often found directly adjacent to large and medium-sized stations, while the distance is generally slightly longer for small stations.

Broadly speaking, it is more common to have a high proportion of elderly people living close to small and medium-sized stations than to large stations—where a younger population is more prevalent. Regardless of the size of the station, people live in households with a slightly lower economic standard than the rest of the population. A higher proportion of people with a foreign background than average is more common, especially close to medium-sized and large stations, but rarer in areas close to the small stations.

4.2.1.1 The Characteristics of the Stations

Out of the 47 stations, nearly all have Real-Time Information (RTI) for train departures and arrivals centrally located in the station (e.g., a departure

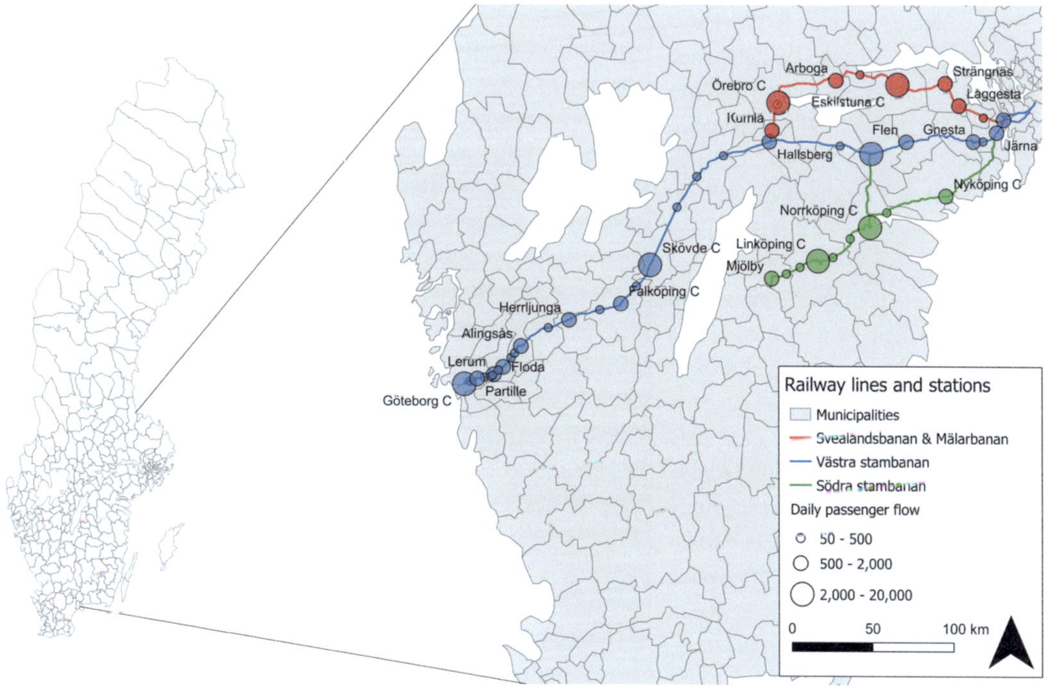

Fig. 4.2 The study area. (*Source* Authors)

board), and almost as many also provide RTI on the platforms. However, for buses, only a third of the stations have centrally located RTI, while nearly two-thirds have it at the bus stops (Fig. 4.3).

In close to 30% of the stations, there are no possibilities to buy train tickets from station personnel or from a ticket machine. These are often small commuter stations where tickets are primarily purchased online via the app of the train operator or directly on the train from the ticket inspector. While this may not be an issue for frequent travellers, it can cause confusion for occasional users, tourists, elderly passengers, or individuals with disabilities. This issue is exacerbated by the fact that nearly half of all the stations lack clear/visible information about the train operator and instructions on how to buy tickets. The availability of ticket sales varies across different lines. For instance, all stations on the red and green lines offer ticket purchasing options, while only half of the stations on the blue line provide this service. The largest variation is in the presence of ticket machines, which are available at only 34 per cent of the stations on the blue line but at all stations on the red and green lines. Over-the-counter ticket sales, however, are similarly common across all three lines (blue: 45%, red: 44%, green: 44%).

The presence of station staff and services also varies significantly. Nearly half (49%) of all stations do not have a kiosk, 43% lack a restaurant or café, and only one station has a host. Many of these stations, particularly smaller ones, do not have any operational personnel, making them more vulnerable to disorder. The red line generally has more facilities, with a higher prevalence of kiosks, cafés, restaurants and waiting rooms, which contribute to social control, but could also create crime-generating opportunities, such as those linked to alcohol sales in restaurants. On the red line, 89% of stations have waiting rooms, compared to just under half of the stations on the blue and green lines. Similarly, free toilets are more common on the red line (67%) than on the blue (41%) and green (33%) lines.

Other services such as ATMs (17%), Wi-Fi (36%), and luggage lockers (19%) are rare across all stations. Toilets, while generally available, are

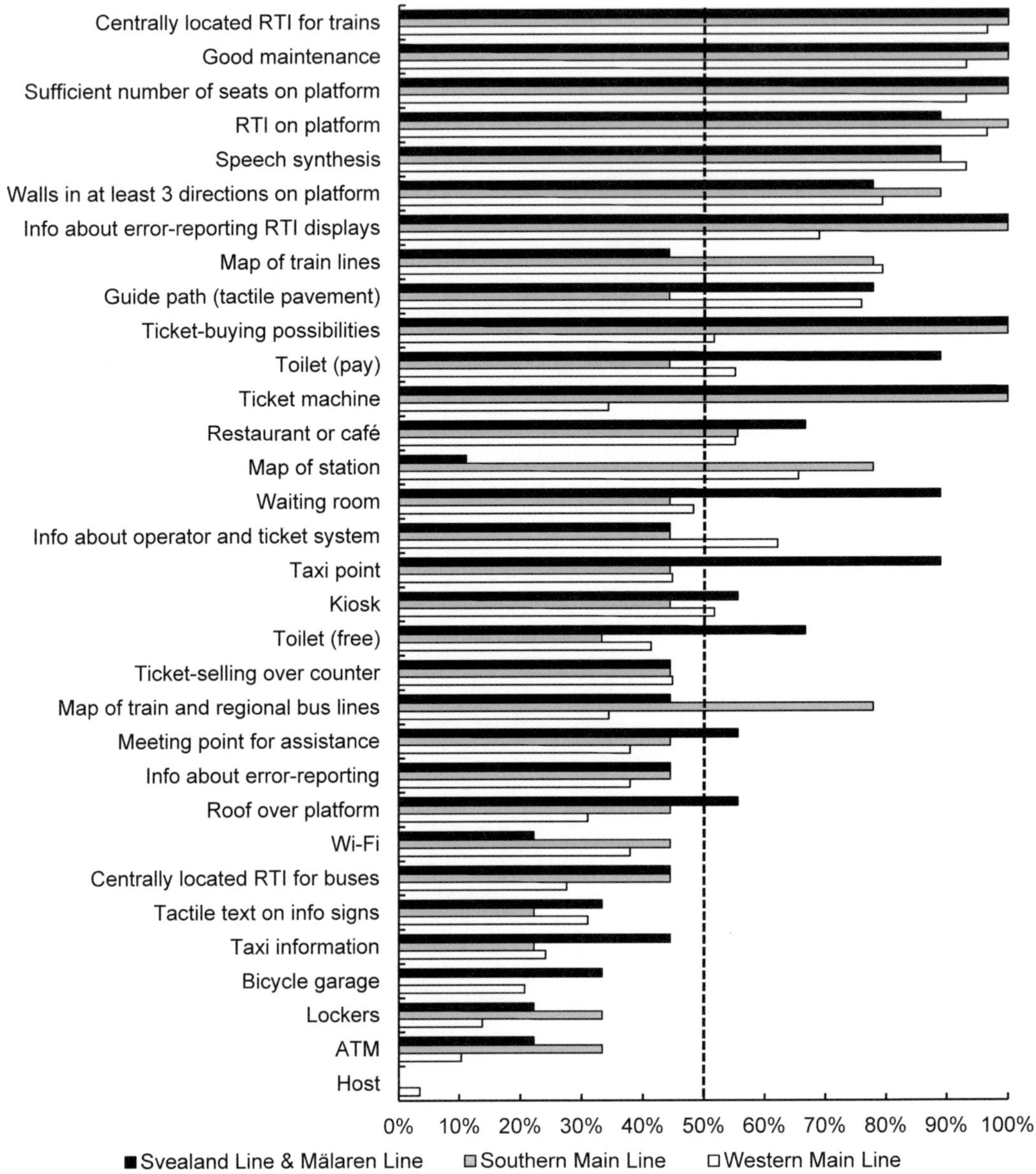

Fig. 4.3 The characteristics of the stations. (*Data source* Fieldwork data collection 2021)

missing in 40% of stations, and in several cases, using them requires a fee. Bicycle racks, which provide secure parking and reduce the risk of theft, are absent along the green line and relatively uncommon on the blue (21%) and red (31%) lines, but more common at larger stations.

Weather protection on the platform is another issue. A significant 62% of stations lack sufficient platform roofing, and 45% do not have a waiting room, meaning that passengers often have little to no shelter while waiting for the train. Seating availability, however, is high, with 96% of stations providing an adequate number of seats on the platform. The lack of weather protection raises pressing concerns in the Swedish context, where snow, wind, and prolonged periods of darkness are the norm rather than an exception.

Adaptations for people with disabilities vary according to station size, but are relatively evenly distributed across the different lines, probably due to standardisation guidelines, although larger stations are significantly better equipped. The four main accessibility features analysed were: (1) meeting points for travellers needing assistance, (2) speech synthesis for audio announcements, (3) tactile text on information signs, and (4) guide paths (tactile paving). More than half of the stations lack a meeting point, which can make travel more difficult for those requiring assistance. Speech synthesis for visually impaired travellers is available at all but four stations (92%). However, tactile text on signage is far less common (30%), and while guide paths are present at a majority of stations (70%), their availability is much less common on the green line.

The size of a station is a key factor influencing its facilities. Larger stations generally offer more services, whereas smaller ones lack many amenities. As seen in Fig. 4.3, small stations on the blue line are particularly deficient in ticket sales, many of which serve Gothenburg's commuter rail network. While frequent travellers may be accustomed to mobile ticketing, the absence of clear purchasing options can create difficulties for occasional users, older passengers, or those with disabilities.

The same trend applies to other station services. Cafés and restaurants are present at all large stations, 71% of medium-sized stations, and just 35% of small ones. Kiosks, waiting rooms, and ATMs follow a similar pattern. Wi-Fi is available at about a third of all stations but is much more common at large stations, found in approximately half of the medium-sized stations, and only 9% of the small stations.

Platforms can range from open-air spaces to those with waiting shelters that provide seating. Platform connections vary from simple ground-level crossings to stair or ramp connections. The availability and clarity of information also differ, with some stations having centrally located, easily accessible displays, while others offer more limited signage. Tunnels can be narrow and dark or spacious, well-lit, and even designed with vibrant colours to create a more welcoming atmosphere. Figure 4.4 illustrates examples of platforms, entries/exits, lounges with different designs, transit and surrounding areas, and tunnels.

4.2.2 Data

A multiple-source database was created using both primary and secondary sources as described below:

4.2.2.1 Police-Recorded Data

Crime and public disorder data were gathered from the Swedish Police Authority headquarters from 2019 to 2021 through X–Y coordinates of all types of offences by date. It was delivered in 28 different datasets, one for each municipality. The datasets were first quality-inspected. Some discrepancies were identified, and new datasets were obtained. The Swedish crime codes (BRÅ, Swedish National Council for Crime Prevention 2021) were used to guide the selection of offences by crime types). The types of crimes included were: assault, sexual offences, threatening behaviour, theft, drug-related offences, robbery, weapon offences, vandalism, violation of privacy, harassment and murder or manslaughter.

Although coordinates were available, it was not possible to know where each incident occurred (at the station, on the train, or at other nearby premises). Crime records were selected for the stations and their surrounding areas by creating buffer zones in GIS. These zones included a 100-m buffer around the stations and a 200-m buffer for the surrounding areas. Using raw data on crime events at and around the stations, we calculated crime rates per passenger based on the estimated passenger flow. Additionally, we calculated crime rates per inhabitant using demographic data to estimate the living population around each station using DeSO areas as a reference. DeSO (Demografiska statistikområden) are small statistical areas defined by Statistics Sweden (SCB) for demographic and socioeconomic analysis. Each DeSO contains approximately 700–2,700 inhabitants in urban areas and 200–1,000 inhabitants in rural areas (Statistics Sweden, 2018).

Fig. 4.4 The station as a system: platform, entry/exit, lounge, travel information and surrounding areas, and underpasses/pedestrian tunnels

4.2.2.2 Data on the Design and Functional Aspects of the Stations

The environment of the railway stations follows some common basic standards (e.g., platforms and illumination). However, as described previously, they are far from being homogeneous, which potentially impacts the vulnerability of the stations to crime, the extent of disorder and the safety perceptions of users. In order to assess these differences, we conducted a systematic and detailed "inspection" of all stations (including photographic documentation), as well as a check on the surrounding areas in winter 2021 and at some selected station in the summer of 2022. Stations were inspected on a weekday.

With the help of a field protocol, a number of variables describing the attributes of the station were determined. Among these were for example whether the station had Real-Time Information (RTI) for trains and for buses, and where the RTI signs were located (on the platform, inside the station, at the bus stops). Moreover, if travellers were able to buy train tickets at the station (over the counter or from a ticket machine), if there was a kiosk, restaurant, ATM, toilet, or waiting room, as well as a sufficient number of seats and weather protection on the platforms. There are also numerous variables concerning people with disabilities, such as speech synthesis, tactile text on signs, and guide paths (tactile pavement).

4.2.2.3 Land-Use Data

Using Open Street Maps and the Open Data Portal, we selected a number of land-use variables associated with particular criminogenic conditions and/or social control in an area. Examples are bars, restaurants, car parks, green areas/parks, secondary schools, the presence of toilets, petrol stations, and the presence of police stations.

4.2.2.4 Demographic and Socioeconomic Data

Data from Statistics Sweden was obtained to create variables on the resident population, young male population, unemployment, household composition, foreign population, and average income per DeSO. Statistical areas (DeSOs) were used as the smallest unit of analysis for the demographic and socioeconomic data for all municipalities (N = 981 DeSO areas), with a population of 1,810 on average and a standard deviation of 482, following county and municipal boundaries. Digital boundaries are available as open spatial data from Statistics Sweden (SCB) at the DeSO level.

Data from the fieldwork inspection (protocol) was entered into spreadsheets and then imported to GIS together with data on land use, crime, and demographic and socio-economic data of the population according to DeSO areas for all municipalities included in the study area. This database allowed us to track the neighbourhood qualities of the railway rides. In order to obtain information about riders' safety perceptions, two major quasi-identical surveys were conducted as described below:

4.2.2.5 Survey with Riders

The survey "To promote safe stations" consisted of a set of 22 questions and was specially designed to capture the importance of the physical and social environment of the stations in the Swedish context (Appendix 1). The survey was distributed using two methods: a printed questionnaire handed out on board the trains and at the stations and a web-based survey hosted on a platform (using Crowdsignal in WordPress). The web survey was promoted by means of posters in the stations (Fig. 4.5), email lists, social media, local Facebook groups, and web pages of the relevant municipalities in the study area. Researchers participated in radio programmes that promoted the research project and encouraged people to answer the survey. The traditional paper survey was carried out during May 2022 with a supplementary session in August 2022, while the web-based survey was active from May to November 2022.

The survey consisted of four parts: introductory questions about frequency of use, where the point of departure of the trip is, the most frequent times of travelling, a main section on victimisation and safety perceptions according to different types of

Fig. 4.5 The two types of posters displayed at the stations, featuring a QR code that directs users to the website for the web-based survey

trips and station environments, and according to the type of victimisation. The survey also included questions about the impact of crime and fear and recommendations for improving safety conditions along the trip, as well as a final section on the background information of the riders, such as gender, age, income, ethnical background, sexual orientation, and disabilities.

The sample was calculated to ensure that the results could be statistically broken down by railway line, station size, gender, age and time of day. A total of 4,893 survey responses were collected (but with a varying N depending on the question). The survey yielded approximately 850 responses from the Svealandsbanan and Mälarbanan (red line), around 2,000 responses from the Western Main Line (blue line), and about 1,000 responses from the Southern Main Line (green line). It was also important to get a sufficient number of responses from each station size category: 800 responses were gathered from small stations 1,750 responses from medium-sized stations, and 1,350 responses from large stations. Table 4.1 shows the total responses for individual characteristics, station size, railway lines, victimisation, and fear of crime. The sample consists mostly of women (61%) and Swedish-born individuals (88%), with the majority aged between 30 and 59 years of age (56%) and fewer elderly people (15% are 60 years of age or more). A significant portion has higher education (59%) and most travel by train less than once a month (32%). Fear of crime was reported by 35% of respondents, while 14.6% had been victims of crime in transit the last five years.

The results from the two surveys (paper and web-based) revealed significant differences in reported victimisation and fear, as well as response rates, which led to a comparative investigation of these two distribution methods. The article

Table 4.1 Description of a sample from "To promote safe stations-surveys"

Gender identity (N = 3,402)	Woman = 61.0% Man = 34.2% LGBTQI+ /non-binary/other = 4.8%
Age (N = 2,793)	Under 30 years = 29.2% 30–59 years = 56.0% 60 years or more = 14.8%
Ethnicity (N = 3,364)	Swedish-born = 88.3% Foreign-born = 11.7%
Income (N = 2,714)	Low income (1–249 KSEK) = 21.1% Medium–low income (250–499 KSEK) = 48.4% Medium–high income (500–749 KSEK) = 23.1% High income (more than 750 KSEK) = 7.4%
Education (N = 3,399)	Elementary school education = 9.0% Upper-secondary school education = 32.2% Higher education = 58.7%
Household (N = 3,374)	Live alone = 23.2% Lone parent = 5.8% Live with partner = 29.8% Live with partner and children = 26.2% Live with parents/other adults = 15.0%
Disability (N = 3,292)	Minor disability = 8.7% Major disability = 1.3%
Travel frequency (N = 4,690)	4–7 days/week = 27.4% 1–3 days/week = 18.3% 1–3 days/month = 20.1% Less than once a month = 32.2% Never = 1.9%
Travel times (N = 4,075)	Morning (06–09) = 54.5% Daytime (09–17) = 64.1% Evening (17–22) = 51.8% Nighttime (22–06) = 9.3%
Miscellaneous (N = 3,479)	Transit captive = 36.6% Comfortable travelling with strangers = 56.7% Safe in neighbourhood = 85.3% Trust in police = 76.9%
Station size (N = 3,912)	Large station = 34.9% Medium-sized station = 44.5% Small station = 20.7%
Train line (N = 3,912)	Green line = 25.6% Blue line = 52.3% Red line = 22.1%
Victimisation (N = 3,964)	Victim of crime = 14.6% Victim of crime (incl. aggressive panhandling) = 24.3% Witness to crime = 34.0%
Fear of crime (N = 3,536)	Fear of crime = 35.0% Fear of crime (train) = 15.3% Fear of crime (station) = 24.9% Fear of crime (way to/from station) = 24.8%

entitled *"Comparing Responses from a Paper-Based Survey with a Web-Based Survey in Environmental Criminology"* (Ceccato et al. 2024) examines the methodological implications of each approach. In summary, the web-based survey revealed that respondents were more open about their experiences with victimisation, fears, and precautions compared to those who completed the paper survey. While response and completion rates were slightly higher in the paper survey, it also faced its challenges, such as respondents' ability to skip questions. Additionally, altering the order of response alternatives (from being fixed in the paper survey to being randomised in the web-based survey) had no significant effect on the results. This finding positively supports the reliability and generalisability of the survey results.

4.2.2.6 Survey with Individuals with Allergies

A separate survey was conducted with a small sample of members of the Swedish Asthma and Allergy Association (N = 66) which was advertised digitally via social media, to investigate victimisation and safety perceptions of this group. Asthma and allergies are common worldwide (Pawankar et al. 2013). Moreover, respondents reporting asthma/allergies had, in our earlier analyses declared themselves as feeling unsafe during their trip compared to travellers without disability. This version of the Web-Based Survey, was supplemented with a response alternative on one of the questions addressing safety, namely "allergens", to study the potential importance of the problem. See Chap. 11 for this study.

4.2.2.7 Interviews

Interviews with representatives of train operators and the public in Sweden and abroad (Denmark, and Holland) were conducted. A representative from an expert from IUC (International Union of Railways, a worldwide railway organisation) was interviewed. In Sweden, 10 interviews were carried out during the autumn of 2021 (12 October–4 November) with experts and key people in public transport in Sweden, including people with a role in safety and security. The people interviewed included representatives from Trafikverket (the Swedish Transport Administration) and various county transport companies, Jernhusen AB (is a state-owned company that owns and manages station buildings) and Travellers Forum (Resenärerna). In selecting interviewees, it was important to include representatives from different actors with different interests in public transport. The initial plan was also to interview representatives from other private or semi-private actors, but some of these declined to participate in the project. The interviews were conducted digitally via Zoom with a semi-structured setup and each interview lasted between 30 and 60 min. The interviews, both Swedish and international, always began with a question about the person's role in the organisation they worked in, and examples of their tasks and areas of responsibility. Questions were then asked about what safety and security-related problems they see, how they work with crime prevention measures, which other actors they cooperate with in this connection and how the distribution of responsibilities is, how they adapt train traffic to the safety needs of different passenger groups (for example the elderly, women, people with disabilities, LGBTQI+ people), and what obstacles and opportunities they see as regards improvement when it comes to dealing with today's problems. Different follow-up questions and discussions were added depending on the interviewee's answers. In conclusion, a question was always asked about what they see as important aspects for a station to function from a safety and security perspective. Read more about the findings from these interviews in Chap. 8.

4.2.3 Methods

As explained in Chap. 1, this study took a systemic look at transit settings, exploring safety as a measure of risk at stations and immediate environments, as experienced by different groups of travellers, in particular those with special mobility needs (individuals with a disability) and with regard to the views of those involved in providing railway services. Using three railway lines, we aimed to examine the nature, type, settings, and extent of victimisation at railway stations according to gender and other individual characteristics, assess individuals' safety needs, and later, make recommendations that can respond to such needs.

4.2.3.1 Temporal Analysis of Crime Data

After extracting the crime data within 100 m of each station using buffers in GIS, the temporal characteristics of these incidents could be analysed. The temporal information in the dataset contained data for the date (N = 5,234) and the time (N = 4,172) of the crimes. Sometimes, there was an exact date and time (N = 863 and 741), but if this did not exist, the start date/time of the interval in which the crime happened was used (N = 3,456 and 2,554). If there was no start date/time but only an end date/time of the crime, this information was used instead (N = 915 and 877). A new column combining the three temporal aspects in the above order had to be created.

In order to be able to run the temporal analyses, the dataset was organised using standard spreadsheets. To start with, all dates starting from 1 January to 31 December 2021 were listed. Then, with a PivotTable it was possible to obtain the count for each crime type per date. The data in the PivotTable was then matched to the corresponding date in the list with all dates using formulas and entering 0 values when no incident happened.

Subsequently, the dates could be transformed to weekdays by changing the format, as well as to seasons depending on the month element of the date (winter = December-February, spring = March–May, summer = June–August, and autumn = September–November). The dataset was now prepared in such a way that an analysis of variance could be conducted (one-way ANOVA test). This type of analysis is used to determine whether there are any statistically significant differences between the means of *three or more* independent (unrelated) groups. In this case, the groups were seasons and the days of the week. An independent sample t-test was performed if there were less than three groups, for example, in the comparison between weekdays and weekends. This method is a statistical technique used to analyse the mean values of *two* independent groups.

For the analysis of time, the data was structured in the same way but replacing the dates with the 24-h clock from 00:00 to 23:59. Since the ANOVA test requires three or more groups, the 24-h clock was divided into four categories: (1) morning = 06:00–12:00, (2) afternoon = 12:00–18:00, (3) evening = 18:00–00:00, and (4) night = 00:00–06:00.

4.2.3.2 Spatial Analysis by Station

For the analysis of spatial patterns, we had two different strategies. First, to take into account the land uses closest to stations, we used the "distance to the nearest hub (points)" function in Geographical Information Systems (GIS), with the option to select a pair of X-Y coordinates and choose the facility (e.g. bar, transportation hub) closest to each station. Second, a similar strategy was used to calculate the number of crimes around the stations after crime codes were selected and the crime data were mapped using GIS. All crime counts for each station and each crime category could be exported into a spreadsheet where rates based on these counts were calculated. These rates considered the number of passengers passing through the station and population using area interpolation procedures in GIS. This method was also used to determine other demographic information around the station, such as the number of people in different age categories, with foreign background, education levels, type of household and ownership, income, etc. Counts and rates by passengers and population of the closest DeSOs were calculated and compared to each other. This pre-analysis was essential to study the distribution of each offence according to stations and lines.

4.2.3.3 Modelling Crime and Safety Perceptions

Hypothesis testing was also used to test the importance of physical and social environment on sexual victimisation and/or safety perceptions using regression models. More specifically, we modelled victimisation/safety perceptions as a function of travellers' individual characteristics, the frequency of use of public transportation, and the environmental covariates of transportation nodes.

For the logistic regression models, a dichotomous dependent variable was created for each outcome—"victimised" and "poor safety perception"—coded as 1 and the rest as 0. A level of significance was set at 5%, and statistically significant results were noted with their respective p-values. The basis for the analysis was formed with two sets of models: one for victimisation and another for safety perceptions. Independent variables included individual passenger characteristics as well as attributes of the railway station and its surroundings. Individual attributes included, for example, gender, sexual orientation, country of birth, travel times, and frequency of travelling by train. Among the station's characteristics, the models included variables that characterise the station (such as the presence of a restaurant or café, toilets, CCTV, and illumination), as well as the surroundings of the station (such as whether the station is perceived as "isolated", whether there is a presence of open drug markets or the existence of tunnels). See the individual chapters for other modeling strategies.

4.2.3.4 Methods Used in Cross-Sectional Chapters

In addition to the analysis of crime and safety perceptions at the stations, seven cross-cutting themes of relevance are reported in the book. Each of these chapters had an independent, standalone theoretical background and an empirical analysis, followed by results, conclusions and recommendations for both research and practice. These themes are covered in Part II and III in this book.

4.2.3.5 Drawing Conclusions and Lessons Learnt for Research and Practice

In the final chapters, we identify commonalities, differences, and lessons from our results in each chapter, and conclude with implications for research and practice. Using the Swedish case, this chapter focuses on the importance of the stations' environments and contexts for crime in transit and travellers' safety perceptions. Drawing from the empirical findings and the international literature on transit safety, environmental criminology, and urban planning, the authors propose a set of recommendations to improve transit safety in the last chapter.

4.3 Research Challenges

As with many other research projects of this scope, we encountered numerous challenges when collecting and analysing the data. This study was no exception, facing the following challenges.

Difficulty Ensuring the Minimum Sample Size Per Station in Rural Contexts In all but six stations (Gårdsjö: N = 1, Jonsered: N = 16, Örebro södra: N = 20, Töreboda: N = 21, Kolmården: N = 25, Stenstorp: N = 27), researchers were able to gather the minimum requested sample size of 30 answers per station. This did not mean these stations were excluded from the analysis; however, the limited sample sizes from these rural and less-frequented stations posed challenges to ensuring representativeness. Consequently, while responses from these stations were still valuable, the low counts prevented reliable, detailed comparisons at the station level.

Difficulty in Ensuring the Minimum Sample from Ethnical Minorities and Genders Obtaining a representative sample across stations, ethnic backgrounds, and gender identities presents significant challenges. While it was technically possible to segment responses according to station type and ethnic background, securing a sufficient number of responses from riders with foreign backgrounds and from LGBTQI+ individuals was difficult. This limitation made it challenging to ensure that minority groups were adequately represented, particularly at smaller stations where fewer responses were collected overall. Achieving a gender-balanced sample at smaller stations was difficult. This limited the robustness of station-level comparisons but did not affect the overall conclusions about these groups' experiences in the transport environment, which are discussed later in the book.

4.4 Concluding Remarks

A key challenge in assessing safety conditions within public transport is reliance on police data, which often fails to capture minor incidents that still contribute to feelings of insecurity. As police data is not a reliable indicator of transit safety in its current form, alternative data sources and more comprehensive reporting systems are necessary to better evaluate safety conditions (during the trip, at the stations and on the way to them) and inform decision-making.

This study combined multiple data sources aiming at better understand safety in station environments, including police-recorded statistics, survey data, land-use mapping, station attributes, and demographic and socioeconomic indicators. By combining all this information, we explored the connections between the physical environment, social conditions, and safety, as well as perceptions of safety. Methodologically, the survey results indicate that response patterns vary depending on the mode of data collection. Participants in online surveys reported higher levels of victimisation and fear compared to those responding via paper surveys, indicating that the survey format may influence responses. This study faced challenges in ensuring a minimum sample size at rural stations, limiting detailed comparison at the station level. Additionally, obtaining a representative sample from ethnic minorities and gender groups, particularly at smaller stations, proved difficult but did not impact the overall conclusions.

To explore the feasibility of developing and implementing an incident-reporting platform as part of a future evidence-based decision support system, it is essential to understand how such a system would function, from the individual rider reporting an incident to the recipients and, ultimately, how the information can be used to enhance the quality of the infrastructure. Achieving this requires understanding the organisational aspects, including the barriers that prevent the system from being implemented and the reasons for the lack of coordination between stakeholders. We would need to investigate further the available technical solutions for an effective incident-reporting system which currently is not available at national level in Sweden.

References

Ceccato V, Gaudelet N, Graf G (2022a) Crime and safety in transit environments: a systematic review of the English and the French literature, 1970–2020. Public Transport 14(1):105–153. https://doi.org/10.1007/s12469-021-00265-1

Ceccato V, Sundling C, Gliori G (2022b) Stationsmiljön, brott och resenärers trygghet - Delrapport 1. Säkerhet och trygghet forskningsgrupp (STF), Institutionen för samhällsplanering och miljö, KTH Kungliga Tekniska Högskolan, Stockholm, 49 sidor

Ceccato V, Gliori G, Näsman P, Sundling C (2024) Comparing responses from a paper-based survey with a web-based survey in environmental criminology. Crime Prevent Community Saf 26(2):216–243

Clarke RV (1995) Situational crime prevention. Crime Justice 19:91–150

Demografiska statistikområden (DeSO) (2018) Statistics Sweden. Sweden. https://www.scb.se/hitta-statistik/regional-statistik-och-kartor/regionala-indelningar/deso---demografiska-statistikomraden/

Moher D, Shamseer L, Clarke M, Ghersi D, Liberati A, Petticrew M, Prisma-P Group (2015) Preferred reporting items for systematic review and meta-analysis protocols (PRISMA-P) 2015 statement. Syst Rev 4:1–9

Pawankar R, Canonica GW, Holgate ST, Lockey RF, Blaiss M (2013) WAO white book on allergy: update 2013. World Allergy Organization 248

Sundling C, Ceccato V (2022) The impact of rail-based stations on passengers' safety perceptions. A systematic review of international evidence. Transp Res Part F: Traffic Psychol Behav 86:99–120. https://doi.org/10.1016/j.trf.2022.02.011

Statistics Sweden (2020) Folkmängden per region efter ålder och kön. Retrieved from: https://www.statistikdatabasen.scb.se/pxweb/sv/ssd/START__BE__BE0101__BE0101Y/FolkmDesoAldKonN/

Steven PL (2014) Crime prevention: approaches, practices, and evaluations. Anderson

Swedish Transport Administration (2009) Number of passengers, Annual Average Daily Traffic (AADT) 2009

Swedish National Council for Crime Prevention, BRÅ (2021) Klassificering av brott: Riktlinjer för antalsräkning av brott (Version 9.2). https://bra.se/download/18.45e4b8e192705389a34e65/1729519891132/2021_Klassificering_av_brott_v%209_2.pdf

Part II
Stations, Places and Contexts

The Impact of the Environment on the Safety of Stations

5

Abstract

The aim of this chapter is to examine the influence of characteristics of both the physical and social environment on victimisation and safety perceptions of those individuals who use railway stations. For this investigation, we examine the influence of physical and social environment characteristics in different transportation settings after controlling for the individual characteristics of the travellers. We created a geodatabase across the study area to execute the analysis. Findings show that environmental factors have a role to play in shaping both risks of victimisation and safety perceptions in train stations and their surrounding areas. Poor illumination, lack of CCTV cameras, inadequate station maintenance, and the presence of drug-related activities contribute to heightened vulnerability and insecurity for passengers. Well-maintained stations with proper lighting, clear signage, and social amenities such as cafés enhance feelings of safety by fostering social interaction and surveillance. Addressing these environmental factors through targeted interventions can significantly improve transit safety, particularly for vulnerable groups such as women, LGBTQI + individuals, young passengers, nighttime travellers, and those with mobility limitations.

Keywords

Railbound · Stations · Transit environment · Illumination · Staff · CCTV · Maintenance · Tunnels

5.1 Introduction

Creating safe and inclusive transit environments is crucial for building sustainable cities and communities where individuals can access public transportation without fear. Safety affects individuals' well-being and influences their mobility choices and, consequently, the sustainability of transportation systems. If transit environments are perceived as unsafe or criminogenic, individuals may be deterred from using public transport, leading to increased reliance on less sustainable transportation options. According to two recent literature reviews (Ceccato et al. 2022; Sundling and Ceccato 2022), illumination and the design of transit environments are key factors, but the social environment also affects safety conditions in these environments. Guardians, attendants, and particularly place managers, such as station staff, oversee the area and its users, contributing to maintaining public order and ensuring safety (Eck 2019; Felson 1995). This evidence is primarily dominated by North American and Western European studies, which often reflect contexts and

V. Ceccato et al., *Transit Safety for a Sustainable City*, Sustainable Development Goals Series,
https://doi.org/10.1007/978-3-031-97427-4_5

environments that may differ significantly from those found in Nordic countries.

Therefore, this chapter examines the influence of physical and social environment characteristics in railway settings on the likelihood of riders' victimisation and perceptions of poor safety. Through fieldwork inspections and analyses of primary and secondary data, we seek to understand what makes a station (un)safe for different types of riders, thereby proposing strategies to improve transit safety and consequently, promoting sustainable transportation choices.

Victimisation in this study covers a range of incidents, including theft, violence, harassment, sexual crimes, and other forms of aggressive behaviour. These incidents are common in transit environments, particularly in and around stations, and significantly impact riders' perceived safety (Ceccato et al. 2022). Safety perceptions in this study refer to the fear of crime and other anxieties expressed by railway riders at stations and during their trips, which can vary over time. As discussed previously, low safety perceptions are influenced by various environmental factors and riders' characteristics (Sundling and Ceccato 2022).

This chapter is composed of several parts: the introduction, followed by the theoretical framework, research design, and a brief description of the study area. It also includes a section on data and methods, an analysis of victimisation characteristics and safety perceptions, a discussion of the results, and concludes with recommendations.

5.2 The Importance of the Station's Design and Environment

The design and environment of stations are crucial for ensuring the safety of passengers. A station's physical layout and overall environment can significantly influence passengers' perceptions of safety and their likelihood of becoming victims of crime. According to two recent systematic literature reviews, well-designed and well-lit stations have been shown to have lower crime rates and higher safety perceptions among passengers (Ceccato et al. 2022; Sundling and Ceccato 2022). The research shows that the design of a station can, for example, affect the flow of passengers and the visibility of potential criminal activities. Stations with open layouts and clear sightlines tend to have better surveillance and fewer hiding spots for potential criminals. In contrast, stations with closed or maze-like layouts may provide more opportunities for criminal activities to go unnoticed. Similarly, the presence of security features such as CCTV cameras, lighting, and security personnel can enhance passengers' sense of safety and deter potential criminals, especially in combination with other safety measures. Moreover, the social environment of the station, including the presence of staff and other passengers, also plays a crucial role in ensuring public order and safety. Station staff, including ticket agents and businesses, can act as guardians and provide a sense of security for passengers. However, the presence of other passengers can contribute to a false sense of safety, as the "bystander effect" suggests that a person witnessing an incident is less likely to intervene if others are present.

Previous evidence from research raises the question of visibility inside stations as an important factor, which can include wide-open design, which straightening sightlines provide, and avoiding obscured areas, corners, hiding spots, and enclosed spaces (e.g. Ceccato et al. 2011). In addition, plants should not become obstacles and impede the field of vision (Mohamed and Stanek 2020; Crime-Concern 2004; Diec et al. 2010). Visibility demands good illumination. In one seminal study by Richards (1980), 93% of respondents suggest lighting improvements to improve station safety. Stations that are well-lit and well-maintained contribute to the overall feeling of safety among metro users in Washington, DC (La Vigne 1996). Research also indicates that good lighting increases women's perceived safety more than men's, and results indicate that the effect of lighting on safety is dependent on type and height (Sundling and Ceccato 2022). Direct glare has a negative impact on perceived safety, as do different exit and entrance brightness levels. Brightness increases perceived safety until a certain threshold (Burnett and Pang 2004).

Surveillance is often highly valued by passengers, and in particular the presence of staff. It is, however, important that staff are easily available, either physically or electronically (Sundling and Ceccato 2022). Likewise, guards and police patrols, again, if they are available, often affect perceived safety in a positive way (Ceccato 2013; Sundling 2020). Moreover, remote surveillance, such as CCTV cameras, is often found to increase perceived safety. However, a prerequisite is that they are monitored by a member of staff (Thomas et al. 2006). Immediate help must be available. Otherwise, they are less valued in regard to perceived safety. Help buttons, emergency alarms, and safety apps (Libardo and Nocera 2012; McCarthy et al. 2016; Debrincat et al. 2017) may be other ways of getting in touch with staff when feeling unsafe. Real-time information (RTI) (Abenoza et al. 2018; Newton 2016) also increases safety perceptions at transit environments.

The impact of management and maintenance on safety perceptions is well documented. According to the "broken window theory" (Wilson and Kelling 1982), see Sect. 2.1, lack of maintenance may attract crime. Therefore, cleanliness contributes to safety, and safety is decreased when places are degraded (Loukaitou-Sideris 2014), especially for casual users (Vanier and de Jubainville 2017). Lack of maintenance may indicate other underlying problems (Eck 2019) which can affect both the perceived and actual safety of transit environments. When stations or transit areas are poorly maintained, it can signal to passengers that the space is neglected or unsafe, leading to increased fear of crime and decreased usage. In the next section, we discuss both individual and environmental factors that affect transit safety.

5.3 Factors Affecting Safety Perceptions

Safety perceptions are also a result of the environment, of what one sees and perceives with one's other senses, an individual's emotional reactions to a place at a particular moment in time and/or memories and associations that a particular setting brings to the surface (Wilhelmsson et al. 2022). For example, exposure to certain criminogenic places may lead to perceptions of lower safety levels. However, more importantly, there seems to be the presence of visual cues in the environment that, if unrepaired or missing, encourage crime and trigger low safety perceptions (Wilson and Kelling 1982). Signs of disorder such as litter, vandalism and loitering, and poor surveillance are thought to trigger fear of crime (Lewis and Maxfield 1980) and are indicators of more serious crimes.

Well-lit stations with clear sightlines and visible security measures, like CCTV cameras and security staff, are generally seen as safer (Sundling and Ceccato 2022). However, recent evidence shows that the colour of the light (especially blue and orange filters) seems to affect safety negatively in metro stations (Ceccato et al. 2025). Features like convex mirrors are said to increase visibility (Crime-Concern 2004) while tools such as pass systems, emergency alarms, crime-resistant materials, and real-time information (RTI) can reduce crime and may increase safety perceptions. High-quality CCTV, context-dependent surveillance, and body-worn cameras have also been proven to improve security and reduce assault incidents at stations (Debrincat et al. 2016).

The social environment, including the presence of staff and other passengers, also contributes to safety. Stations with visible staff and higher foot traffic are perceived as safer, while isolated, empty stations can increase feelings of vulnerability (Sundling and Ceccato 2022). Studies show that platforms near cafés or shops tend to be seen as safer, while secluded stations are perceived as more dangerous (Libardo and Nocera 2012). However, the presence of unruly or disruptive individuals can decrease safety perceptions (Kennedy 2008). Research also indicates that moderate crowds, particularly for women, contribute to perceptions of safety. This has been found in different parts of the world, such as India (Gopal and Shin 2019) as areas with more people tend to experience lower crime rates (Noble 2015).

However, too crowded places can induce anxiety as offenders can act anonymously. A study in Taiwan found crowding to be the most important reason for anxiety when travelling (Cheng 2010). Regular cleaning and frequent services, especially in the evenings, also help establish a safer environment (Ceccato et al. 2022).

Safety perceptions vary over time. Peak hours in transit environments can be perceived as unsafe for women (Mitra-Sarkar and Partheeban 2011) and for people with disabilities (Wayland et al. 2020). However, in a study by Yavuz and Welch (2010), peak-hour travelling was safer for men while this was not found to be true for women. Much of the research compares day and night travel. It is usually found that nighttime travel feels more unsafe than daytime travel, especially for women (e.g., Ceccato & Loukaitou-Sideris 2022). Also, at night, waiting times are generally longer, and longer waiting times have been found to increase the lack of safety (Fan et al. 2016). The same is true for travel time: the longer the travel time, the more people usually feel unsafe (Vilalta 2011).

5.4 Location and Area Characteristics

The location and area characteristics of railway stations can significantly influence safety perceptions and the occurrence of crime (Ceccato et al. 2022). These features include the socioeconomic status of the neighbouring area, the availability of amenities and commercial establishments, and the general urban setting where the station is located. Criminological literature has shown evidence that regions with higher poverty rates, unemployment, and deprivation may be more prone to crime, such as theft, vandalism, and assault. These areas may also have higher levels of drug and alcohol abuse, which can contribute to anti-social behaviour and criminal activity (Ceccato 2013).

The presence of amenities and businesses around stations can also influence safety perception and security. For example, stations near shopping centres, restaurants, and entertainment centres may experience higher levels of pedestrian traffic and social activity, contributing to a safer environment (Coppola and Silvestri 2020), but it may attract other types of crimes, such as pickpocketing. Conversely, stations in more isolated or industrial areas may have fewer amenities and less pedestrian traffic, making them more vulnerable to crime (Ceccato and Uittenbogaard 2014).

Stations in densely populated urban areas, characterised by high levels of pedestrian activity, often exhibit lower crime rates and enhanced safety due to the "eyes on the street" effect (Ceccato et al. 2024). This effect suggests that areas with more people present are less likely to be targeted by criminal activities, as there is a greater chance for potential offenders of being noticed. In remote or rural settings stations may experience a higher crime rate and lower levels of safety due to limited surveillance and social activity. These areas may also have fewer resources for law enforcement and social support services, which may contribute to higher criminogenic conditions.

In summary, international research indicates that factors such as lighting, maintanence, station design, crowd density, and security measures all directly impact how safe people feel at a station. For this case study, we drew on this research on transit safety to explore the following research questions:

- How does the station environment, as well as the journey to the station, affect victimisation?
- How are people's perceptions of safety influenced by the station environment and surrounding areas?
- What changes or recommendations are needed to improve safety of travellers in railway environments?

5.5 Study Area Research Design

The study covers 47 stations across three railway lines in central and southern Sweden, serving 28 municipalities with a population of 1.78 million (Statistics Sweden 2020). With 10.4 million people, Sweden has 87% of its population in urban areas. The study area is in the southern part of the country, with a higher population density. The second-largest city in Sweden, with about 580,000 inhabitants, Gothenburg, is part of the study area but not the capital. Stations are categorised by passenger flow: seven with over 2,000 daily passengers, 18 with fewer than 500, and 22 with 500–2,000 covering various contexts. To assess station environments, we systematically inspected all stations, including photographic documentation, and checked their surroundings in winter 2021, with a selected group inspected in summer 2022.

After approval from the Swedish Ethical Review Authority, data collection was carried from May to September 2022. We combined responses from two quasi-identical surveys, one print and one online, with 47 questions—the "To Promote Safe Stations" survey aimed to capture the importance of station environments. Posters and cards promoting the survey were displayed during fieldwork in the summer of 2022 and distributed in Facebook groups. Researchers also participated in radio programmes. The total sample was 4,893 riders ($N_{printsurvey} = 2{,}180$ and $N_{onlinesurvey} = 2{,}713$), but they were not representative of all stations. We also excluded 784 responses because riders did not travel within the study area, and we could not link the attributes of the stations with their responses. Additionally, we interviewed ten representatives of train operators and public officers between October and November 2021, and this information was used in the recommendations section.

5.6 Methods

The survey consisted of four parts: introductory questions about frequency of use, trip departure location, most frequent times of the trip, a main section on victimisation and safety perceptions of different types of trips and station environments, and according to the type of victimisation. The survey also included questions about the impact of crime and fear, recommendations for improving safety conditions along the trip, and a final section on the background information of the riders, such as gender, age, income, ethnic background, sexual orientation, and disabilities. The questions used in this analysis were:

- Victimisation: "In the past 5 years, have you been victimised by crime on the train, at the station, or on the way to/from the station?" (a list of crime types)
- Safety perceptions: "When travelling by train, do you often feel afraid of being exposed to the following?" (a list of crime types). Further questions are presented in the Appendix.

The data from the survey were transferred from the web-based platform via spreadsheet to a standard statistical software package, where the analyses were conducted. We used Geographical Information Systems (GIS) to map relevant land use and include variables that indicate the demographic and socioeconomic contexts of the stations, such as income and age. Three separate databases were combined into a large database. Given the large number of variables and aiming to obtain parsimonious models, we used exploratory data analysis and hypothesis testing (e.g., frequencies, cross-tables, correlation).

For the logistic regression models, a dichotomous dependent variable was created ("being victimised" and "poor safety perceptions" = 1, 0 = Otherwise). A 5% significance level was applied, and p-values were reported for significant results. Two sets of models were used to analyse victimisation and safety perceptions, with independent variables covering passengers' individual attributes, station characteristics, and surrounding areas. For an extended/complementary version of this study, see Ceccato et al. (2024).

5.7 Results

The environment of a station plays a significant role in shaping safety perceptions, even after accounting for individual, station-specific, and contextual factors. Positive elements, such as well-maintained facilities, restaurants, cafés, and adequate lighting, create a more welcoming and secure atmosphere, contributing to an overall sense of safety. These features provide practical benefits and promote visibility, social interaction, and a feeling of vitality, all of which can deter crime and reduce fear. On the other hand, stations with isolated areas, such as tunnels or poorly lit spots, are often perceived as unsafe. These spaces, lacking in social presence or proper maintenance, can be seen as vulnerable to criminal activities and heighten feelings of insecurity among travellers (Fig. 5.1). Before we further discuss these findings in detail, we start by describing the main types of safety problems and when and where they happen.

Vandalism was the most common crime around stations (42%) reported to the police, followed by property crimes, mainly theft (33%). Open drug markets and drug or weapon-related offences were also significant (21%). Crime patterns varied by season, day, and time, peaking in summer and autumn, with lower levels in winter. Most incidents occurred in the evening (20:00–00:00), particularly on Fridays and Saturdays.

Of the surveyed travellers, 14% (N = 519) reported victimisation while travelling by train in the past five years. Nearly half of the incidents occurred on the way to the station (47%), one-third at the station (34%), and the fewest on the train (19%). The most frequently reported crimes in the survey were stalking (6%), sexual harassment (5%), and property crimes (5%). However, if considering aggressive panhandling, this is the most commonly reported incident by respondents (14%). The modelling results for respondents' victimisation and safety perceptions are presented in Tables 5.1 and 5.2.

Thus, we see that the police data identify vandalism as the most prevalent crime, focusing on property-related offences around stations while survey data shows aggressive panhandling as the most common issue reported by travellers, with a significant emphasis on personal victimisation such as stalking and sexual harassment, which are not as frequently noted in police data. These differences suggest that personal safety concerns might be under-reported in official crime statistics, with travellers more likely to report victimisation and harassment through surveys than through formal police recording channels.

5.7.1 Victimisation

Table 5.1 shows the results of individuals' victimisation (a) in the station (b) and on the path to the station (c). Drug use or open drug markets around stations, stations located in isolated areas, lack of staff at stations, and the presence of restaurants or cafés were significant factors affecting victimisation. Isolated stations increased the odds of victimisation, highlighting the importance of station proximity to other facilities and populated areas. Drug-related activities significantly increased the risk of victimisation, as did insufficient staff presence at stations. On the other hand, having a restaurant or café in the station reduced the risk of victimisation. Individuals' characteristics, such as age and gender, were also important in explaining the likelihood of victimisation. Victimisation was significantly higher among individuals aged 18 years or younger and women. Respondents with reduced mobility were also more likely to be victimised, as were those travelling at night and those travelling more frequently. Riders who did not use cars, who were transit-dependent, were also more likely to be victimised.

5.7.2 Safety Perceptions

Table 5.2 indicates the results of respondents' poor safety perceptions (a) at the station (b) and on the way to the station (c). Previous victimisation affected riders' safety perceptions, in other words, respondents previously victimised in transit were more likely to feel unsafe than those who were not victimised. LGBTQI + /non-binary/ other riders were more likely to feel unsafe at the station and on the way to or from the train. Women also felt more unsafe, particularly on the way to the station, where they were more likely to be victimised. Stations with tunnels were perceived as being more unsafe, as were those stations where riders experienced a lack of staff. Drug use or open drug markets around stations doubled the odds of

Fig. 5.1 The environment of a station influences safety, even after accounting for individual, station, and contextual factors. Restaurants, cafés, good lighting, and maintenance positively contribute to the safety conditions of the station. In contrast, tunnels and isolated stations are seen to negatively impact safety

riders feeling unsafe at the station. Overall, the station environment has a more substantial influence on travellers' safety perceptions than on their actual victimisation.

5.8 Discussion

The importance of good management and maintenance in transit safety is well documented, with poor cleaning practices and a lack of staff contributing to reduced safety perceptions and an increased likelihood of victimisation at stations. Victimisation was found to be less dependent on station environmental conditions compared to individuals' safety perceptions. Poor safety perceptions, often associated with visual signs such as poor maintenance, is more prevalent than victimisation. Poor safety perceptions may stem from various factors, including aggressive panhandling, open drug markets, or station vandalism. Interestingly, while many incidents occurred on platforms, trains, or on the way to stations, riders felt often unsafe in the tunnels leading to stations and in peripheral areas such as bus terminals and car parks. These isolated spaces can make riders feel that they are easy targets for criminals, even if they have not been victimised there. The literature often identifies long corridors with sharp corners and restricted sightlines as unsafe places, indicating that impaired visibility is closely linked to declared poor safety perceptions.

The study's findings show the complexity of transit environments and their impact on safety. There is no single physical or social characteristic that can solve safety issues, but safe stations share common features such as good natural surveillance, illumination, and the presence of people, both staff and passengers. Riders' victimisation and safety perceptions are influenced by gender and other individual factors, emphasising the need for a nuanced approach to transit safety. Specific groups, such as women, LGTBQI + individuals, and those with reduced mobility, have unique safety needs that must be addressed, as discussed in other chapters of this book.

The study suggests that future research should explore the interaction effects between individual characteristics and station conditions, particularly a lack of staff and poor maintenance. Moreover, as travel services become more digitalised and automated, tailored policies are needed to address the safety needs of specific groups of transit

Table 5.1 Logistic regression results of Y = Victimisation (a), in the station (b) and on the path to the station (c)

Y = Victimised = 1, otherwise, 0		Victimisation (a) N = 313				Victimisation at the station (b) N = 195				Victimisation on the way to the station (c) N = 231			
		OR	C.I. 95%		Sig.	OR	C.I. 95%		Sig.	OR	C.I. 95%		Sig.
Individual characteristics	Woman	**1.547**	**1.166**	**2.053**	**0.003**	1.222	0.870	1.718	0.247	1.547	1.166	2.053	0.061
	Young (under 18)	**1.696**	**1.095**	**2.627**	**0.018**	1.636	0.954	2.807	0.074	1.696	1.095	2.627	0.235
	LGTBQI + / Non-binary/ Other	1.368	0.756	2.476	0.301	0.645	0.276	1.508	0.312	1.368	0.756	2.476	0.463
	Foreign born	1.125	0.740	1.709	0.581	1.054	0.615	1.807	0.847	1.125	0.74	1.709	0.672
	Frequent traveller	**1.363**	**1.042**	**1.782**	**0.024**	**1.701**	**1.231**	**2.350**	**0.001**	**1.363**	**1.042**	**1.782**	**0.002**
	Travel during nighttime	**1.621**	**1.252**	**2.098**	**0.000**	**1.728**	**1.248**	**2.392**	**0.001**	**1.621**	**1.252**	**2.098**	**0.000**
	University education	**0.701**	**0.537**	**0.917**	**0.009**	**0.657**	**0.472**	**0.916**	**0.013**	**0.701**	**0.537**	**0.917**	**0.003**
	Reduced mobility	**1.776**	**1.103**	**2.857**	**0.018**	**2.275**	**1.324**	**3.908**	**0.003**	1.776	1.103	2.857	0.203
	Transit captive	**1.293**	**1.004**	**1.665**	**0.047**	1.277	0.934	1.746	0.126	1.293	1.004	1.665	0.213
Station's characteristics	Restaurant/ café	0.851	0.575	1.261	0.423	**0.532**	**0.322**	**0.880**	**0.014**	0.851	0.575	1.261	0.249
	Toilet (pay)	1.082	0.725	1.615	0.700	1.605	0.954	2.700	0.075	1.082	0.725	1.615	0.253
	Poor illumination	0.843	0.641	1.109	0.223	0.993	0.711	1.388	0.969	0.843	0.641	1.109	0.509
	No. CCTV	0.919	0.785	1.074	0.288	1.055	0.870	1.279	0.584	0.919	0.785	1.074	0.192
	No. operators	0.996	0.893	1.110	0.938	0.926	0.811	1.057	0.254	0.996	0.893	1.110	0.615
	Lack of staff	1.189	0.919	1.539	0.187	**1.787**	**1.288**	**2.480**	**0.001**	1.189	0.919	1.539	0.135
	Poor maintenance	1.133	0.849	1.511	0.396	1.208	0.858	1.701	0.278	1.133	0.849	1.511	0.358
Station's location & area characteristics	Tunnel	1.377	0.986	1.925	0.061	1.207	0.790	1.845	0.385	1.377	0.986	1.925	0.175
	Isolated station	**1.334**	**1.012**	**1.759**	**0.041**	1.259	0.898	1.766	0.181	1.334	1.012	1.759	0.106
	Drug use/ selling at station	**1.573**	**1.203**	**2.057**	**0.001**	**2.062**	**1.485**	**2.864**	**0.000**	**1.573**	**1.203**	**2.057**	**0.000**

OR Odds ratio; *CI* Confidence interval; *Sig.* *p*-value; statistically significant at the 5% level at most
Source Adapted from Ceccato et al. (2024), p. 9

riders. These policies should prioritise the safety of women, LGTBQI + individuals, those who travel at night, and those with reduced physical mobility.

Safety conditions in transit nodes differ from those experienced or perceived on the way to them. Fixing victimisation does not necessarily solve issues related to poor safety perceptions. The lack of a central system for recording incidents means that interventions by safety experts and planners are not evidence-based and this is a problem because resources may be misallocated and priorities set without a clear understanding of where and what types of incidents are most prevalent. Without systematic data, it is challenging to develop targeted strategies to improve

Table 5.2 Logistic regression results of Y = Poor safety perception (a), at the station (b), and on the way to the station (c)

Y = Poor safety perception = 1, otherwise, 0		Poor safety perception (a) N = 871				Poor safety perception at the station (b) N = 637				Poor safety perception on the way to the station (c) N = 648			
		OR	C.I. 95%		Sig.	OR	C.I. 95%		Sig.	OR	C.I. 95%		Sig.
Individual characteristics	Woman	**2.075**	**1.664**	**2.587**	**0.000**	**1.847**	**1.451**	**2.352**	**0.000**	**2.502**	**1.961**	**3.193**	**0.000**
	Young (under 18)	0.912	0.610	1.365	0.655	1.047	0.689	1.590	0.829	0.976	0.646	1.474	0.909
	LGTBQI + / Non-binary/ Other	**3.103**	**1.920**	**5.014**	**0.000**	**3.396**	**2.076**	**5.554**	**0.000**	**3.274**	**2.000**	**5.360**	**0.000**
	Foreign born	0.896	0.637	1.260	0.528	0.847	0.582	1.233	0.386	1.063	0.744	1.519	0.738
	Frequent traveller	**1.342**	**1.078**	**1.671**	**0.008**	1.180	0.934	1.492	0.165	1.247	0.991	1.569	0.060
	Travel during nighttime	**1.590**	**1.304**	**1.938**	**0.000**	**1.420**	**1.146**	**1.759**	**0.001**	**1.692**	**1.369**	**2.091**	**0.000**
	University education	**0.659**	**0.532**	**0.815**	**0.000**	**0.703**	**0.561**	**0.883**	**0.002**	**0.597**	**0.478**	**0.747**	**0.000**
	Reduced mobility	1.199	0.775	1.857	0.415	1.230	0.782	1.937	0.370	0.998	0.633	1.572	0.992
	Victim to crime	**4.985**	**3.721**	**6.677**	**0.000**	**4.656**	**3.538**	**6.129**	**0.000**	**3.416**	**2.608**	**4.473**	**0.000**
	Transit captive	0.993	0.809	1.219	0.946	0.993	0.797	1.237	0.953	1.071	0.864	1.329	0.531
Station's characteristics	Restaurant/ café	**0.603**	**0.442**	**0.824**	**0.001**	**0.641**	**0.459**	**0.897**	**0.009**	**0.607**	**0.437**	**0.841**	**0.003**
	Toilet (pay)	1.045	0.763	1.431	0.786	1.159	0.822	1.633	0.399	0.925	0.665	1.288	0.646
	Poor illumination	1.103	0.889	1.367	0.373	1.060	0.844	1.332	0.617	**1.263**	**1.010**	**1.580**	**0.041**
	No. CCTV	**0.837**	**0.739**	**0.947**	**0.005**	**0.829**	**0.727**	**0.946**	**0.005**	0.911	0.799	1.038	0.162
	No. operators	**1.116**	**1.025**	**1.217**	**0.012**	**1.135**	**1.035**	**1.244**	**0.007**	1.000	0.967	1.159	0.220
	Lack of staff	**2.477**	**2.023**	**3.033**	**0.000**	**2.454**	**1.973**	**3.052**	**0.000**	**2.036**	**1.644**	**2.521**	**0.000**
	Poor maintenance	**1.452**	**1.151**	**1.832**	**0.002**	**1.578**	**1.240**	**2.008**	**0.000**	**1.288**	**1.012**	**1.640**	**0.039**
Station's location & area characteristics	Tunnel	**2.030**	**1.562**	**2.637**	**0.000**	**2.118**	**1.583**	**2.833**	**0.000**	**2.205**	**1.655**	**2.937**	**0.000**
	Isolated station	1.060	0.848	1.325	0.608	1.127	0.891	1.424	0.318	1.227	0.973	1.545	0.081
	Drug use/ selling at station	**1.852**	**1.496**	**2.292**	**0.000**	**2.024**	**1.618**	**2.533**	**0.000**	**1.752**	**1.402**	**2.190**	**0.000**

OR Odds ratio *CI* Confidence interval; *Sig. p*-value; statistically significant at the 5% level at most
Source Adapted from Ceccato et al. (2024), p. 10

transit safety. A centralised system for recording incidents is essential for planning the safety of train riders in a better way and ensuring a more sustainable future.

5.9 Conclusions and Recommendations

Police data highlights vandalism as the most common crime problem, while survey data reveal that aggressive panhandling and personal victimisation (e.g., stalking, sexual harassment) as being the most problematic issues. This suggests that personal safety concerns may be under-reported in official crime statistics, as passengers are more likely to disclose such incidents in surveys than to the police. A unified reporting system is crucial to capturing these unrecorded incidents. Without it, these problems remain invisible in official data, preventing targeted interventions. This lack of information leaves those working with transit safety operating in the dark, unable to address key risks effectively.

The analysis shows that individual characteristics and environmental factors influence the risk of victimisation at train stations and on the way to them. Women, young individuals (under 18), LGBTQI + /Non-binary/Other individuals, frequent passengers, nighttime travellers, and those with reduced mobility are at higher risk in both contexts. However, environmental conditions play a crucial role, with poor illumination and station maintenance increasing the risk of victimisation inside stations. At the same time, other environmental factors (e.g, the lack of CCTV cameras and the presence of drug-related activities) seem to further elevate risks on the way to the station. These findings indicate that while individual vulnerabilities matter, improving station and journey conditions, such as enhancing lighting, maintenance, and surveillance, can significantly reduce victimisation and improve safety perceptions.

Environmental factors significantly shape safety perceptions at and on the way to train stations. Poor illumination, the absence of CCTV cameras, and a lack of staff contribute to a heightened sense of insecurity in both contexts. Additionally, stations with isolated areas or tunnels are associated with lower safety perceptions, as these features limit natural surveillance and increase vulnerability. The presence of drug use or selling spots near stations further exacerbates safety concerns, reinforcing the importance of addressing social disorder in transit environments. However, specific amenities can enhance perceived safety, such as stations with cafés or restaurants foster a sense of security by increasing social actvity and natural surveillance. These findings emphasise the critical role of well-maintained, well-lit, and actively monitored station environments in improving passengers' safety perceptions and confidence in public transportation. What travellers see (and do not see) affects their feelings of safety, along with comfortable waiting areas and clear signage, all of which help create a safer and reassuring transit experience. Safety interventions should prioritise the most vulnerable groups in order to create safer transit environments. This would ensure that women, LGBTQI + individuals, young passengers, nighttime travellers, and those with mobility limitations feel safer using public transportation.

Safety perceptions in railway environments depend on passenger characteristics and environmental features. Crime-free stations may feel unsafe due to poor maintenance or a lack of staff. Research suggests that these interactions shape perceived safety, highlighting the need to further explore how passenger profiles influence safety concerns. Some of these unanswered questions are further investigated in the following chapters of this book.

The absence of a unified incident reporting system means that current interventions by safety experts and planners lack a solid evidence base. While some transit operators have developed their databases, these are often limited to specific stations and rarely shared across networks, making it difficult to generalise findings and implement effective solutions nationwide. As our study shows, relying solely on police statistics to understand station-related crime can be misleading, as only serious offences tend to be reported. Minor incidents of incivilities or public disturbances—while not classified as crimes—significantly impact people's sense of safety and often go unnoticed. In other rail-bound transit systems, between 80 and 90% of recorded incidents reported by travellers and staff are not crimes but still affect the overall perception of safety. A well-structured and centralised incident-reporting system is essential for informed decision-making, better safety planning, and the long-term sustainability of public transportation.

While reducing crime in transit is essential, policymakers must also address travellers' declared poor safety perceptions through continuous safety surveys, community engagement, better communication with stakeholders, and environmental improvements. Well-lit spaces, visible staff presence, and clear signage can enhance passengers' safety perceptions, reducing anxiety even in stations with low crime rates.

Additionally, improving stakeholder coordination from national to local levels is fundamental for effective safety interventions. Establishing accurate, updated databases on passenger victimisation and safety concerns would provide valuable insights for evidence-based policies. Tailoring prevention measures to the specific needs of different passenger groups—such as women, LGBTQI + individuals, and night-time travellers—ensures that safety initiatives are inclusive and sustainable. By integrating these approaches, policymakers can create safer, more welcoming transit environments that encourage mobility for all.

References

Abenoza RF, Ceccato V, Susilo YO, Cats O (2018) Individual, travel, and bus stop characteristics influencing travelers' safety perceptions. Transp Res Rec 2672(8):19–28

Burnett J, Pang A (2004) Design and performance of pedestrian subway lighting systems. Tunn Undergr Space Technol 19:619–628

Ceccato V (2013) Moving safely: crime and perceived safety in Stockholm's subway stations. Lexington, Plymouth

Ceccato V, Loukaitou-Sideris A (2022) Fear of sexual harassment and its impact on safety perceptions in transit environments: a global perspective. Violence against Women 28(1):26–48

Ceccato V, Uittenbogaard AC (2014) Space-time dynamics of crime in transport nodes. Ann Assoc Am Geogr 104(1):131–150

Ceccato V, Uittenbogaard A, Bamzar R (2011) Security in Stockholm's underground stations: the importance of environmental attributes and context. Secur J. https://doi.org/10.1057/sj.2011.32

Ceccato V, Gaudelet N, Graf G (2022) Crime and safety in transit environments: a systematic review of the English and the French literature, 1970–2020. Public Transport 14(1):105–153

Ceccato V, Sundling C, Gliori G (2024) What makes a railway station safe and for whom? The impact of transit environments on passengers' victimisation and safety perceptions. Eur Transp Res Rev 16(1):21

Ceccato V, Besenecker U, Ariel B, Eizadi H, Höglund M, Näsman P (2025) Assessing safety perceptions and lighting conditions in a metro station. J Exp Criminol 1–27. https://doi.org/10.1007/s11292-025-09683-8

Cheng Y-H (2010) Exploring passenger anxiety associated with train travel. Transportation 37:875–896

Coppola P, Silvestri F (2020) Assessing travelers' safety and security perception in railway stations. Case Stud Transport Policy 8(4):1127–1136

Crime Concern (2004) People's perceptions of personal security and their about crime on public transport: research findings

Debrincat M, Dupart C, Laurent C (2016) Étude sur le harcèlement sexiste et les violences sexuelles faites aux femmes dans les transports publics. FNAUT, Paris

Debrincat M, Dupart S, Moggio C (2017) Harcèlement sexiste dans les transports collectifs routiers et les pôles d'échanges multimodaux: L'analyse de la FNAUT. FNAUT, Paris, France

Diec J, Coxon S, De Bono A (2010) Designing a public train station shelter to minimise anti-social behaviour and crime in Melbourne's metropolitan rail environment

Eck JE (2019) Place managers and crime places. In: Oxford Research Encyclopedia of Criminology and Criminal Justice

Fan Y, Guthrie A, Levinson D (2016) Waiting time perceptions at transit stops and stations: effects of basic amenities, gender, and security. Transp Res Part a: Policy Pract 88:251–264

Felson M (1995) Those who discourage crime. Crime Place 4(3):53–66

Gopal K, Shin EJ (2019) The impacts of rail transit on the lives and travel experiences of women in the developing world: evidence from the Delhi Metro. Cities 88:6–75. https://www.scopus.com/inward/record.uri?eid=2-s2.0-85059830906&doi=10.1016%2fj.cities.2019.01.008&partnerID=40&md5=b8df07810bc8278b1a4358b39ef4414a

Kennedy, D (2008) Personal security in public transport travel in New Zealand: Problems, issues & solutions. Land Transport New Zealand Research Report, 344.

La Vigne NG (1996) Safe transport: security by design on the Washington Metro. Preventing Mass Transit Crime 6:163–197
Lewis DA, Maxfield MG (1980) Fear in the neighborhoods: an investigation of the impact of crime. J Res Crime Delinq 17(2):160–189. https://doi.org/10.1177/002242788001700203
Libardo A, Nocera S (2012) Exploring the perceived security in transit: the Venetian students' perspective. WIT Trans Ecol Environ 155:943–954
Loukaitou-Sideris A (2014) Fear and safety in transit environments from the women's perspective. Secur J 27(2):242–256
McCarthy OT, Caulfield B, O'Mahony M (2016) How transport users perceive personal safety apps. Transport Res f: Traffic Psychol Behav 43:166–182
Mitra-Sarkar S, Partheeban P (2011) Abandon all hope, ye who enter here: understanding the problem of "eve teasing" in Chennai, India. In: Transportation Research Board Conference Proceedings, vol 2, issue 46
Mohamed AA, Stanek D (2020) The influence of street network configuration on sexual harassment patterns in Cairo. Cities 98:102583
Noble J (2015) L'insécurité personnelle des jeunes dans les transports en commun franciliens: une étude dispositionnelle du sentiment d'insécurité Université Sorbonne Paris Cité]
Newton, A (2016). Transport technology and crime. In: The handbook of technology, crime and justice. Routledge, Abingdon 2017, pp 281–294
Richards LJI (1980) Passenger value structure model. Final Report DOT Contract TSC-1314. Dunlap and Associates, Inc., Darien
Statistics Sweden (2020) Folkmängden per region efter ålder och kön. Retrieved from: https://www.statistikdatabasen.scb.se/pxweb/sv/ssd/START__BE__BE0101__BE0101Y/FolkmDesoAldKonN/
Sundling C (2020) Huddinge, Sweden. In: Transit Crime and Sexual Violence in Cities. Routledge, pp 167–175
Sundling C, Ceccato V (2022) The impact of rail-based stations on passengers' safety perceptions. A systematic review of international evidence. Transport Res f: Traffic Psychol Behav 86:99–120
Thomas LJ, Rhind DJ, Robinson KJ (2006) Rail passenger perceptions of risk and safety and priorities for improvement. Cogn Technol Work 8:67–75
Vanier C, De Jubainville HD (2017) Feeling unsafe in public transportation: a profile analysis of female users in the Parisian region. Crime Prevent Community Saf 19(3–4):251–263
Vilalta CJ (2011) Fear of crime in public transport: research in Mexico City. Crime Prevent Community Saf 13:171–186
Wayland S, Newland J, Gill-Atkinson L, Vaughan C, Emerson E, Llewellyn G (2020) I had every right to be there: discriminatory acts towards young people with disabilities on public transport. Disabil Soc 1–24
Wilhelmsson M, Ceccato V, Gerell M (2022) What effect does gun-related violence have on the attractiveness of a residential area? The case of Stockholm, Sweden. J Eur Real Estate Res 15(1):39–57
Wilson JQ, Kelling GL (1982) Broken windows. Atl Mon 249:29–38
Yavuz N, Welch EW (2010) Addressing fear of crime in public space: gender differences in reaction to safety measures in train transit. Urban Stud 47(12):2491–2515

6 Does Socio-economic Context Influence Transit Safety During Transit?

Abstract

This Chapter examines how the characteristics of individuals' residential areas and the conditions surrounding transit stations shape their experiences of victimisation and perceptions of safety. By analysing the influence of individual and socioeconomic factors, it seeks to identify disparities in safety perceptions and victimisation risks along the journey—from home to the departure station and onward to the destination—using southern Sweden as an example based on 3,407 survey responses. The findings suggest that the environment surrounding the station where a trip originates has a greater influence on travellers' safety perceptions than the conditions at the destination station or their place of residence. These results highlight the need for enhanced stakeholder collaboration to address travellers' safety concerns at different stages of their journey and to better understand how different groups adapt their routines to mitigate potential risks.

Keywords

Whole journey approach · Place of residence · Stations · Survey

6.1 Introduction

Transit safety is context-dependent, meaning that individuals' perceptions of safety and their likelihood of experiencing victimisation are influenced by both personal factors (such as gender, age, previous experiences, and socioeconomic background) and situational factors (such as the quality of the physical social environment or the time of day or the year). However, traditional safety measures often focus on isolated segments of a trip, such as just the station or a specific segments of the journey. This fragmented approach to the trip overlooks the broader context of how safety risks and perceptions evolve throughout the transit experience. When the whole journey is not considered, critical aspects in which individuals become more vulnerable may be missed—such as the journey from home to the station—leading to incomplete knowledge about what happens to travellers. By adopting a whole journey perspective (Loukaitou-Sideris 2012; Natarajan et al. 2017), travellers' needs can be better addressed at every stage of their trip and ensure a more comprehensive approach to reducing victimisation risks and enhancing safety perceptions.

Ceccato et al. (2021) suggest that transit safety along the trip is gendered and may differ from different types of transportation models—buses or trains. For instance, they show that exposure to high-crime environments on the way to the metro significantly reduces young people's safety perceptions (but not on the way to bus stops),

V. Ceccato et al., *Transit Safety for a Sustainable City*, Sustainable Development Goals Series,
https://doi.org/10.1007/978-3-031-97427-4_6

especially for men. These results are consistent with previous studies in the US that showed that students become victims of crimes along the trip, affecting, therefore, their safety perceptions on the way to or from the station/bus stops (Wiebe et al. 2014).

This Chapter explores how residential area characteristics and station conditions influence individuals' victimisation experiences and safety perceptions. Using data from 3,407 survey responses in southern Sweden, it examines how individual and socioeconomic factors shape safety perceptions and victimisation risks from home to the station of destination. Using descriptive statistics (t-test and correlation), we compare victimisation, for example, to determine whether there is a statistically significant difference between them in the place of residence, then at the station of origin, and at the station of destination. Correlation measures are used to indicate the strength and direction of the relationship between the socioeconomic variables in different locations along the trip, indicating whether and how they change together. Finally, we use logistic regression models to assess socioeconomic factors shaping safety perceptions and victimisation risks from home to the destination station after controlling travellers' personal characteristics.

The Chapter starts with a theoretical background which is followed by a description of the case study with results showing the impact of the station and resident contexts on victimisation and perceived safety. The Chapter finishes with a discussion of the results and conclusions.

6.2 Theoretical Background

Victimisation and the fear of crime are deeply influenced by both the characteristics of the individual (Chapters 10, 11, 12, and 13) and environmental factors (Chapter 5), including the broader neighbourhood conditions. Research on this phenomenon has highlighted how factors such as social vulnerability and community cohesion in a station's surrounding areas contribute to travellers' victimisation and increased levels of fear (see e.g., Ceccato 2013).

Neighbourhood conditions are crucial in shaping individuals' perceptions of safety. While higher crime rates are often linked to increased fear, research has shown that fear is not solely determined by crime rates or the type of crime in an area. For example, Lewis and Maxfield (1980) found that residents of neighbourhoods with lower crime rates can still experience higher levels of fear, primarily due to perceptions of disorder or "incivility". This suggests that crime is not always the direct cause of fear; rather, factors such as socio-economic disadvantage and lack of community cohesion can intensify feelings of insecurity. In fact, neighbourhoods marked by disorder or economic inequality often foster greater levels of fear, even when crime rates are low (Pain 2000; Rader et al. 2012).

At the macro level, larger societal factors such as income inequality and social protection play significant roles in shaping fear of crime. Wilkinson and Pickett (2010) argue that income inequality is a powerful predictor of various social problems, including fear of crime. They propose that societies with greater income inequality foster a sense of social division, where status competition becomes more intense and those at the bottom of the social ladder are perceived as more likely to engage in criminal behaviour. Other studies have also found a strong association between income inequality and fear of crime (Hummelsheim et al. 2011; Vieno et al. 2013), supporting the idea that inequality intensifies fears of violence and crime by signalling social instability.

Additionally, factors such as economic disadvantages, immigrant concentrations, and social capital have been identified as contributing to variations in crime and safety perceptions. For instance, research indicates that greater economic disadvantage, such as high unemployment rates, correlates with an increased fear of crime (Franklin et al. 2008; Roman and Chaflin 2008), although findings can be inconsistent (Visser et al. 2013). Structural factors, including the concentration of immigrants in certain neighbourhoods, have also been linked to heightened fear, particularly when communities feel threatened by perceived cultural or economic differences (Sampson and Raudenbush 1997; Wang 2012). A

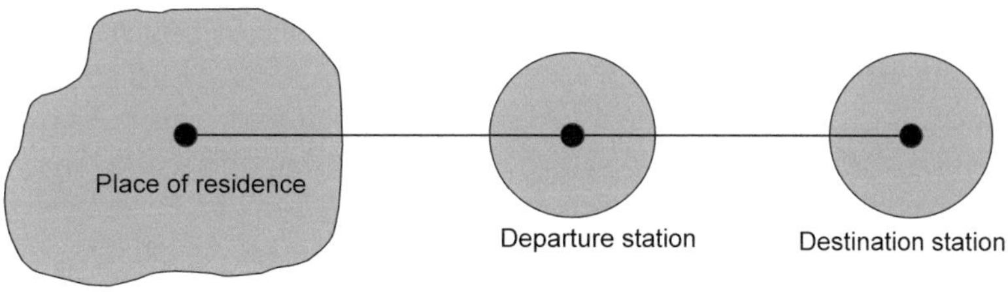

Fig. 6.1 The context of travellers: place of residence, departure station and destination along the trip

lack of social capital, or community ties, further increases fear, as individuals in communities with low trust and cohesion often feel more vulnerable and less secure (Ferguson and Mindel 2007).

When examining fear of crime at the level of train stations, which serve as spaces of transition between neighbourhoods, these neighbourhoods and macro-level factors are particularly important. Train stations are often situated in areas that are socioeconomically diverse, with varying levels of crime and social cohesion. The broader socioeconomic context of the surrounding neighbourhoods—characterised by economic inequality, immigration concentration, or social fragmentation—can influence how safe commuters feel. For example, train stations located in economically disadvantaged areas may experience higher levels of fear, even if actual crime rates are low, because individuals perceive these areas as disorderly or unsafe. As suggested by Brunton-Smith and Sturgis (2011), fear of crime is influenced not only by the direct experience of crime but also by the broader social and economic conditions that individuals encounter in their communities.

Understanding how the socioeconomic context and neighbourhood conditions surrounding train stations affect fear of crime is important when it comes to understanding that fear of crime goes beyond the train stations themselves, with the surrounding socioeconomic conditions playing a key role.

6.3 The Case Study

This Chapter explores how travellers' residential area characteristics and station conditions influence individuals' victimisation experiences and fear of crime (Fig. 6.1). The study aims to address the following questions:

1. How do travellers' places of residence and local socioeconomic contexts influence victimisation and safety perceptions during train travel?
2. Does the context of the stations affect victimisation and safety perception among travellers?
3. How does the context of the destination station influence travellers' victimisation and safety perception during the train journey?

6.3.1 Study Area

The study focuses on 47 stations located across three railway lines in central and southern Sweden, serving 28 municipalities with a combined population of 1.78 million (Statistics Sweden 2020). The second-largest city in Sweden, Gothenburg, with a population of about 580,000, is included in the study area, though the capital, Stockholm, is not. Additionally, this particular study extends beyond the 28 municipalities to consider where people reside, capturing how the local socioeconomic contexts there influence their experiences of victimisation and safety perceptions along the trip (Figure 6.3 in Appendix A).

6.3.2 Data

Data collection was carried out from May to September 2022, following approval from the Swedish Ethical Review Authority. The study used a single survey entitled "To Promote Safe Stations", which consisted of 47 questions. This survey aimed to understand the importance of station environments and was available in both printed and online formats. To encourage participation, posters and cards were displayed at stations during the fieldwork, and the survey was also promoted through Facebook groups and radio programmes. The final sample included 4,893 travellers (2,180 from the print survey and 2,713 from the online survey).

Among these, 3,484 respondents provided their postal codes, which allowed for spatial mapping of their locations. This enabled the integration of socioeconomic data, such as population density, unemployment rates, income levels, ethnic background, and demographic composition, sourced from Statistics Sweden, for further analysis.

6.3.3 Methods

Socioeconomic data at the DeSO level (one of the smallest official units of analysis available nationally in Sweden) was mapped using GIS and linked to the respondent's location of residence. For the analysis around train stations, a 200-m buffer zone was created around each station and intersected with the socioeconomic data to define the station area. Since some station areas overlap multiple DeSO areas, the proportion of each DeSO area within the buffer zone was calculated based on its size. This proportion for each DeSO was then used to estimate each station area's demographics and socioeconomic conditions. Figure 6.2 illustrates an example of two survey respondents, the different residential contexts they live in, their routes to the station, and the contexts of the departure/destination station.

Independent-sample t-tests assessed whether significant differences existed between groups based on selected socioeconomic variables (Appendix B). The groups for comparison were those who had been victimised by crime and those who had not, as well as those who feared crime and those who did not. To assess the magnitude and direction of the difference, Cohen's d was calculated as a measure of effect size, indicating the standardised difference between the two means (Cohen 1988). A positive d means Group 1 has a higher mean than Group 2, while a negative d suggests the opposite. Cohen classified effect sizes as small ($d = 0.2$), medium ($d = 0.5$) and large ($d \geq 0.8$). Additionally, bivariate correlation analysis was used to examine the association between the variables and reinforce the analysis (Appendix C). Later, logistic regression was used to control for travellers' individual characteristics. This method allowed for assessing the independent effect of socioeconomic conditions on fear of crime while accounting for travellers' personal characteristics such as gender, age, previous victimisation, and mobility constraints.

The questions used in the survey were:

- Victimisation: "In the past 5 years, have you been victimised by crime on the train, at the station, or on the way to/from the station?" (a list of crime types)
- Safety perception: "When travelling by train, do you often feel afraid of being exposed to the following?" (a list of crime types with location selections: at the station, on the way to the station, and on the train)

6.4 Results

6.4.1 The Impact of the Traveller's Place of Residence

6.4.1.1 Victimisation

According to the descriptive analyses (Tables 6.8 and 6.9, Appendix B and Table 6.15, Appendix C), victimised individuals tended to live in areas with lower population density, socio-economic conditions, and education levels than non-victimised individuals. While income and unemployment rates showed no significant differences overall, those who had experienced victimisation resided closer to the train stations. The proportion of residents with a foreign background was slightly higher in areas where victimised riders lived. Expanding the definition of victimisation to include aggressive panhandling made differences in the effect of the neighbourhood variables foreign background and unemployment, which became more pronounced, while population density ceased to be a factor.

The logistic regression analysis of residential area characteristics shows that most factors had no apparent impact on victimisation in

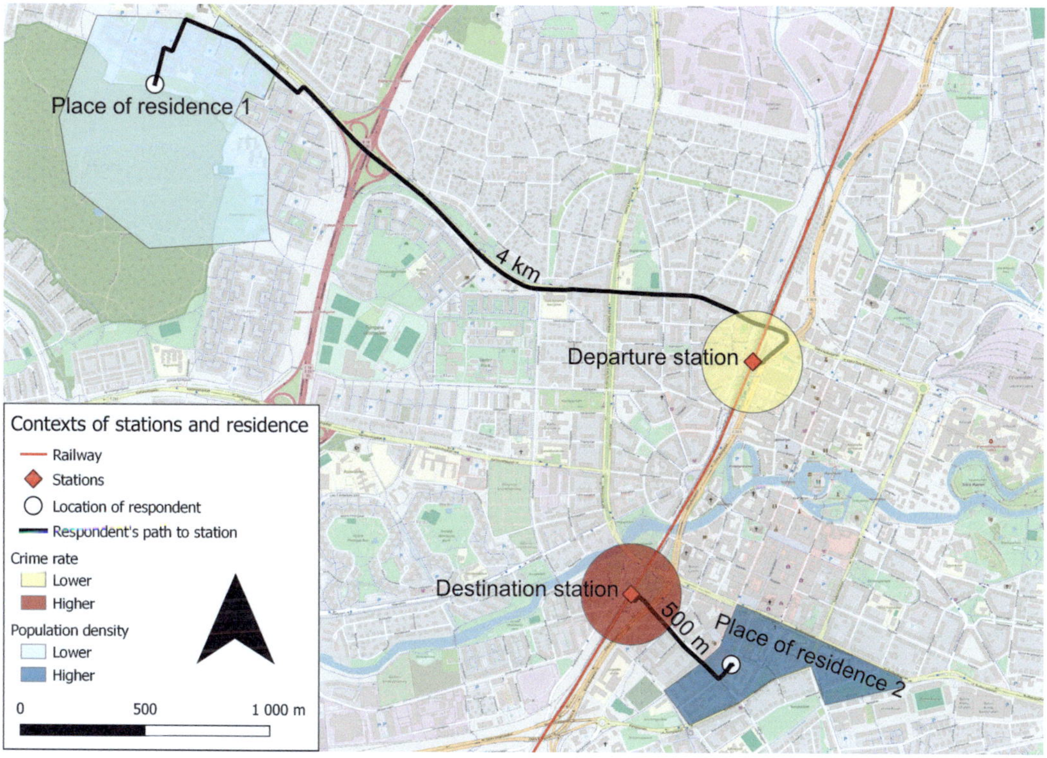

Fig. 6.2 Example of two survey respondents, the different residential contexts they live in, their routes to the station, and the contexts of the departure/destination station

transit environments when controlling for travellers' individual characteristics (note that all models show poor explanatory power). Population size, the proportion of residents with a foreign background, employment rate, housing composition, income levels, and local crime rates did not significantly influence the likelihood of victimisation on the way to the station, at the station, or on the train.

However, two variables stood out. A higher share of university-educated residents in the area of residence was linked to a lower likelihood of victimisation on the way to the station (OR = 0.971, $p = 0.031$) and at the station (OR = 0.963, $p = 0.014$), suggesting a possible protective effect. Additionally, individuals living closer to transit stations faced a slightly higher risk of victimisation on their way to the station (OR = 0.990, $p = 0.039$), though this association was weaker for incidents occurring at the station ($p = 0.096$) or on the train ($p = 0.063$). It is worth noting that drug-related activity at stations strongly increased victimisation across all contexts. The modelling results are presented in Table 6.1.

6.4.1.2 Fear of Crime

The descriptive analyses (Table 6.10, Appendix B and Table 6.15, Appendix C) show that fearful individuals were more likely to live in lower-density areas and closer to the train stations. Socioeconomic factors such as lower economic standards (at risk of poverty) and education levels were associated with higher fear levels, while income and unemployment showed no significant impact. Poor safety perceptions were also linked to housing composition, with individuals living in areas with more single-family housing reporting higher fear levels than those living in areas with primarily multi-family housing. Employment rates were slightly higher in the residential areas of those who declared experiencing fear of crime, but demographic characteristics such as foreign background showed no significant relationship.

Table 6.1 The impact of place of residence on (a) Victimisation on the way to the station (b) Victimisation at the station (c) Victimisation on the train

Y = 1 = Victimisation, 0 = Otherwise		(a) Victimisation on the way to the station (N = 187)				(b) Victimisation at the station (N = 158)				(c) Victimisation on the train (N = 140)			
		OR	95% C.I.for OR		Sig.	OR	95% C.I.for OR		Sig.	OR	95% C.I.for OR		Sig.
			Lower	Upper			Lower	Upper			Lower	Upper	
Individual characteristics	Woman	1.248	0.895	1.739	0.192	1.305	0.908	1.875	0.151	**1.531**	**1.034**	**2.266**	**0.033**
	Young (18 or under)	1.258	0.700	2.258	0.443	1.748	0.974	3.136	0.061	1.252	0.640	2.446	0.512
	Reduced mobility	**1.999**	**1.104**	**3.620**	**0.022**	**3.230**	**1.825**	**5.715**	**0.000**	**3.388**	**1.903**	**6.034**	**0.000**
Area of residence characteristics	Population	1.000	1.000	1.000	0.620	1.000	1.000	1.000	0.761	1.000	1.000	1.000	0.745
	Foreign background (%)	1.002	0.983	1.021	0.826	0.999	0.978	1.021	0.940	1.014	0.992	1.036	0.205
	Higher education (%)	**0.971**	**0.945**	**0.997**	**0.031**	**0.963**	**0.934**	**0.992**	**0.014**	0.980	0.950	1.012	0.213
	Employment rate (%)	0.973	0.924	1.025	0.300	0.981	0.924	1.042	0.537	1.016	0.953	1.083	0.636
	Multi-family housing (%)	0.994	0.986	1.003	0.185	1.005	0.996	1.015	0.288	0.996	0.986	1.006	0.454
	Income (1000 SEK)	1.008	0.998	1.018	0.122	1.010	0.999	1.021	0.076	1.004	0.992	1.015	0.543
	Distance to station (km)	**0.990**	**0.980**	**0.999**	**0.039**	0.991	0.981	1.002	0.096	0.989	0.978	1.001	0.063
	Observed drug use/selling at the station	**2.087**	**1.529**	**2.849**	**0.000**	**2.777**	**1.978**	**3.898**	**0.000**	**1.963**	**1.377**	**2.797**	**0.000**
	Crime rate in area of residence	0.490	0.169	1.420	0.189	0.413	0.127	1.343	0.142	0.771	0.292	2.036	0.600

(a) Nagelkerke = 0.060.
(b) Nagelkerke = 0.084.
(c) Nagelkerke = 0.061.

When controlling for individual characteristics, the analysis of area of residence characteristics reveals limited significant associations with fear of crime across the three contexts. Population density was not a significant predictor of fear in the models. Similarly, the proportion of residents with a foreign background did not show a meaningful association with fear of crime on the way to the station, at the station, or on the train. In addition to this, higher education levels, employment rate, the proportion of multi-family housing, and income levels did not emerge as significant predictors in any of the models, suggesting that these factors have minimal direct influence on perceived safety in transit environments.

The only significant predictor of fear among the area of residence characteristics was the distance to the station, with individuals living closer to the station being more likely to report fear of crime (note that all models show moderate explanatory power). This effect remained consistent across all models, showing significant associations with fear on the way to the station (OR = 0.986, $p < 0.001$), at the station (OR = 0.985, $p < 0.001$), and on the train (OR = 0.992, $p = 0.041$). The crime rate in the respondents' residential area did not show any significant association with fear of crime in any of the models. The modelling results are displayed in Table 6.2.

6.4.2 The Station's Socioeconomic Context and Impact on Safety

6.4.2.1 Victimisation

This section examines how the socioeconomic characteristics of station areas relate to victimisation risks in and around transit environments. The descriptive analyses (Table 6.11, Appendix B and Table 6.16, Appendix C) show that stations in areas where travellers experienced victimisation had lower socioeconomic conditions compared to those without. The median income was also slightly lower. However, housing composition and the proportion of residents with a foreign background showed no significant impact.

Building on these findings, the logistic regression models that account for individual-level factors further highlight which station area characteristics were significantly associated with victimisation risk in transit environments. A higher proportion of multi-family housing was linked to a reduced likelihood of victimisation both on the way to the station (OR = 0.261, $p = 0.004$) and on the train (OR = 0.299, $p = 0.021$). Additionally, income levels were associated negatively with victimisation risk on the way to the station (OR = 0.985, $p = 0.026$), although this effect was not significant at the station or on the train. Drug-related activity at the station emerged as the most influential environmental factor, substantially increasing victimisation risk in all contexts: on the way to the station (OR = 2.159, $p < 0.001$), at the station (OR = 2.492, $p < 0.001$), and on the train (OR = 2.061, $p < 0.001$).

Other socioeconomic characteristics, including population size, the proportion of residents with a foreign background, higher education levels, employment rates, and local crime rates, did not show a significant impact on victimisation in any of the three contexts. The full modelling results for victimisation can be found in Table 6.3.

6.4.2.2 Fear of Crime

After examining victimisation, this section considers how station-area socioeconomic conditions influence travellers' fear of crime. In the descriptive analysis (Table 6.12, Appendix B and Table 6.16, Appendix C), population density around the stations affect travellers' fear of crime, with fearful individuals more likely to travel from densely populated station areas. Education and employment in these areas were also shown to have an impact—stations in areas with fewer university-educated residents and lower employment rates were associated with higher fear levels. Housing type was relevant, as single-family housing correlated with fear, while multi-family housing was associated with lower fear levels. Financial status followed a similar pattern, with fear more common in lower-income areas. Notably, crime rates were lower in station's areas where fear was reported.

Table 6.2 The impact of place of residence on (a) Fear of crime on the way to the station (b) Fear of crime at the station (c) Fear of crime on the train

Y = 1 = Fear of crime, 0 = Otherwise		(a) Fear of crime on the way to the station (N = 528)				(b) Fear of crime at the station (N = 517)				(c) Fear of crime on the train (N = 290)			
		OR	95% C.I. for OR		Sig.	OR	95% C.I. for OR		Sig.	OR	95% C.I. for OR		Sig.
			Lower	Upper			Lower	Upper			Lower	Upper	
Individual characteristics	Woman	**1.860**	**1.464**	**2.364**	**0.000**	**1.493**	**1.175**	**1.897**	**0.001**	**1.519**	**1.136**	**2.032**	**0.005**
	Young (18 or under)	1.185	0.759	1.850	0.454	1.164	0.740	1.832	0.511	1.475	0.895	2.432	0.128
	Victim of crime	**3.595**	**2.702**	**4.782**	**0.000**	**4.613**	**3.451**	**6.165**	**0.000**	**3.485**	**2.572**	**4.723**	**0.000**
	Reduced mobility	1.042	0.633	1.713	0.872	1.409	0.860	2.310	0.173	1.445	0.833	2.506	0.190
Area of residence characteristics	Population	1.000	1.000	1.000	0.100	1.000	1.000	1.000	0.619	1.000	1.000	1.000	0.516
	Foreign background (%)	1.011	0.997	1.025	0.137	0.997	0.983	1.011	0.676	1.013	0.996	1.029	0.129
	Higher education (%)	0.990	0.972	1.009	0.308	0.993	0.975	1.012	0.465	1.001	0.978	1.023	0.957
	Employment rate (%)	1.034	0.994	1.076	0.098	1.016	0.977	1.056	0.435	1.037	0.989	1.087	0.132
	Multi-family housing (%)	0.999	0.993	1.005	0.756	0.997	0.991	1.003	0.307	0.996	0.988	1.003	0.241
	Income (1000 SEK)	1.002	0.995	1.009	0.591	0.998	0.991	1.005	0.580	0.997	0.988	1.006	0.502
	Distance to station (km)	**0.986**	**0.979**	**0.993**	**0.000**	**0.985**	**0.979**	**0.992**	**0.000**	**0.992**	**0.984**	**1.000**	**0.041**
	Observed drug use/selling at the station	**2.221**	**1.776**	**2.778**	**0.000**	**2.505**	**1.997**	**3.142**	**0.000**	**2.324**	**1.781**	**3.034**	**0.000**
	Crime rate in area of residence	0.706	0.405	1.230	0.219	1.081	0.695	1.681	0.729	1.041	0.594	1.822	0.889

(a) Nagelkerke = 0.183.
(b) Nagelkerke = 0.194.
(c) Nagelkerke = 0.135.

Table 6.3 The impact of departure station area on (a) Victimisation on the way to the station (b) Victimisation at the station (c) Victimisation on the train

Y = 1 = Victimisation, 0 = Otherwise		(a) Victimisation on the way to the station (N = 239)				(b) Victimisation at the station (N = 205)				(c) Victimisation on the train (N = 176)			
		OR	95% C.I. for OR		Sig.	OR	95% C.I. for OR		Sig.	OR	95% C.I. for OR		Sig.
			Lower	Upper			Lower	Upper			Lower	Upper	
Individual characteristics	Woman	1.286	0.965	1.715	0.086	1.240	0.911	1.685	0.171	**1.536**	**1.092**	**2.161**	**0.014**
	Young (18 or under)	**1.608**	**1.010**	**2.561**	**0.045**	**1.844**	**1.143**	**2.974**	**0.012**	**1.925**	**1.167**	**3.175**	**0.010**
	Reduced mobility	1.495	0.869	2.572	0.147	**2.309**	**1.377**	**3.870**	**0.002**	**2.432**	**1.415**	**4.178**	**0.001**
Departure station area characteristics	Population	1.000	0.999	1.000	0.675	1.000	1.000	1.001	0.722	1.000	1.000	1.001	0.149
	Foreign background (%)	0.509	0.091	2.835	0.441	0.688	0.111	4.280	0.688	0.171	0.023	1.255	0.082
	Higher education (%)	48.467	0.357	6580.310	0.121	4.422	0.024	807.472	0.576	0.836	0.004	180.407	0.948
	Employment rate (%)	0.931	0.009	100.915	0.975	0.339	0.002	50.974	0.672	0.199	0.001	38.819	0.549
	Multi-family housing (%)	**0.261**	**0.104**	**0.655**	**0.004**	0.513	0.191	1.376	0.185	**0.299**	**0.108**	**0.832**	**0.021**
	Income (1000 SEK)	**0.985**	**0.972**	**0.998**	**0.026**	0.993	0.979	1.007	0.342	0.995	0.981	1.009	0.485
	Observed drug use/selling at the station	**2.159**	**1.641**	**2.841**	**0.000**	**2.492**	**1.856**	**3.346**	**0.000**	**2.061**	**1.500**	**2.830**	**0.000**
	Crime rate in area of residence	1.021	0.972	1.071	0.409	1.026	0.976	1.077	0.317	1.043	0.997	1.091	0.066

(a) Nagelkerke = 0.051.
(b) Nagelkerke = 0.059.
(c) Nagelkerke = 0.063.

In the logistic regression models, when controlling for individual factors, the analysis of station area characteristics shows several factors that influence fear of crime. Population density remained a significant predictor of fear on the way to the station, at the station, and on the train, though the effect was small (OR = 1.001, $p < 0.05$). This suggests that higher population density around the station may contribute to lower levels of safety perception in these contexts. The proportion of residents with a foreign background had a mixed effect, showing a significant negative association with fear at the station (OR = 0.137, $p = 0.002$), indicating that individuals travelling from stations in areas with a higher proportion of foreign-born residents were less likely to report fear in this setting. However, no significant effects were found for fear on the way to the station or on the train.

Higher education levels showed inconsistent results. While education was not a significant predictor of fear on the way to the station or at the station, it was negatively associated with fear on the train (OR = 0.019, $p = 0.043$), suggesting that travellers from stations in areas with a higher proportion of educated residents were less likely to feel unsafe while travelling. The employment rate was a significant protective factor across all models, with higher employment rates around the station being associated with lower declared fear.

Multi-family housing did not have a consistent effect, and income showed only weak associations, with significance found only at the station (OR = 0.988, $p = 0.016$). Denser residential areas may provide a greater sense of safety or increased social control, which helps deter crime. Finally, the crime rate around the station had a small but significant negative impact on fear on the way to the station (OR = 0.948, $p = 0.008$), though no significant effects were found for fear at the station or on the train.

Income levels around the station had a significant negative impact on travellers' fear at the station (OR = 0.988, $p = 0.016$), but no significant effects in the other contexts. The crime rate around the station was negatively associated with fear on the way to the station (OR = 0.948, $p = 0.008$), suggesting that individuals travelling from stations with higher reported crime rates were slightly less likely to report fear in this setting. However, this association was not significant for fear at the station or on the train. The modelling results for travellers' fear of crime are presented in Table 6.4.

6.4.3 Safety to Destination Station

6.4.3.1 Victimisation

According to the descriptive analysis, the socioeconomic characteristics of destination stations showed minimal differences between those where riders experienced victimisation and those where they did not. The population size, proportion of residents with a foreign background, university education levels, housing composition, employment rates, economic standards,[1] and average income all exhibited no significant impact on victimisation.

However, a notable distinction emerged in travel distance, with victims of crime travelling significantly shorter distances on average compared to non-victims. This suggests that those experiencing victimisation may be more likely to travel within shorter commuting ranges. The full results are presented in Table 6.13, Appendix B and Table 6.17, Appendix C.

In the logistic regression analysis (Table 6.5), after adjusting for individual characteristics, no destination station area characteristics emerged as statistically significant predictors of victimisation on the train.

6.4.3.2 Fear of Crime

The descriptive analysis (Table 6.14, Appendix B and Table 6.17, Appendix C) revealed differences in population size and housing composition that emerged between destination stations where riders reported fear of crime and those where they did not. Stations associated with fear of crime

[1] Low economic standard: Share of people with an equivalised disposable income below 60% of the national median. High economic standard: Share of people with an equivalised disposable income above 200% of the national median.

Table 6.4 The impact of departure station area on (a) Fear of crime on the way to the station (b) Fear of crime at the station (c) Fear of crime on the train

Y = 1 = Fear of crime, 0 = Otherwise		(a) Fear of crime on the way to the station (N = 658)				(b) Fear of crime at the station (N = 646)				(c) Fear of crime on the train (N = 375)			
		OR	95% C.I. for OR		Sig.	OR	95% C.I. for OR		Sig.	OR	95% C.I. for OR		Sig.
			Lower	Upper			Lower	Upper			Lower	Upper	
Individual characteristics	Woman	**1.947**	**1.580**	**2.400**	**0.000**	**1.562**	**1.267**	**1.926**	**0.000**	**1.481**	**1.153**	**1.902**	**0.002**
	Young (18 or under)	1.304	0.909	1.871	0.149	**1.450**	**1.004**	**2.094**	**0.047**	**1.859**	**1.257**	**2.748**	**0.002**
	Victim of crime	**3.830**	**2.975**	**4.931**	**0.000**	**4.539**	**3.511**	**5.869**	**0.000**	**3.573**	**2.729**	**4.679**	**0.000**
	Reduced mobility	1.135	0.737	1.750	0.565	1.281	0.832	1.972	0.260	1.514	0.941	2.438	0.088
Departure station area characteristics	Population	**1.001**	**1.000**	**1.001**	**0.000**	**1.000**	**1.000**	**1.001**	**0.040**	**1.001**	**1.000**	**1.001**	**0.000**
	Foreign background (%)	0.423	0.126	1.413	0.162	**0.137**	**0.039**	**0.483**	**0.002**	0.624	0.146	2.665	0.524
	Higher education (%)	0.065	0.003	1.534	0.090	2.147	0.071	65.152	0.661	**0.019**	**0.000**	**0.877**	**0.043**
	Employment rate (%)	**0.412**	**0.219**	**0.775**	**0.006**	**0.251**	**0.130**	**0.486**	**0.000**	**0.415**	**0.192**	**0.895**	**0.025**
	Multi-family housing (%)	0.166	0.006	4.461	0.285	1.695	0.058	49.193	0.759	1.007	0.019	53.631	0.997
	Income (1000 SEK)	1.006	0.997	1.015	0.171	**0.988**	**0.979**	**0.998**	**0.016**	1.007	0.996	1.017	0.219
	Observed drug use/selling at the station	**2.229**	**1.827**	**2.720**	**0.000**	**2.615**	**2.137**	**3.200**	**0.000**	**2.300**	**1.816**	**2.914**	**0.000**
	Crime rate around station	**0.948**	**0.911**	**0.986**	**0.008**	0.963	0.922	1.006	0.093	0.997	0.958	1.036	0.861

(a) Nagelkerke = 0.170.
(b) Nagelkerke = 0.202.
(c) Nagelkerke = 0.137.

Table 6.5 The impact of destination station area on (a) Victimisation on the train and (b) Fear of crime on the train

Y = 1 = Victimised and Fear of crime, 0 = Otherwise		(a) Victimised on the train (N = 104)				(b) Fear of crime on the train (N = 175)			
		OR	95% C.I. for OR		Sig.	OR	95% C.I. for OR		Sig.
			Lower	Upper			Lower	Upper	
Individual characteristics	Woman	1.467	0.939	2.293	0.092	**1.527**	**1.060**	**2.200**	**0.023**
	Young (18 or under)	1.351	0.702	2.598	0.368	1.416	0.830	2.415	0.202
	Victim of crime	–	–	–	–	**3.324**	**2.286**	**4.834**	**0.000**
	Reduced mobility	**3.888**	**1.920**	**7.871**	**0.000**	1.816	0.903	3.652	0.094
Destination station area characteristics	Population	1.000	0.999	1.000	0.334	1.000	0.999	1.001	0.862
	Foreign background (%)	1.749	0.062	48.935	0.742	0.445	0.028	7.184	0.568
	Higher education (%)	13.222	0.004	49,528.249	0.539	89.953	0.094	86,500.316	0.199
	Employment rate (%)	98.919	0.010	981,536.917	0.328	0.115	0.000	185.688	0.566
	Multi-family housing (%)	0.738	0.101	5.383	0.764	1.296	0.249	6.740	0.758
	Income (1000 SEK)	0.987	0.961	1.014	0.355	0.991	0.969	1.013	0.415
	Travel distance	0.993	0.985	1.001	0.097	0.998	0.993	1.004	0.568
	Observed drug use/selling at the station	1.459	0.554	3.843	0.445	0.904	0.396	2.062	0.810
	Crime rate around station	0.988	0.870	1.122	0.853	1.019	0.936	1.109	0.669

(a) Nagelkerke = 0.045.
(b) Nagelkerke = 0.090.

were in areas with a significantly larger population and a slightly higher proportion of multi-family housing. While these differences were statistically significant, the effects were small, suggesting a limited practical impact.

In contrast, other socioeconomic characteristics of destination stations, including the proportion of residents with a foreign background, university education levels, employment rates, economic standards, average income, crime rates, and travel distances, showed no significant variation between the two groups.

After adjusting for individual characteristics, no destination station area factors were found to significantly predict fear of crime on the train (Table 6.5).

6.5 Discussion

Victimisation and fear of crime during train journeys are shaped by the entire travel experience, from the place where a person lives to their departure station, journey, and destination. The results suggest that socioeconomic characteristics of an individual's place of residence and departure station impact victimisation, but that the effect is limited when controlling for travellers' individual factors.

Population density showed different effects on victimisation depending on location (Table 6.6). Lower population density in residential areas was associated with higher victimisation risks, while higher population density at departure stations was linked to increased victimisation. This contrast may be due to differing social dynamics, where crowded environments facilitate anonymity and crime opportunities, while sparsely populated areas lack the protective effects of social control. However, these variables were not significant after controlling for individual factors in the regression models.

At the station of departure, lower economic standards were associated with higher victimisation risks, while higher economic standards at both residential and departure stations were linked to lower victimisation levels. Income and employment rates at departure stations also showed a negative relationship with victimisation, indicating that individuals travelling from wealthier, more economically stable areas were less likely to experience victimisation. This suggests that economic stability is linked to lower crime risks, as areas with higher income and employment rates may have fewer criminogenic conditions and greater social control (Wilkinson and Pickett 2010; Hummelsheim et al. 2011, Vieno et al. 2013). Notably, the distance travelled was associated negatively with victimisation at both residence (even after controlling for individual factors) and destination levels, suggesting that those experiencing victimisation tend to travel shorter distances. This may be because daily commuters and frequent users of the system, who often take shorter trips, are more exposed to the transit environment, increasing their likelihood of encountering situations that lead to victimisation.

Table 6.6 The relationship between the socioeconomic context of the respondents' residential area, departure station and destination station with victimisation

	Residence	Departure station	Destination station
Population density	(–)	(+)	n.s.
Low economic standard	n.s.	(+)	n.s.
High economic standard	(–)	(–)	n.s.
Income	n.s.	–	n.s.
Foreign background	n.s.	n.s.	n.s.
Employment rate	n.s	(–)	n.s.
Distance	–	N/A	(–)
Higher education	–	(–)	n.s.
Multi-family housing	(–)	–	n.s.
Crime rate	n.s.	n.s.	n.s.

n.s. = not significant
In parantheses = not significant after controlling for individual factors

In contrast, the proportion of foreign background residents and crime rates in these areas did not significantly correlate with travellers' victimisation, regardless of location. This suggests that area level demographic composition or crime do not strongly influence travellers' victimisation risk in transit environments.

Fear of crime demonstrated a somewhat different pattern, with a notable influence of population density across all stations' spatial levels (Table 6.7). Higher population density was linked to greater fear of crime at both the departure and destination stations, while lower residential population density was associated with lower fear levels. Similarly to victimisation, this may be due to the differing social dynamics. However, only population density around the departure station remained significant after controlling for individual factors. Economic factors also played a role, with lower economic standards at departure stations linked to greater fear. In comparison, higher economic standards at residential and departure station levels were associated with lower fear. This aligns with research suggesting that socially and economically disadvantaged areas can increase perceptions of poor safety, even in the absence of high-crime rates, which had a negative correlation in the respondent's area of residence and the departure station (Franklin et al. 2008; Sampson and Raudenbush 1997).

Table 6.7 The relationship between the socioeconomic context of the respondents' residential area, departure station and destination station with fear of crime

	Residence	Departure station	Destination station
Population	(–)	+	(+)
Low economic standard	n.s.	(+)	n.s.
High economic standard	(–)	(–)	n.s.
Income	n.s.	–	n.s.
Foreign background	n.s.	–	n.s.
Employment rate	(+)	–	n.s.
Distance	–	N/A	n.s.
Higher education	(–)	(–)	n.s.
Multi-family housing	(–)	(–)	(+)
Crime rate	(–)	(–)	n.s.

n.s. = not significant
In parantheses = not significant after controlling for individual factors

Employment rates showed mixed effects—higher employment at the residential level correlated with increased fear, while higher employment at departure stations was linked to lower fear. Yet, after controlling for individual factors, employment at departure stations was no longer a significant predictor.

Interestingly, multi-family housing was negatively linked to fear of crime at both the area of residence and departure station levels but positively linked to fear at the destination station. This suggests that individuals from less densely populated areas may feel more vulnerable when travelling to and waiting at their departure stations. Conversely, arriving at a destination with a larger population and more multi-family housing may also heighten perceptions of poor safety, possibly due to unfamiliar surroundings, denser crowds, or a perceived higher risk of crime. The relationship between housing types and fear of crime also implies that stations located in areas dominated by single-family housing have higher levels of fear compared to those in areas with more multi-family housing, which could be linked to perceptions of greater surveillance and informal social control (Brunton-Smith and Sturgis 2011). Furthermore, the only significant predictor of fear among the area of residence characteristics was distance to the station, with individuals living closer to the station being more likely to report fear of crime.

In summary, socioeconomic factors increase the fear of crime more than travellers' victimisation in their place of residence. Economic standards and crime rates influence fear, but these effects disappear when travellers' individual factors are controlled for. At the departure station, population density and economic conditions influence travellers' fear of crime and victimisation. Higher population density increases travellers' fear and victimisation, while travellers' lower economic standards increase both risks. However, income and education reduce victimisation risk, while for fear of crime, the effect of financial stability diminishes after controlling for travellers' individual factors. During the trip towards the destination station, travellers' fear of crime is influenced by population density, while victimisation is mainly linked to distance of travel, with longer distances reducing risk. Socioeconomic conditions at the destination station do not significantly influence victimisation.

6.6 Conclusions

Our findings demonstrate that travellers' individual factors remain the most significant predictors of victimisation and fear of crime in transit environments. However, specific socioeconomic characteristics also impact travellers' victimisation and fear of crime. Across the three locations, travellers' place of residence, departure station, and destination station, the departure station has the strongest impact on both fear of crime and victimisation, driven by population density and financial conditions of these areas. Travellers' fear of crime is more influenced by socioeconomic factors at the place of residence, while local conditions at the departure station shape victimisation. The destination station plays a minor role, with longer travel distances reducing travellers' victimisation, while population density affects travellers' fear of crime. The results underscore the need for adaptive safety strategies that balance structural improvements with understanding individual travel behaviours and patterns. They can also help inform strategies to create more inclusive and secure public transport systems, which aligns with Sustainable Development Goal (SDG) 11, precisely target 11.2, which seeks to provide access to safe, affordable, accessible, and sustainable transport systems for all.

Future research should further explore how different traveller groups experience safety along the journey, particularly in relation to gender, age and time of travel, to refine safety measures that are inclusive and effective across diverse populations. This study examines the journey through three key locations, but future research should consider the entire travel experience, including the return journey, to better understand transit safety throughout the trip.

Appendix A: Map

(See Fig. 6.3).

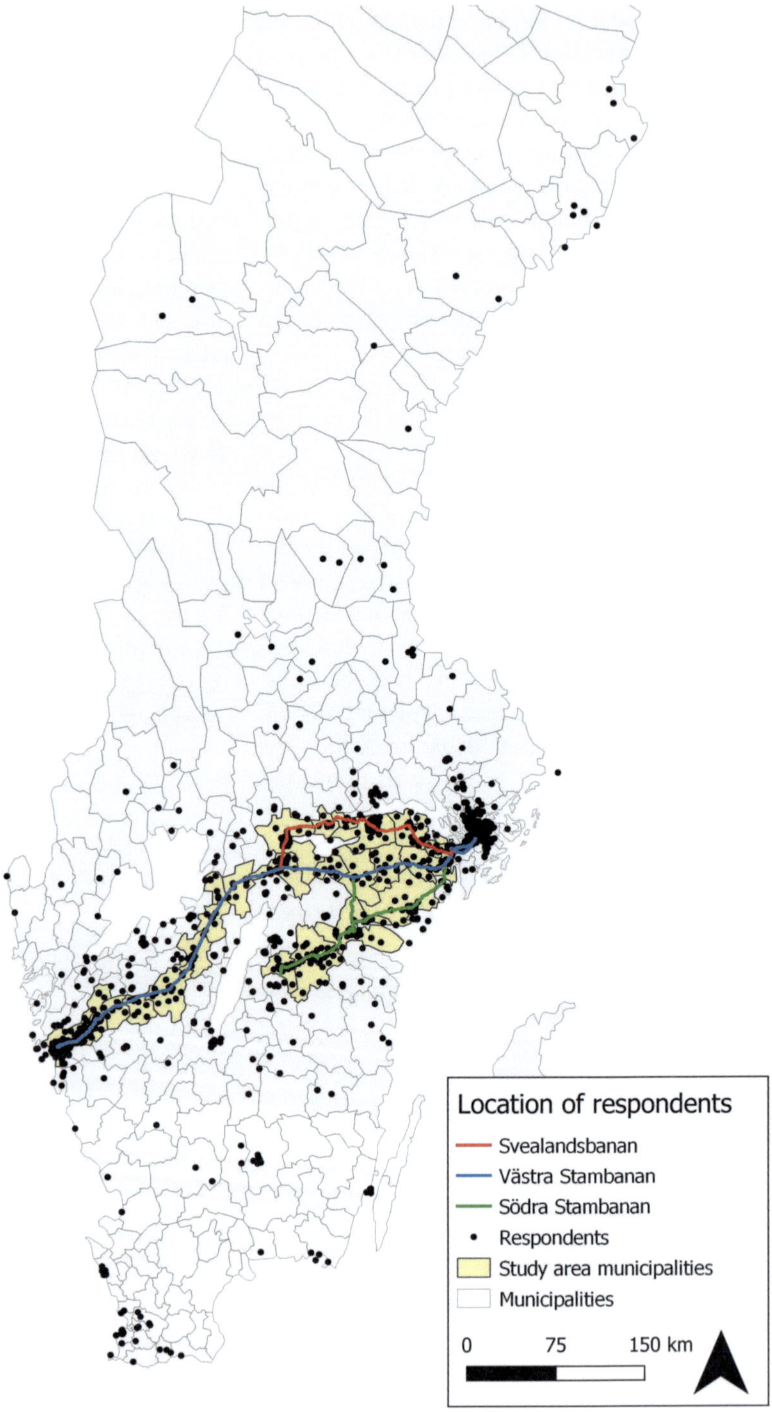

Fig. 6.3 The railway lines and location of respondents

Appendix B: T-tests

(See Tables 6.8, 6.9, 6.10, 6.11, 6.12, 6.13, 6.14).

Table 6.8 Socioeconomic differences in respondents' residing areas by victimisation status

Variable	Victimised mean ± SD (N = 488)	Non-victimised mean ± SD (N = 2,996)	t	p	d
Population density (per km^2)	**2,221 ± 4,717**	**2,736 ± 5,011**	**−2.12**	**0.034**	**−0.10**
Distance to station (km)	**14.1 ± 29.5**	**30.3 ± 67.6**	**−5.2**	**< 0.001**	**−0.25**
Unemployment rate (%)	4.1 ± 3.4	3.9 ± 3.2	1.73	0.084	0.08
Low economic standard (%)	13.2 ± 9.9	12.6 ± 9.5	1.29	0.198	0.06
High economic standard (%)	**5.4 ± 4.3**	**6.2 ± 5.8**	**−2.85**	**0.004**	**−0.14**
Income (1000 SEK)	302.3 ± 49.5	304.7 ± 51.2	−0.97	0.335	−0.05
Foreign background (%)	18.7 ± 12.0	17.9 ± 11.7	1.68	0.093	0.08
Young men (15–24 years) (%)	5.9 ± 1.7	6.0 ± 1.9	−1.14	0.254	−0.06
Employment rate (%)	82.1 ± 7.7	82.2 ± 7.5	−0.30	0.761	−0.02
University education (%)	**27.2 ± 11.7**	**29.4 ± 12.2**	**−3.68**	**< 0.001**	**−0.18**
Multi-family housing (%)	38.9 ± 32.7	42.1 ± 34.5	−1.90	0.058	−0.09

Table 6.9 Socioeconomic differences in respondents' residing areas by victimisation status (including aggressive panhandling)

Variable	Victimised incl. aggressive panhandling mean ± SD (N = 833)	Non-victimised incl. aggressive panhandling mean ± SD (N = 2,651)	t	p	d
Population density (per km^2)	2,780 ± 5,130	2,627 ± 4,822	0.78	0.218	0.03
Distance to station (km)	**18.2 ± 44.9**	**31.1 ± 68.5**	**−5.1**	**< 0.001**	**−0.2**
Unemployment rate (%)	**4.1 ± 3.3**	**3.9 ± 3.2**	**1.82**	**0.035**	**0.07**
Low economic standard (%)	13.0 ± 9.5	12.6 ± 9.6	1.19	0.118	0.05
High economic standard (%)	5.9 ± 5.6	6.1 ± 5.7	−0.90	0.370	−0.04
Foreign background (%)	**25.1 ± 16.1**	**23.53 ± 15.5**	**2.55**	**0.005**	**0.10**
Young men (15–24 years) (%)	6.0 ± 1.7	6.0 ± 2.0	−1.25	0.105	−0.05
Employment rate (%)	81.9 ± 7.5	82.3 ± 7.5	−1.10	0.135	−0.04
University education (%)	**28.2± 12.2**	**29.3 ± 12.2**	**−2.28**	**0.023**	**−0.09**
Multi-family housing (%)	42.6 ± 34.0	41.4 ± 34.3	0.88	0.378	0.04

Table 6.10 Socioeconomic differences in respondents' residing areas by fear status

Variable	Fear of crime mean ± SD (N = 1,091)	No fear mean ± SD (N = 2,040)	t	p	d
Population density (per km^2)	**2,506 ± 4,688**	**2,908 ± 5,283**	**−2.11**	**0.035**	**−0.08**
Distance to station (km)	**15.8 ± 37.6**	**37.0 ± 74.1**	**−8.8**	**< 0.001**	**−0.33**
Unemployment rate (%)	4.0 ± 3.4	3.93 ± 3.2	0.15	0.882	0.01
Low economic standard (%)	12.7 ± 9.5	12.91 ± 9.8	−0.71	0.478	−0.03
High economic standard (%)	**5.6 ± 5.0**	**6.4 ± 6.1**	**−3.28**	**0.001**	**−0.12**
Income (1000 SEK)	303.0 ± 48.4	304.90 ± 53.1	−0.98	0.328	−0.04
Foreign background (%)	23.7 ± 15.7	24.3 ± 15.8	−1.05	0.295	−0.04
Young men (15–24) (%)	6.0 ± 1.7	6.1 ± 2.1	−0.90	0.366	−0.03
Employment rate (%)	**82.4 ± 7.4**	**81.8 ± 7.7**	**2.02**	**0.044**	**0.08**
University education (%)	**28.2 ± 11.7**	**29.8 ± 12.6**	**−3.53**	**< 0.001**	**−0.13**
Multi-family housing (%)	**40.2 ± 33.2**	**43.6 ± 35.2**	**−2.65**	**0.008**	**−0.10**

Table 6.11 Socioeconomic differences by victimisation status at the departure stations and its surroundings (200 m around)

Variable	Victimised mean ± SD (N = 496)	Non-victimised mean ± SD (N = 3,079)	t	p	d
Population	394 ± 488	473 ± 451	0.96	0.337	0.05
Foreign background (%)	28.2 ± 16.5	27.7 ± 15.1	0.79	0.432	0.15
University education (%)	**24.8 ± 9.2**	**26.2 ± 9.0**	**−3.15**	**< 0.001**	**−0.15**
Single-family housing (%)	32.8 ± 29.4	30.7 ± 29.2	1.51	0.131	0.07
Multi-family housing (%)	59.1 ± 27.8	61.1 ± 27.9	−1.53	0.127	−0.07
Employed (%)	**47.3 ± 6.2**	**48.2 ± 6.1**	**−3.03**	**0.002**	**−0.15**
High economic standard (%)	**4.7 ± 2.8**	**5.3 ± 3.0**	**−4.17**	**< 0.001**	**−0.20**
Low economic standard (%)	**18.5 ± 10.7**	**17.1 ± 9.9**	**2.92**	**0.004**	**0.14**
Average income (1000 SEK)	**305.2 ± 51.7**	**311.8± 47.4**	**−2.82**	**0.005**	**−0.14**
Crime rate (per inhabitant)	2.5 ± 4.0	2.3 ± 3.4	1.39	0.166	0.07

Table 6.12 Socioeconomic differences by fear status at the departure stations and its surroundings (200 m around)

Variable	Fear of crime (Station) mean ± SD (N = 776)	No fear mean (Station) ± SD (N = 2,420)	t	p	d
Population	**420 ± 499**	**356 ± 430**	**3.46**	**< 0.001**	**0.14**
Foreign background (%)	27.8 ± 16.3	27.7 ± 14.7	0.18	0.858	0.01
University education (%)	**24.6 ± 8.9**	**26.7 ± 9.1**	**−5.73**	**< 0.001**	**−0.24**
Single-family housing (%)	**32.8 ± 29.8**	**29.7 ± 28.6**	**2.57**	**0.010**	**0.11**
Multi-family housing (%)	**59.6 ± 27.8**	**61.9 ± 27.7**	**−2.02**	**0.044**	**−0.08**
Employed (%)	**47.2 ± 6.1**	**48.5 ± 6.1**	**−5.28**	**< 0.001**	**−0.22**
High economic standard (%)	**4.6 ± 2.8**	**5.5 ± 3.0**	**−7.55**	**< 0.001**	**−0.31**
Low economic standard (%)	**18.4 ± 10.4**	**16.8 ± 9.8**	**3.95**	**< 0.001**	**0.16**
Average income (1000 SEK)	**303.1 ± 49.3**	**314.1 ± 46.5**	**−5.65**	**< 0.001**	**−0.23**
Crime rate (per inhabitant)	**1.8 ± 2.4**	**2.5 ± 3.8**	**−4.57**	**< 0.001**	**−0.19**

Table 6.13 Socioeconomic differences by victimisation status at the destination stations and its surroundings (200 m around)

Variable	Victim of crime (Train) mean ± SD (N = 215)	Victim of crime (Train) mean ± SD (N = 2,495)	t	p	d
Population	482 ± 398	464 ± 394	0.62	0.533	0.04
Foreign background (%)	29.2 ± 13.6	27.8 ± 13.3	1.51	0.130	0.11
University education (%)	30.6 ± 9.8	30.4 ± 9.8	0.25	0.802	0.02
Multi-family housing (%)	73.0 ± 23.1	71.8 ± 24.8	0.71	0.479	0.05
Employment rate (%)	79.8 ± 7.1	79.6 ± 6.8	0.34	0.736	0.02
High economic standard (%)	6.3 ± 3.2	6.5 ± 3.2	−0.93	0.353	−0.07
Low economic standard (%)	16.2 ± 9.1	15.71 ± 8.8	0.77	0.441	0.06
Average income (1000 SEK)	317.1 ± 41.2	319.3 ± 41.4	−0.76	0.448	−0.05
Crime rate (per inhabitant)	2.2 ± 2.4	2.3 ± 3.2	−0.22	0.826	−0.02
Travel distance (km)	**46.5 ± 45.7**	**63.5 ± 64.2**	**−3.76**	**< 0.001**	**−0.2**

Table 6.14 Socioeconomic differences by fear status at the destination stations and its surroundings (200 m around)

Variable	Fear of crime (Train) mean ± SD (N = 337)	No fear (Train) mean ± SD (N = 2,091)	t	p	d
Population	**514 ± 399**	**450 ± 390**	**2.80**	**0.006**	**0.16**
Foreign background (%)	28.3 ± 13.0	27.8 ± 13.3	0.57	0.560	0.03
University education (%)	30.7 ± 10.3	30.3 ± 9.6	0.70	0.505	0.04
Multi-family housing (%)	**74.2 ± 23.2**	**71.3 ± 24.9**	**2.03**	**0.043**	**0.12**
Employment rate (%)	79.8 ± 7.0	79.6 ± 6.7	0.68	0.511	0.03
High economic standard (%)	6.5 ± 3.3	6.4 ± 3.2	0.41	0.694	0.02
Low economic standard (%)	16.1 ± 8.8	15.6 ± 8.8	0.96	0.338	0.06
Average income (1000 SEK)	316.8 ± 41.6	319.7 ± 40.9	−1.19	0.235	−0.07
Crime rate (per inhabitant)	2.08 ± 2.60	2.30 ± 3.27	−1.40	0.163	−0.07
Travel distance (km)	59.4 ± 61.3	62.2 ± 61.3	−0.82	0.413	−0.05

Appendix C: Correlation Analysis

(See Tables 6.15, 6.16, 6.17).

Table 6.15 Bivariate Pearson correlation analysis of victimisation, fear, and socioeconomic context of respondent's residence area

Variables	Victim of crime	Victim to crime (incl. aggressive panhandling)	Victim of crime (Way)	Victim of crime (Station)	Victim of crime (Train)	Fear of crime	Fear of crime (Way)	Fear of crime (Station)	Fear of crime (Train)	Pop density	Low economic std	Income	Foreign back-ground	Employed	Distance to station	University education	High economic std	Multi-family housing	Crime rate in residence area
Victim of crime	1	**0.720** (<0.001)	**0.805** (<0.001)	**0.777** (<0.001)	**0.722** (<0.001)	**0.279** (<0.001)	**0.248** (<0.001)	**0.281** (<0.001)	**0.241** (<0.001)	**−0.036** (0.034)	0.022 (0.198)	−0.016 (0.335)	0.028 (0.093)	−0.005 (0.761)	**−0.088** (<0.001)	**−0.062** (<0.001)	**−0.048** (0.004)	−0.032 (0.058)	−0.013 (0.509)
Victim to crime (incl. aggressive panhandling)	**0.720** (<0.001)	1	**0.580** (<0.001)	**0.559** (<0.001)	**0.519** (<0.001)	**0.298** (<0.001)	**0.255** (<0.001)	**0.289** (<0.001)	**0.240** (<0.001)	0.013 (0.437)	0.020 (0.236)	−0.006 (0.734)	**0.043** (0.011)	−0.019 (0.270)	**−0.086** (<0.001)	**−0.039** (0.023)	−0.015 (0.370)	0.015 (0.378)	**0.047** (0.012)
Victim of crime (Way)	**0.805** (<0.001)	**0.580** (<0.001)	1	**0.707** (<0.001)	**0.577** (<0.001)	**0.275** (<0.001)	**0.268** (<0.001)	**0.287** (<0.001)	**0.207** (<0.001)	**−0.034** (0.042)	0.020 (0.232)	−0.018 (0.289)	0.021 (0.226)	−0.004 (0.836)	**−0.065** (<0.001)	**−0.058** (0.001)	**−0.044** (0.010)	**−0.044** (0.009)	−0.020 (0.285)
Victim of crime (Station)	**0.777** (<0.001)	**0.559** (<0.001)	**0.707** (<0.001)	1	**0.637** (<0.001)	**0.279** (<0.001)	**0.260** (<0.001)	**0.291** (<0.001)	**0.265** (<0.001)	−0.018 (0.301)	0.008 (0.651)	−0.005 (0.758)	0.022 (0.194)	0.007 (0.693)	**−0.055** (0.002)	**−0.049** (0.004)	−0.032 (0.062)	−0.012 (0.475)	0.005 (0.772)
Victim of crime (Train)	**0.722** (<0.001)	**0.519** (<0.001)	**0.577** (<0.001)	**0.637** (<0.001)	1	**0.226** (<0.001)	**0.202** (<0.001)	**0.232** (<0.001)	**0.304** (<0.001)	−0.006 (0.706)	0.003 (0.856)	0.009 (0.598)	0.028 (0.099)	0.008 (0.625)	**−0.054** (0.003)	**−0.036** (0.033)	−0.024 (0.151)	−0.023 (0.181)	−0.023 (0.231)
Fear of crime	**0.279** (<0.001)	**0.298** (<0.001)	**0.275** (<0.001)	**0.279** (<0.001)	**0.226** (<0.001)	1	**0.782** (<0.001)	**0.789** (<0.001)	**0.570** (<0.001)	**−0.038** (0.035)	−0.013 (0.478)	−0.017 (0.328)	−0.019 (0.295)	**0.036** (0.044)	**−0.156** (<0.001)	**−0.063** (<0.001)	**−0.059** (0.001)	**−0.047** (0.008)	**−0.046** (0.021)
Fear of crime (Way)	**0.248** (<0.001)	**0.255** (<0.001)	**0.268** (<0.001)	**0.260** (<0.001)	**0.202** (<0.001)	**0.782** (<0.001)	1	**0.600** (<0.001)	**0.413** (<0.001)	−0.032 (0.071)	−0.038 (0.035)	0.004 (0.808)	−0.024 (0.172)	**0.054** (0.002)	**−0.123** (<0.001)	**−0.035** (0.049)	**−0.054** (0.003)	**−0.046** (0.010)	**−0.049** (0.015)
Fear of crime (Station)	**0.281** (<0.001)	**0.289** (<0.001)	**0.287** (<0.001)	**0.291** (<0.001)	**0.232** (<0.001)	**0.789** (<0.001)	**0.600** (<0.001)	1	**0.509** (<0.001)	**−0.048** (0.007)	0.004 (0.837)	−0.032 (0.073)	−0.022 (0.228)	0.020 (0.275)	**−0.082** (<0.001)	**−0.074** (<0.001)	**−0.064** (<0.001)	**−0.050** (0.006)	−0.038 (0.059)
Fear of crime (Train)	**0.241** (<0.001)	**0.240** (<0.001)	**0.207** (<0.001)	**0.265** (<0.001)	**0.304** (<0.001)	**0.570** (<0.001)	**0.413** (<0.001)	**0.509** (<0.001)	1	0.005 (0.797)	0.003 (0.887)	−0.008 (0.642)	0.034 (0.060)	0.006 (0.753)	**0.111** (<0.001)	−0.034 (0.058)	−0.029 (0.103)	−0.009 (0.612)	−0.013 (0.528)
Pop density	**−0.036** (0.034)	0.013 (0.437)	**−0.034** (0.042)	−0.018 (0.301)	−0.006 (0.706)	**−0.038** (0.035)	−0.032 (0.071)	**−0.048** (0.007)	0.005 (0.797)	1	**0.194** (<0.001)	0.016 (0.345)	**0.280** (<0.001)	**−0.267** (<0.001)	**−0.056** 0.001	**0.333** (<0.001)	**0.228** (<0.001)	**0.592** (<0.001)	**0.123** (<0.001)
Low economic std	0.022 (0.198)	0.020 (0.236)	0.020 (0.232)	0.008 (0.651)	0.003 (0.856)	−0.013 (0.478)	**−0.038** (0.035)	0.004 (0.837)	0.003 (0.887)	**0.194** (<0.001)	1	**−0.792** (<0.001)	**0.754** (<0.001)	**−0.926** (<0.001)	**0.124** (<0.001)	**−0.446** (<0.001)	**−0.444** (<0.001)	**0.543** (<0.001)	**0.168** (<0.001)
Income	−0.016 (0.335)	−0.006 (0.734)	−0.018 (0.289)	−0.005 (0.758)	0.009 (0.598)	−0.017 (0.328)	0.004 (0.808)	−0.032 (0.073)	−0.008 (0.642)	0.016 (0.345)	**−0.792** (<0.001)	1	**−0.496** (<0.001)	**0.725** (<0.001)	−0.014 0.406	**0.686** (<0.001)	**0.773** (<0.001)	**−0.374** (<0.001)	**−0.134** (<0.001)
Foreign background	0.028 (0.093)	**0.043** (0.011)	0.021 (0.226)	0.022 (0.194)	0.028 (0.099)	−0.019 (0.295)	−0.024 (0.172)	−0.022 (0.228)	0.034 (0.060)	**0.280** (<0.001)	**0.754** (<0.001)	**−0.496** (<0.001)	1	**−0.751** (<0.001)	**−0.052** 0.002	**−0.214** (<0.001)	**−0.282** (<0.001)	**0.591** (<0.001)	**0.241** (<0.001)

(continued)

Table 6.15 (continued)

Variables	Victim of crime	Victim to crime (incl aggressive panhandling)	Victim of crime (Way)	Victim of crime (Station)	Victim of crime (Train)	Fear of crime	Fear of crime (Way)	Fear of crime (Station)	Fear of crime (Train)	Pop density	Low economic std	Income	Foreign back-ground	Employed	Distance to station	University education	High economic std	Multi-family housing	Crime rate in residence area
Employed	−0.005 (0.761)	−0.019 (0.270)	−0.004 (0.836)	0.007 (0.693)	0.008 (0.625)	**0.036** (0.044)	**0.054** (0.002)	0.020 (0.275)	0.006 (0.753)	**−0.267** (<0.001)	**−0.926** (<0.001)	**0.725** (<0.001)	**−0.751** (<0.001)	1		0.027 (0.097)	**−0.140** (<0.001)	**0.150** (<0.001)	**−0.219** (<0.001)
Distance to station	**−0.088** (<0.001)	**−0.086** (<0.001)	**−0.065** (<0.001)	**−0.055** (0.002)	**−0.054** (0.003)	**−0.156** (<0.001)	**−0.130** (<0.001)	**−0.123** (<0.001)	**−0.075** (<0.001)	**0.111** (<0.001)	**−0.045** 0.007	**0.124** (<0.001)	−0.014 0.406	0.013 0.440	1	**0.287** (<0.001)	**0.301** (<0.001)	**−0.567** (<0.001)	**0.073** (<0.001)
University education	**−0.062** (<0.001)	**−0.039** (0.023)	**−0.058** (0.001)	**−0.049** (0.004)	**−0.036** (0.033)	**−0.063** (<0.001)	**−0.035** (0.049)	**−0.074** (<0.001)	−0.034 (0.058)	−0.034 (0.058)	**−0.435** (<0.001)	**−0.446** (<0.001)	**0.686** (<0.001)	**−0.214** (<0.001)	0.027 (0.097)	1	**0.658** (<0.001)	**0.076** (<0.001)	0.032 (0.088)
High economic std	**−0.048** (0.004)	−0.015 (0.370)	**−0.044** (0.010)	−0.032 (0.062)	−0.024 (0.151)	**−0.059** (0.001)	**−0.054** (0.003)	**−0.064** (<0.001)	−0.029 (0.103)	−0.029 (0.103)	**−0.370** (<0.001)	**−0.444** (<0.001)	**0.773** (<0.001)	**−0.274** (<0.001)	**−0.140** (<0.001)	**0.658** (<0.001)	1	**−0.072** (<0.001)	**0.045** (0.015)
Multi−family housing	−0.032 (0.058)	0.015 (0.378)	**−0.044** (0.009)	−0.012 (0.475)	−0.023 (0.181)	**−0.047** (0.008)	**−0.046** (0.010)	**−0.050** (0.006)	−0.009 (0.612)	−0.009 (0.612)	**0.513** (<0.001)	**0.543** (<0.001)	**−0.374** (<0.001)	**0.612** (<0.001)	**0.150** (<0.001)	**0.076** (<0.001)	**−0.072** (<0.001)	1	**−0.319** (<0.001)
Crime rate in residence area	−0.013 (0.509)	**0.047** (0.012)	−0.020 (0.285)	0.005 (0.772)	−0.023 (0.231)	**−0.046** (0.021)	**−0.049** (0.015)	−0.038 (0.059)	−0.013 (0.528)	**0.123** (<0.001)	**0.168** (<0.001)	**−0.134** (<0.001)	**0.241** (<0.001)	**−0.219** (<0.001)	**0.073** (<0.001)	0.032 (0.088)	**0.045** (0.015)	**−0.319** (<0.001)	1

Table 6.16 Bivariate Pearson correlation analysis of victimisation, fear and socioeconomic context of departure station

	Victim to crime	Victim to crime (incl. aggressive panhandling)	Victim to crime (Way to station)	Victim to crime (Station)	Victim to crime (Train)	Fear of crime	Fear of crime (Way)	Fear of crime (Station)	Fear of crime (Train)	Population	Foreign background	University education	Multi-family housing	Employed	High economic std	Low economic std	Income	Crime rate around station
Victim to crime	1	0.729 (<0.001)	**0.805** (<0.001)	0.767 (<0.001)	**0.706** (<0.001)	**0.281** (<0.001)	**0.246** (<0.001)	**0.277** (<0.001)	**0.239** (<0.001)	0.016 (0.337)	0.013 (0.432)	**−0.053** (0.002)	−0.026 (0.127)	**−0.051** (0.002)	**−0.070** (<0.001)	**0.049** (0.004)	**−0.047** (0.005)	0.023 (0.166)
Victim to crime (incl. aggressive panhandling)	0.729 (<0.001)	1	**0.580** (<0.001)	0.559 (<0.001)	**0.514** (<0.001)	**0.299** (<0.001)	**0.254** (<0.001)	**0.284** (<0.001)	**0.243** (<0.001)	0.018 (0.281)	0.022 (0.183)	**−0.075** (<0.001)	**−0.050** (0.003)	**−0.068** (<0.001)	**−0.095** (<0.001)	**0.053** (0.001)	**−0.052** (0.002)	**0.042** (0.012)
Victim to crime (Way to station)	**0.805** (<0.001)	**0.580** (<0.001)	1	**0.707** (<0.001)	**0.577** (<0.001)	**0.275** (<0.001)	**0.268** (<0.001)	**0.287** (<0.001)	**0.207** (<0.001)	**−0.034** (0.042)	0.020 (0.232)	−0.018 (0.289)	0.021 (0.226)	−0.004 (0.836)	**−0.065** (<0.001)	**−0.058** (0.001)	**−0.044** (0.010)	**−0.044** (0.009)
Victim to crime (Station)	**0.767 (<0.001)**	**0.559 (<0.001)**	**0.707** (<0.001)	**1**	**0.620** (<0.001)	**0.278** (<0.001)	**0.259** (<0.001)	**0.289** (<0.001)	**0.260** (<0.001)	**0.033** (0.049)	0.031 (0.062)	**−0.056** (0.001)	0.007 (0.675)	**−0.055** (0.001)	**−0.070** (<0.001)	**0.066** (<0.001)	**−0.065** (<0.001)	−0.007 (0.694)
Victim to crime (Train)	**0.706 (<0.001)**	**0.514 (<0.001)**	**0.577** (<0.001)	**0.620 (<0.001)**	1	**0.226 (<0.001)**	**0.201 (<0.001)**	**0.224 (<0.001)**	**0.302 (<0.001)**	−0.007 (0.667)	−0.011 (0.495)	**−0.050 (0.003)**	**−0.043 (0.010)**	**−0.033 (0.050)**	**−0.055 (0.001)**	0.019 (0.251)	−0.023 (0.171)	**0.038 (0.023)**
Fear of crime	**0.281** (<0.001)	**0.299** (<0.001)	**0.275** (<0.001)	**0.278** (<0.001)	**0.226** (<0.001)	1	**0.783** (<0.001)	**0.786** (<0.001)	**0.579** (<0.001)	**0.040** (0.025)	−0.008 (0.642)	**−0.086** (<0.001)	**−0.062** (<0.001)	**−0.081** (<0.001)	**−0.114** (<0.001)	**0.040** (0.023)	**−0.065** (<0.001)	**−0.064** (<0.001)
Fear of crime (Way)	**0.246** (<0.001)	**0.254** (<0.001)	**0.268** (<0.001)	**0.259** (<0.001)	**0.201** (<0.001)	**0.783** (<0.001)	1	**0.603** (<0.001)	**0.426** (<0.001)	**0.041** (0.021)	−0.026 (0.149)	**−0.048** (0.006)	**−0.066** (<0.001)	**−0.050** (0.005)	**-0.081** (<0.001)	0.003 (0.873)	−0.029 (0.106)	**−0.079** (<0.001)
Fear of crime (Station)	**0.277** (<0.001)	**0.284** (<0.001)	**0.287** (<0.001)	**0.289** (<0.001)	**0.224** (<0.001)	**0.786** (<0.001)	**0.603** (<0.001)	1	**0.517** (<0.001)	**0.061** (<0.001)	0.003 (0.858)	**−0.101** (<0.001)	**-0.036** (0.044)	**−0.093** (<0.001)	**−0.132** (<0.001)	**0.070** (<0.001)	**−0.099** (<0.001)	**−0.081** (<0.001)
Fear of crime (Train)	**0.239** (<0.001)	**0.243** (<0.001)	**0.207** (<0.001)	**0.260** (<0.001)	**0.302** (<0.001)	**0.579** (<0.001)	**0.426** (<0.001)	**0.517** (<0.001)	1	0.032 (0.071)	−0.009 (0.608)	**−0.053** (0.003)	**−0.048** (0.006)	**-0.036** (0.039)	**−0.073** (<0.001)	0.016 (0.360)	−0.030 (0.086)	−0.015 (0.402)
Population	0.016 (0.337)	0.018 (0.281)	**−0.034** (0.042)	**0.033** 0.049	−0.007 (0.667)	**0.040** (0.025)	**0.041** (0.021)	**0.061** (<0.001)	0.032 (0.071)	1	**0.625** (<0.001)	**−0.049** (0.002)	**0.619** (<0.001)	**−0.177** (<0.001)	**−0.228** (<0.001)	**0.539** (<0.001)	**−0.571** (<0.001)	**−0.284** (<0.001)
Foreign background	0.013 (0.432)	0.022 (0.183)	0.020 (0.232)	0.031 (0.062)	−0.011 (0.495)	−0.008 (0.642)	−0.026 (0.149)	0.003 (0.858)	−0.009 (0.608)	**0.625** (<0.001)	**1**	**−0.319** (<0.001)	**0.534** (<0.001)	**−0.510** (<0.001)	**−0.359** (<0.001)	**0.816** (<0.001)	**−0.719** (<0.001)	−0.012 (0.455)
University education	**−0.053** (0.002)	**−0.075** (<0.001)	−0.018 (0.289)	**−0.056** (0.001)	**−0.050** (0.003)	**−0.086** (<0.001)	**−0.048** (0.006)	**−0.101** (<0.001)	**−0.053** (0.003)	**−0.049** (0.002)	**−0.319** (<0.001)	1	**0.067** (<0.001)	**0.770** (<0.001)	**0.881** (<0.001)	**−0.682** (<0.001)	**0.759** (<0.001)	**0.046** (0.004)
Single-family housing	0.025 (0.131)	**0.045** (0.007)	0.021 (0.226)	−0.004 0.791	**0.040** (0.017)	**0.073** (<0.001)	**0.076** (<0.001)	**0.045** (0.010)	**0.051** (0.004)	**−0.644** (<0.001)	**−0.532** (<0.001)	**−0.064** (<0.001)	**−0.988** (<0.001)	**−0.038** (0.016)	**−0.105** (<0.001)	**−0.438** (<0.001)	**0.456** (<0.001)	**0.175** (<0.001)

(continued)

Table 6.16 (continued)

	Victim to crime	Victim to crime (inc. aggressive panhandling)	Victim to crime (Way to station)	Victim to crime (Station)	Victim to crime (Train)	Fear of crime	Fear of crime (Way)	Fear of crime (Station)	Fear of crime (Train)	Population	Foreign background	University education	Multi-family housing	Employed	High economic std	Low economic std	Income	Crime rate around station
Multi-family housing	−0.026 (0.127)	**−0.050** (0.003)	−0.004 (0.336)	0.007 0.675	**−0.043** (0.010)	**−0.062** (<0.001)	**−0.066** (<0.001)	**−0.036** (0.044)	**−0.048** (0.006)	**0.619** (<0.001)	**0.534** (<0.001)	**0.067** (<0.001)	1	**0.038** (0.018)	**0.114** (<0.001)	**0.440** (<0.001)	**−0.453** (<0.001)	**−0.209** (<0.001)
Employed	**−0.051** (0.002)	**−0.068** (<0.001)	**−0.065** (<0.001)	**−0.055** (0.001)	**−0.033** (0.050)	**−0.081** (<0.001)	**−0.050** (0.005)	**−0.093** (<0.001)	**−0.036** (0.039)	**−0.177** (<0.001)	**−0.510** (<0.001)	**0.779** (<0.001)	**0.038** (0.018)	1	**0.729** (<0.001)	**−0.594** (<0.001)	**0.750** (<0.001)	**0.057** (0.001)
High economic std	**−0.070** (<0.001)	**−0.095** (<0.001)	**−0.058** (0.001)	**−0.070** (<0.001)	**−0.055** (<0.001)	**−0.114** (<0.001)	**−0.081** (<0.001)	**−0.132** (<0.001)	**−0.073** (<0.001)	**−0.228** (<0.001)	**−0.359** (<0.001)	**0.881** (<0.001)	**0.114** (<0.001)	**0.729** (<0.001)	1	**−0.711** (<0.001)	**0.833** (<0.001)	**0.050** (0.003)
Low economic std	**0.049** (0.004)	**0.053** (0.001)	**−0.044** (0.010)	**0.066** (<0.001)	0.019 (0.251)	**0.040** (0.023)	0.003 (0.873)	**0.070** (<0.001)	0.016 (0.360)	**0.539** (<0.001)	**0.816** (<0.001)	**−0.682** (<0.001)	**0.440** (<0.001)	**−0.594** (<0.001)	**−0.711** (<0.001)	1	**−0.711** (<0.001)	**0.152** (<0.001)
Income	**−0.047** (0.005)	**−0.052** (0.002)	**−0.044** (0.009)	**−0.062** (<0.001)	−0.023 (0.171)	**−0.065** (<0.001)	−0.029 (0.106)	**−0.099** (<0.001)	−0.030 (0.086)	**−0.571** (<0.001)	**−0.719** (<0.001)	**0.759** (<0.001)	**−0.453** (<0.001)	**0.750** (<0.001)	**0.833** (<0.001)	**−0.711** (<0.001)	1	**0.088** (<0.001)
Crime rate around station	0.023 (0.166)	**0.042** (0.012)	−0.020 (0.285)	−0.007 0.694	**0.038** (0.023)	**−0.064** (<0.001)	**−0.079** (<0.001)	**−0.081** (<0.001)	−0.015 (0.402)	**−0.284** (<0.001)	−0.012 (0.455)	**0.046** (0.004)	**−0.209** (<0.001)	**0.057** (0.001)	**0.050** (0.003)	**0.152** (<0.001)	**0.088** (<0.001)	1

Table 6.17 Bivariate Pearson correlation analysis of victimisation, fear and socioeconomic context of destination station

	Victim to crime	Victim to crime (incl. aggressive panhandling)	Victim to crime (Station)	Victim to crime (Train)	Fear of crime	Fear of crime (Way)	Fear of crime (Station)	Fear of crime (Train)	Population	Foreign background	University education	Multi-family housing	Employment rate	High economic std	Low economic std	Income	Crime rate around station	Travel Distance
Victim to crime	1	**0.729 (<0.001)**	**0.767 (<0.001)**	**0.706 (<0.001)**	**0.281 (<0.001)**	**0.246 (<0.001)**	**0.277 (<0.001)**	**0.239 (<0.001)**	0.032 (0.092)	0.023 (0.231)	0.001 (0.966)	0.023 (0.232)	−0.005 (0.801)	−0.015 (0.436)	0.028 (0.147)	−0.023 (0.228)	0.012 (0.547)	**−0.070 (<0.001)**
Victim to crime (incl. aggressive panhandling)	**0.729 (<0.001)**	1	**0.559 (<0.001)**	**0.514 (<0.001)**	**0.299 (<0.001)**	**0.254 (<0.001)**	**0.284 (<0.001)**	**0.243 (<0.001)**	0.034 (0.074)	0.025 (0.200)	0.001 (0.949)	0.031 (0.101)	0.005 (0.785)	−0.019 (0.318)	0.024 (0.212)	−0.018 (0.346)	**0.038 (0.046)**	**−0.057 (0.003)**
Victim to crime (Station)	**0.767 (<0.001)**	**0.559 (<0.001)**	1	**0.620 (<0.001)**	**0.278 (<0.001)**	**0.259 (<0.001)**	**0.289 (<0.001)**	**0.260 (<0.001)**	0.030 (0.114)	0.017 (0.364)	0.003 (0.886)	0.026 (0.175)	−0.014 (0.474)	−0.007 (0.702)	0.030 (0.123)	−0.025 (0.186)	−0.002 (0.915)	**−0.046 (0.016)**
Victim to crime (Train)	**0.706 (<0.001)**	**0.5140 (<0.001)**	**0.620 (<0.001)**	1	**0.226 (<0.001)**	**0.201 (<0.001)**	**0.224 (<0.001)**	**0.302 (<0.001)**	0.012 (0.533)	0.029 (0.130)	0.005 (0.802)	0.014 (0.479)	−0.005 (0.789)	−0.018 (0.353)	0.015 (0.441)	−0.015 (0.448)	−0.004 (0.826)	**−0.072 (<0.001)**
Fear of crime	**0.281 (<0.001)**	**0.299 (<0.001)**	**0.278 (<0.001)**	**0.226 (<0.001)**	1	**0.783 (<0.001)**	**0.786 (<0.001)**	**0.579 (<0.001)**	**0.123 (<0.001)**	−0.007 (0.725)	**0.077 (<0.001)**	**0.105 (<0.001)**	**0.052 (0.010)**	**0.044 (0.030)**	0.003 (0.878)	−0.009 (0.642)	**−0.062 (0.002)**	−0.021 (0.294)
Fear of crime (Way)	**0.246 (<0.001)**	**0.254 (<0.001)**	**0.259 (<0.001)**	**0.201 (<0.001)**	**0.783 (<0.001)**	1	**0.603 (<0.001)**	**0.426 (<0.001)**	**0.085 (<0.001)**	−0.034 (0.092)	**0.101 (<0.001)**	**0.085 (<0.001)**	**0.081 (<0.001)**	**0.070 (0.001)**	−0.039 (0.055)	0.033 (0.101)	**−0.055 (0.007)**	**−0.069 (0.001)**
Fear of crime (Station)	**0.277 (<0.001)**	**0.284 (<0.001)**	**0.289 (<0.001)**	**0.224 (<0.001)**	**0.786 (<0.001)**	**0.603 (<0.001)**	1	**0.517 (<0.001)**	**0.101 (<0.001)**	−0.017 (0.390)	**0.071 (<0.001)**	**0.099 (<0.001)**	**0.062 (0.002)**	**0.046 (0.025)**	−0.003 (0.885)	−0.004 (0.831)	**−0.059 (0.004)**	−0.001 (0.969)
Fear of crime (Train)	**0.239 (<0.001)**	**0.243 (<0.001)**	**0.260 (<0.001)**	**0.302 (<0.001)**	**0.579 (<0.001)**	**0.426 (<0.001)**	**0.517 (<0.001)**	1	**0.057** (0.005)	0.012 (0.566)	0.014 (0.485)	**0.041 (0.043)**	0.003 (0.878)	0.008 (0.684)	0.020 (0.336)	−0.024 (0.229)	−0.024 (0.236)	−0.017 (0.412)
Population	0.032 (0.092)	0.034 (0.074)	0.030 (0.114)	0.012 (0.533)	**0.123 (<0.001)**	**0.085 (<0.001)**	**0.101 (<0.001)**	**0.057** (0.005)	1	**0.364 (<0.001)**	**0.133 (<0.001)**	**0.626 (<0.001)**	**0.042 (0.021)**	−0.011 (0.550)	**0.338 (<0.001)**	**−0.388 (<0.001)**	**−0.322 (<0.001)**	**0.045 (0.038)**
Foreign background	0.023 (0.231)	0.025 (0.200)	0.017 (0.364)	0.029 (0.130)	−0.007 (0.725)	−0.034 (0.092)	−0.017 (0.390)	0.012 (0.566)	**0.364 (<0.001)**	1	**−0.279 (<0.001)**	**0.378 (<0.001)**	**−0.470 (<0.001)**	**−0.326 (<0.001)**	**0.760 (<0.001)**	**−0.633 (<0.001)**	**0.093 (<0.001)**	**0.088 (<0.001)**
University education	0.019 (0.333)	0.019 (0.312)	0.017 (0.367)	0.007 (0.716)	**−0.039 (0.054)**	**−0.071 (<0.001)**	**−0.040 (0.049)**	−0.002 (0.936)	**0.088 (<0.001)**	**0.679 (<0.001)**	**−0.816 (<0.001)**	**−0.052 (0.004)**	**−0.874 (<0.001)**	**−0.782 (<0.001)**	**0.914 (<0.001)**	**−0.864 (<0.001)**	−0.036 (0.051)	0.038 (0.083)
Multi-family housing	0.001 (0.966)	0.001 (0.949)	0.003 (0.886)	0.005 (0.802)	**0.077 (<0.001)**	**0.101 (<0.001)**	**0.071 (<0.001)**	0.014 (0.485)	**0.133 (<0.001)**	**−0.279 (<0.001)**	1	**0.340 (<0.001)**	**0.801 (<0.001)**	**0.909 (<0.001)**	**−0.683 (<0.001)**	**0.774 (<0.001)**	**0.077 (<0.001)**	−0.014 (0.522)
Employment rate	0.023 (0.232)	0.031 (0.101)	0.026 (0.175)	0.014 (0.479)	**0.105 (<0.001)**	**0.085 (<0.001)**	**0.099 (<0.001)**	**0.041 (0.043)**	**0.626 (<0.001)**	**0.378 (<0.001)**	**0.340 (<0.001)**	1	**0.293 (<0.001)**	**0.355 (<0.001)**	**0.200 (<0.001)**	**−0.197 (<0.001)**	**−0.202 (<0.001)**	**0.120 (<0.001)**

(continued)

Table 6.17 (continued)

	Victim to crime	Victim to crime (incl. aggressive panhandling)	Victim to crime (Station)	Victim to crime (Train)	Fear of crime	Fear of crime (Way)	Fear of crime (Station)	Fear of crime (Train)	Population	Foreign background	University education	Multi-family housing	Employment rate	High economic std	Low economic std	Income	Crime rate around station	Travel Distance
High economic std	−0.005 (0.801)	0.005 (0.785)	−0.014 (0.474)	−0.005 (0.789)	**0.052** **(0.010)**	**0.081** **(<0.001)**	**0.062** **(0.002)**	0.003 (0.878)	**0.042** **(0.021)**	**−0.470** **(<0.001)**	**0.801** **(<0.001)**	**0.293** **(<0.001)**	1	**0.734** **(<0.001)**	**−0.784** **(<0.001)**	**0.751** **(<0.001)**	**0.158** **(<0.001)**	0.016 (0.473)
Low economic std	−0.015 (0.436)	−0.019 (0.318)	−0.007 (0.702)	−0.018 (0.353)	**0.044** **(0.030)**	**0.070** **(0.001)**	**0.046** **(0.025)**	0.008 (0.684)	−0.011 (0.550)	**−0.326** **(<0.001)**	**0.909** **(<0.001)**	**0.355** **(<0.001)**	**0.734** **(<0.001)**	1	**−0.698** **(<0.001)**	**0.771** **(<0.001)**	**0.040** **(0.028)**	−0.027 (0.216)
Income	0.028 (0.147)	0.024 (0.212)	0.030 (0.123)	0.015 (0.441)	0.003 (0.878)	−0.039 (0.055)	−0.003 (0.885)	0.020 (0.335)	**0.338** **(<0.001)**	**0.760** **(<0.001)**	**−0.683** **(<0.001)**	**0.200** **(<0.001)**	**−0.784** **(<0.001)**	**−0.698** **(<0.001)**	1	**−0.944** **(<0.001)**	**−0.157** **(<0.001)**	**0.081** **(<0.001)**
Dest_Crime rate	−0.022 (0.261)	−0.016 (0.416)	−0.029 (0.134)	−0.014 (0.455)	−0.018 0.376)	0.029 (0.156)	−0.007 (0.737)	−0.028 (0.174)	**−0.411** **(<0.001)**	**−0.636** **(<0.001)**	**0.736** **(<0.001)**	**−0.222** **(<0.001)**	**0.792** **(<0.001)**	**0.714** **(<0.001)**	**−0.942** **(<0.001)**	**0.984** **(<0.001)**	**0.303** **(<0.001)**	**−0.082** **(<0.001)**
Travel Distance	−0.023 (0.228)	−0.018 (0.346)	−0.025 (0.186)	−0.015 (0.448)	−0.009 (0.642)	0.033 (0.101)	−0.004 (0.831)	−0.024 (0.229)	**−0.388** **(<0.001)**	**−0.633** **(<0.001)**	**0.774** **(<0.001)**	**−0.197** **(<0.001)**	**0.751** **(<0.001)**	**0.771** **(<0.001)**	**−0.944** **(<0.001)**	1	**0.278** **(<0.001)**	**−0.078** **(<0.001)**
Crime rate around station	0.012 (0.547)	**0.038** **(0.046)**	−0.002 (0.915)	−0.004 (0.826)	**−0.062** **(0.002)**	**−0.055** **(0.007)**	**−0.059** **(0.004)**	−0.024 (0.236)	**−0.322** **(<0.001)**	**0.093** **(<0.001)**	**0.077** **(<0.001)**	**−0.202** **(<0.001)**	**0.158** **(<0.001)**	**0.040** **(0.028)**	**−0.157** **(<0.001)**	**0.278** **(<0.001)**	1	0.037 (0.090)
Travel Distance	**−0.070** **(<0.001)**	**−0.057** **(0.003)**	**−0.046** **(0.016)**	**−0.072** **(<0.001)**	−0.021 (0.294)	**−0.069** **(0.001)**	−0.001 (0.969)	−0.017 (0.412)	**0.045** **(0.038)**	**0.088** **(<0.001)**	−0.014 (0.522)	**0.120** **(<0.001)**	0.016 (0.473)	−0.027 (0.216)	**0.081** **(<0.001)**	**−0.078** **(<0.001)**	0.037 (0.090)	1

References

Brunton-Smith IR (2011) Do neighbourhoods generate fear of crime? In: Royal Statistical Society Local Group

Ceccato V (2013) Moving safely: crime and perceived safety in Stockholm's subway stations. Lexington, Plymouth

Ceccato V, Langefors L, Näsman P (2021) Young people's victimization and safety perceptions along the trip. Nordic J Criminol 22(1):106–125. https://doi.org/10.1080/2578983X.2021.1882744

Cohen J (1988) Statistical power analysis for the behavioral sciences, 2nd edn. Routledge

Ferguson KM, Mindel CH (2007) Modeling fear of crime in Dallas neighborhoods: a test of social capital theory. Crime Delinq 53(2):322–349

Franklin TW, Franklin CA, Fearn NE (2008) A multilevel analysis of the vulnerability, disorder, and social integration models of fear of crime. Social Justice Res 21:204–227

Hummelsheim D, Hirtenlehner H, Jackson J, Oberwittler D (2011) Social insecurities and fear of crime: a cross-national study on the impact of welfare state policies on crime-related anxieties. Eur Sociol Rev 27(3):327–345

Lewis D, Maxfield M (1980) Fear in the neighborhoods: an investigation of the impact of crime. J Res Crime Delinq 17:160–189

Loukaitou-Sideris A (2012) Safe on the move: the importance of the built environment. In: The urban fabric of crime and fear. Springer, pp 85–110

Natarajan M, Schmuhl M, Sudula S, Mandala M (2017) Sexual victimization of college students in public transport environments: a whole journey approach. Crime Prevent Community Saf 19(3):168–182

Pain R (2000) Place, social relations and the fear of crime: a review. Prog Hum Geogr 24(3):365–387

Rader NE, Cossman JS, Porter JR (2012) Fear of crime and vulnerability: using a national sample of Americans to examine two competing paradigms. J Crim Just 40(2):134–141

Roman CG, Chalfin A (2008) Fear of walking outdoors: a multilevel ecologic analysis of crime and disorder. Am J Prev Med 34(4):306–312

Sampson RJ, Raudenbush SW, Earls F (1997) Neighborhoods and violent crime: a multilevel study of collective efficacy. Science 277(5328):918–924

Statistics Sweden (2020) Folkmängden per region efter ålder och kön. Retrieved from: https://www.statistikdatabasen.scb.se/pxweb/sv/ssd/START__BE__BE0101__BE0101Y/FolkmDesoAldKonN/

Vieno A, Roccato M, Russo S (2013) Is fear of crime mainly social and economic insecurity in disguise? A multilevel multinational analysis. J Community Appl Social Psychol 23(6):519–535

Visser M, Scholte M, Scheepers P (2013) Fear of crime and feelings of unsafety in European countries: macro and micro explanations in cross-national perspective. Sociol Q 54(2):278–301

Wang X (2012) Undocumented immigrants as perceived criminal threat: a test of the minority threat perspective. Criminology 50(3):743–776

Wiebe DJ, Richmond TS, Poster J, Guo W, Allison PD, Branas CC (2014) Adolescents' fears of violence in transit environments during daily activities. Secur J 27(2):226–241. https://doi.org/10.1057/sj.2014.8

Wilkinson R, Pickett K (2010) The spirit level. Why equality is better for everyone. Penguin UK

7 Safe Havens or Isolated Spots? Crime and Safety in Areas on the Rural–Urban Continuum

Abstract

Research on rural transit safety is limited. Rural stations contribute to connectivity, economic sustainability, and social cohesion in rural areas, but how safe are they? We start this chapter by presenting stories from rural areas worldwide to illustrate the potential of stations to serve as lifelines for remote areas. Then, we provide an in-depth analysis of how railway stations are affected by crime and how passengers perceive the environment of these stations when travelling during the day and night, using southern Sweden as an example based on 3,407 survey responses. Desolation and anonymity are issues that are associated with more rural stations, lack of shelter and access to personnel make these transit environments less safe, especially from the perspective of vulnerable groups. Findings show that individuals from rural stations reported higher victimisation rates than those in urban stations, with significantly more incidents involving threats, hate crimes, and violence but no significant differences for robbery, theft, sexual harassment, or stalking. Fear of crime was also more prevalent among rural individuals than those living in urban municipalities. The disparities in victimisation and fear raise questions about equity and justice, as rural residents may face insufficient safety measures, underreporting, or less effective interventions to address crime and fear in these areas. Targeted strategies are needed to ensure a fairer distribution of safety, from remote rural areas to larger cities.

Keywords

Stations · Rural · Urban · Survey · Crime · Safety · Women · Disability · Rural–urban continuum

7.1 Introduction

All over the world, limited public transport in rural areas disproportionately affects rural populations (Gray et al. 2001; Kenyon et al. 2002; McGuire et al. 2022; Velaga et al. 2012), making them heavily reliant on automobiles. Limited access to public transportation is a major constraint in rural areas, where most of the world's poor people live. According to the World Bank (2017), in rural Africa, for example, limited connectivity due to poor transport infrastructure affects over 450 million people, or 70% of the rural population. In urban areas, rapid population growth, expected to reach two billion by 2045, is outpacing the development of public transport, restricting access to economic and social opportunities. This imbalance between urban and rural areas affects everyone, but in particular vulnerable groups, such as children, women, older adults, individuals with disabilities, and the mobility-impaired living in

V. Ceccato et al., *Transit Safety for a Sustainable City*, Sustainable Development Goals Series,
https://doi.org/10.1007/978-3-031-97427-4_7

remote areas all over the world. The risk of victimisation, or the inability to use public transport while feeling entirely safe (whether on public transport, walking to or from a transit facility or stop, or waiting at a bus, transit stop, or station platform) can decrease the use of public transit when it is available.

A recent study in Sweden reveals that young women in rural areas may adopt different precautionary measures compared to their counterparts in urban areas (Ceccato et al. 2024b). Building on this initial study and dataset, this chapter reports on patterns of victimisation and safety perceptions of individuals living in areas across the rural-urban continuum, using responses from a survey from southern Sweden as a case study.

We use the concept of "mobility justice" here to call for fair and equitable transportation for all (Sheller 2018) to discuss temporal and spatial patterns of victimisation and safety perceptions in transit environments with a particular focus on women and disabled individuals. The concept of mobility justice also embraces the importance of areas across the rural-urban continuum, which refers to the spectrum of places that range from rural to urban, highlighting a gradual transition rather than a strict divide between these categories. Mobility, and consequently victimisation and fear, vary across the rural-urban continuum due to differences in urbanisation, lifestyles, population density, infrastructure, and land use, spanning remote rural areas, small towns, suburbs, and densely populated urban centres. This continuum highlights the complex interplay of social, economic, and environmental factors across different geographical scales in a globalised world, revealing the fluid and interconnected nature of places that sometimes show conventional signs of an "urban" nature and sometimes that of a "rural" nature (Ceccato and Abraham 2022).

We examine strategies to sustain and enhance the viability of rural stations, offering relevant examples from the study area to engage an international audience interested in both rural development and safety issues (see e. g. Hale and Harkness 2022; Harkness and White 2021). This chapter also connects to the SDGs by promoting sustainable communities (SDG 11), building resilient infrastructure (SDG 9), ensuring peace and justice (SDG 16), reducing inequalities (SDG 10), and supporting economic growth (SDG 8). It highlights the role of safe, accessible rural transportation in fostering equitable rural development, social cohesion, and sustainability.

This chapter examines crime and fear in rural transit systems, focusing on victimisation and safety perceptions across rural-urban contexts. Section 7.1 introduces the aim and objectives, followed by Sect. 7.2, which discusses rural transit conditions and their impact on crime and transit safety. Section 7.3 outlines the case study, data, and methods. Section 7.4 introduces victimisation patterns, while Sect. 7.5 explores patterns of fear of crime. Section 7.6 identifies victimisation and safety perception determinants, with subsections on women's safety and safety precautions. Finally, Sect. 7.7 summarises findings, discusses policy implications, and suggests future research. This study aims to provide information on equitable safety measures and bridge the rural-urban divide by addressing rural transit safety.

7.2 Situational Conditions of Crime and Fear in Rural Transit Contexts

Crime and safety in rural areas deserve special attention. First, crime is not exclusive to urban settings; it is shaped by its environment, with certain crimes unique to rural contexts. Influenced by local and global factors, rural crime challenges law enforcement and threatens rural sustainability (Barclay et al. 2004; Donnermeyer 2016; Mawby 2015). Second, low crime rates in rural areas often mask underlying crime-related issues that affect communities (Yarwood 2001). Third, police efforts and crime prevention should be tailored to the safety needs of those living in the most sparsely populated parts of the country (Harkness 2020; Hogg and Carrington 2006; Weisheit et al. 2005). Lastly, rural criminology plays a vital role in bridging natural and social sciences, offering interdisciplinary insights that support the sustainable development of rural areas and aligning with the United Nations Sustainable Development Goals (SDGs) (UN 2015). Recognising and addressing these issues is essential for scholars, policymakers, and society.

Crime and safety in rural areas can be approached through a situational lens. This means focusing on the specific environmental, social, and temporal factors that create crime opportunities (Clarke 1992). This approach examines how variations in daily routines, the presence or absence of capable guardians, and the movement of potential offenders and suitable targets contribute to patterns of victimisation (Cohen and Felson 1979) and people's safety perceptions in these environments. By analysing these dynamics, we can better understand how specific contexts influence the risk of crime and plan for more context-sensitive safety interventions (Aransiola and Ceccato 2020).

In comparison to urban areas, people living in rural areas often face a lack of reliable public transportation, compelling individuals to rely on personal vehicles or to walk long distances (Ceccato 2015). The great distances between residences and the sparse population density in rural areas can exacerbate a sense of isolation and fear. Moreover, rural areas often experience fewer police patrols, which can amplify the fear of crime. This perception stems from a sense of inadequate protection and the belief that reporting incidents may be less impactful or yield limited results (Abraham and Ceccato 2022).

Desolation and anonymity are significant challenges often associated with transit nodes, bus stops and rural train stations. The lack of shelter, limited access to personnel, and inadequate infrastructure contribute to transit environments that feel unwelcoming and unsafe. Without the presence of staff or visible security measures, these spaces can become particularly intimidating, especially during late hours or in adverse weather conditions. This issue is further amplified in regions such as Scandinavian countries, where long, dark winters exacerbate feelings of isolation and vulnerability. The combination of harsh weather, extended periods of darkness, and sparse station facilities can deter passengers, particularly women and younger people, from using rail-bound transportation.

In contrast, urban areas, while offering better access to public transportation and a better police presence, may still present challenges to women's safety, such as crowded or poorly maintained environments. What remains unclear is whether people's safety needs differ significantly between rural and urban contexts or if they share similar underlying concerns.

The literature highlights the impact of factors in transport nodes such as restaurants, CCTV, lighting, staff, and surveillance on safety. These factors include the extent of empty space in stations, open environments, kiosks, underpasses, size, lighting, body-worn cameras, and maintenance. As suggested in the initial chapters of this book, in train stations, safety is influenced not only by physical design but also by human activities and the presence or absence of staff. They are shaped by their surroundings, including the socioeconomic conditions of the local population, which influence crime rates and travellers' safety perceptions both at and en route to the station. However, the international literature lacks clarity on whether these findings apply to railway stations in smaller municipalities. Research on rural areas is limited. A recent literature review on perceived safety concluded that most studies address large cities such as London, Madrid, or Delhi, with only a small share of focus on small towns and villages (Sundling and Ceccato 2022) and there is an overall limited understanding of which crimes occur within station environments and their adjacent areas (Ceccato et al. 2024c).

According to the World Bank report from 2017, a lack of safety in the public transport system limits labour market participation and the job search radius of the poor, particularly for women, but improving nighttime lighting reduces the incidence of crime around rapid transit bus stops (World Bank 2017). One reason is that in rural communities, the close-knit nature of social networks may discourage people, especially women, from reporting crime due to fears of social stigma, ostracism, retaliation, or the spread of rumours, particularly in cases of domestic violence (DeKeseredy 2020).

For this case study, we followed the recent strand of Western research on transit safety and set out to investigate the following research questions.

- How do travellers' experiences of victimisation and their perceptions of safety vary across the rural-urban continuum?
- What factors are most associated with victimisation and safety perceptions in transit in areas across the rural-urban continuum?
- What are the changes/recommendations needed for improving individuals' safety?

7.3 The Case Study

The study area comprises 47 rail-bound stations distributed over three railway lines stretching through central and southern Sweden, serving 28 municipalities with a total population of 1.78 million inhabitants (Statistics Sweden 2020). We have adopted the definition of "rural municipality" ("landsbygdskommun") provided by the Swedish Agency for Economic and Regional Growth (Tillväxtverket 2021), which identifies municipalities based on population density and distance to urban centres (Fig. 7.1). According to this definition, municipalities are categorised as rural if most of their population lives in rural areas and are not close to larger urban centres. In the study area, nine municipalities are classified as rural municipalities, containing ten stations, while the remaining municipalities fall under mixed or metropolitan categories. Rural municipalities are characterised by smaller populations and a lack of proximity to major urban centres, while metropolitan municipalities have higher population densities and tend to have a greater concentration of services and public transportation options. Mixed municipalities, on the other hand, combine rural and urban features, offering a blend of access to urban amenities and rural surroundings.

Altogether, 64,000 passengers travel on these railway lines annually (Swedish Transport Administration 2009). Most stations are situated in small towns with populations under 15,000, but many of these towns are near larger cities or municipalities, enabling commuting to urban areas with populations between 15,000 and 37,499. Medium-sized urban areas along the routes include Södertälje, Eskilstuna, Örebro, Norrköping, and Linköping, while Gothenburg is the only city classified as a major urban centre with over 37,500 inhabitants. The stations studied primarily function as long-distance railway stations, though some also serve as commuter hubs. Table 7.1 shows the distribution of respondents, stations and municipalities by municipality type, from urban to rural.

7.4 Methods for Analysis

Cross-tables were used to identify patterns of fear of crime, victimisation, and safety precautions across the three municipality types: rural, mixed, and urban. The Chi-square test was applied to determine whether the observed differences in the variables across rural, mixed, and urban municipalities were statistically significant.

In addition to descriptive analyses, logistic regression models were conducted to assess the likelihood of victimisation and fear across different municipality types. These models allowed us to explore the contributing factors, how they differ across rural, mixed, and urban municipalities, and their respective impacts on victimisation and fear.

Prior to conducting the regression analysis, a bivariate correlation analysis was carried out to ensure that highly correlated independent variables were not included in the models. A range of variables was tested before selecting the final set included in this chapter. These included, for example, ethnicity, income, and the number of operators at the station. However, none of these variables significantly impacted either fear or victimisation and were excluded from the final models. Additionally, the variable "unsafe in the neighbourhood" was excluded to avoid circular

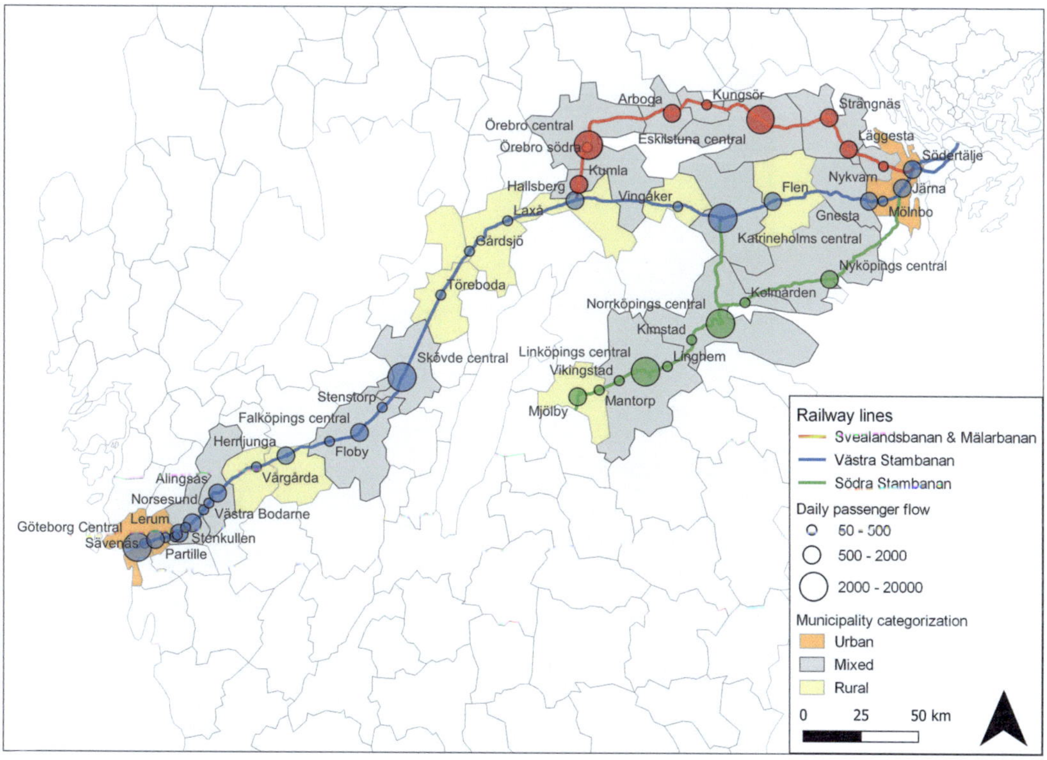

Fig. 7.1 Map of the study area with rural municipalities marked. (*Source* Authors)

Table 7.1 Distribution of respondents, stations, and municipalities by municipality types

	Municipalities	Stations	Respondents	Respondents/ Stations	Average no. of passengers daily	Average no. of operators
Rural	9	10	609	61	645	2.3
Mixed	16	30	2,726	91	1,142	2.0
Urban	3	7	578	83	3,329	2.3
Total	28	47	3,913	83	1,362	2.1

reasoning, as it directly describes safety perceptions.

7.5 Results

7.5.1 Victimisation Across the Rural-Urban Continuum

Individuals using stations in rural municipalities reported higher victimisation levels than those living in municipalities with stations classified as urban stations, but note that this relationship is significant at the 10% level. Specifically, 16% of respondents from rural stations reported being victims, in relation to 12% from urban stations (Chi-square = 3.16, df = 1, $p = 0.075$). Incidents involving threats or hate crimes were significantly more common in rural stations (6%) in comparison to urban areas (3%; Chi-square = 5.59, df = 1, $p = 0.018$). Similarly, victimisation involving violence was more prevalent in rural stations (4% compared to 2%; Chi-square = 4.77, df = 2, $p = 0.029$). However, no statistically significant differences were found in the reports of robbery, theft, sexual harassment, or stalking between stations in

(a)

(b)

Fig. 7.2 Examples of rural stations in (a) summer and (b) winter. (*Source* Authors (2022))

urban and rural municipalities. Figure 7.2 illustrates stations in smaller municipalities in the summer (a) and winter (b).

7.5.2 Fear of Crime Across the Rural-Urban Continuum

Poor safety perceptions (fear of crime) were generally higher among respondents from stations in rural municipalities than in urban ones. For instance, 38% of rural respondents reported a fear of crime, compared to 30% in urban areas (Chi-square = 5.64, df = 1, p = 0.018). Fear of specific crimes, such as property crimes, especially robbery, was much higher in rural stations (29%) compared to urban areas (21%; Chi-square = 6.69, df = 1, p = 0.010). Fear of assault was particularly pronounced, with more than a quarter of rural respondents expressing concern compared to 16% in urban areas (Chi-square = 16.02, df = 1, $p < 0.001$). Similarly, rural travellers also reported a greater fear of being victims of threats or hate crimes (21%) in comparison to urban travellers

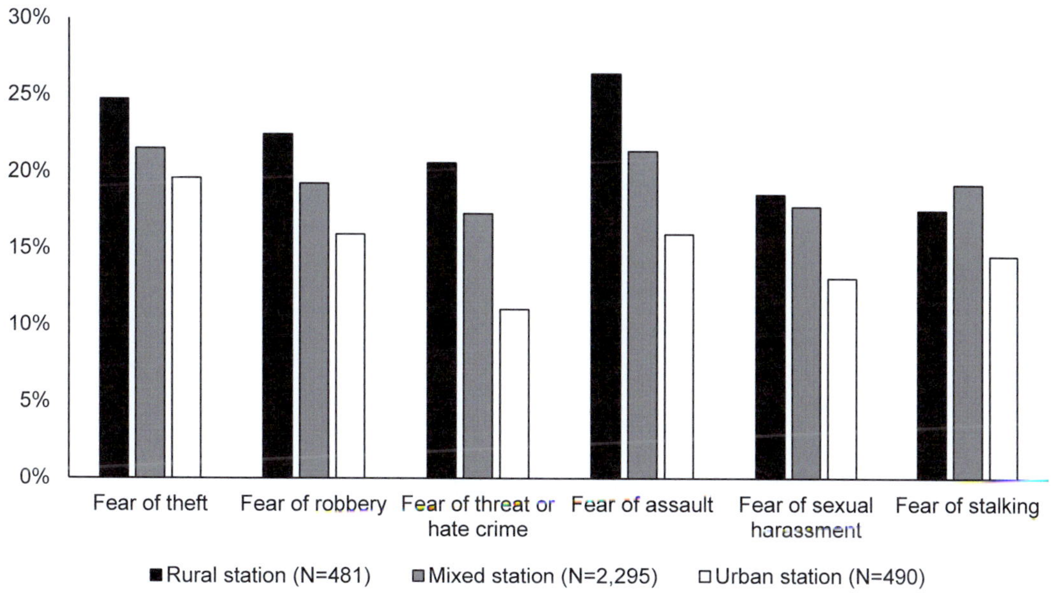

Fig. 7.3 Travellers' fear of crime categorised by crime type

(11.0%; Chi-square = 16.72, df = 1, $p < 0.001$), as well as sexual harassment (18% compared to 13%; Chi-square = 5.42, df = 1, $p = 0.020$). However, the fear of theft and stalking does not show a significant difference between the two types of stations (Fig. 7.3).

7.5.2.1 Spatial and Temporal Patterns

Safety perceptions vary significantly between respondents from urban and rural municipalities, both temporally and spatially. Rural respondents are more likely to feel unsafe in specific environments, especially at stations. While there were no notable differences in safety perceptions onboard the train or during the journey to and from the station, rural respondents reported much higher levels of fear at the station. Nearly a third of rural respondents expressed fear of crime at the station, compared to only 18% of urban travellers (Chi-square = 18.65, df = 1, $p < 0.001$). Similarly, 27% of rural travellers felt unsafe in the waiting rooms, while just 13.5% of urban travellers reported the same (Chi-square = 21.32, df = 1, $p < 0.001$). Although many rural stations lack toilets, those that do have them are not perceived as safe. Our findings show that poor safety perceptions in station toilets was higher in rural areas (19.5% compared to 13.0%; Chi-square = 4.24, df = 1, $p = 0.040$).

Additionally, rural respondents were more likely to feel unsafe during the day, with 40.9% declaring to *sometimes or always* feel unsafe compared to 37% in urban areas (Chi-square = 8.91, df = 1, $p = 0.003$). This difference was statistically significant for safety perceptions onboard the train, at the station, and during the journey to and from the station during the daytime. However, poor safety perceptions at night did not show significant differences between the two groups, suggesting that daytime fears are a more notable issue in stations areas in rural municipalities.

7.5.3 Determinants of Victimisation and Safety Perceptions Across the Rural-Urban Continuum

To better capture the rural–urban continuum, we break down the stations' contexts into three types of municipalities rather than two. Table 7.2 presents the results of logistic regression models analysing victimisation in transit across three geographical contexts: rural areas (N = 360), mixed areas (N = 1,618), and urban areas (N = 318). Drug use/selling at stations is a key risk factor in rural and mixed areas, while LGBTQI+ individuals face heightened vulnerability in mixed areas. The variable young age traveller strongly predicts victimisation in rural areas, whereas reduced mobility and tunnel usage are significant challenges in mixed areas.

For safety perceptions (Table 7.3), across all contexts, being a woman, identifying as LGBTQI+ , and previous victimisation emerge as significant predictors of fear in transit, with the highest odds observed in urban areas. Rural areas emphasise personal safety concerns, mixed areas highlight social vulnerabilities and accessibility challenges, and urban areas reveal heightened environmental and nighttime travel risks. Table 7.3 shows the models for individual safety perceptions in transit.

In rural areas, victimisation in transit appears to be influenced by several key factors. Young individuals (18 or under) are 4.7 times as likely to be victimised as other groups. Additionally, the presence of drug use/selling at stations doubles the odds of victimisation. A lack of staff also plays a significant role, emphasising the importance of the presence of staff in mitigating victimisation risks. Other factors, such as gender, reduced mobility, and tunnel usage, do not show significant effects, suggesting a unique victimisation profile for rural transit environments. However, gender affects perceived safety. In rural areas (N = 362), fear in transit is strongly associated with being a woman, travelling only at night, and being a victim of crime. Lack of staff and poor maintenance also contribute significantly to safety perceptions. Drug use or the selling of drugs at stations is another critical factor. Individuals with a university education are less likely to experience fear. These findings highlight the importance of addressing personal safety and environmental conditions in rural transit contexts.

In mixed areas, women are significantly more likely to report victimisation, indicating gender-specific vulnerabilities in these contexts. Similarly, individuals identifying as LGBTQI+ , non-binary, or other gender identities face higher odds of victimisation. Individuals with reduced mobility are 2.3 as likely to be victimised as the rest of the sample, suggesting the need for a revision of the transit infrastructure with accessible transit design. Transit environments such as tunnels and stations with drug-related activity are particularly worrying. In mixed areas, women are almost twice as likely to report fear in transit, while LGBTQI+ individuals face significantly heightened fear. Travelling at night and being a victim of crime are key predictors. Environmental factors, such as a lack of staff and the presence of tunnels, further elevate transit-related fear. Poor maintenance and drug-related activities also play significant roles.

Urban areas exhibit distinct victimisation patterns compared to rural and mixed contexts. Travellers' university education is a protective factor, significantly lowering the odds of victimisation. In contrast, other variables, including gender, reduced mobility, and tunnel usage, do not yield statistically significant results. In urban areas (N = 330), women and LGBTQI+ individuals are significantly more likely to report fear in transit. Key environmental factors include lack of staff and the use of tunnels. Fear is also strongly associated with previous victimisation. Nighttime travel significantly increases the odds of fear, while poor maintenance and drug-related activities do not show significant effects.

7.5.4 Transit Safety of Women in Rural-Urban Contexts

Although rural women feel safer than their urban counterparts, previous research in this study area shows they are more often willing to take precautionary measures before taking a trip (e.g., travelling with someone in the evening or avoiding certain stations). Chi-square analysis and logistic

Table 7.2 Determinants of victimisation and safety perceptions across the rural-urban continuum

	(a) Victimisation in transit in rural areas (N = 360)				(b) Victimisation in transit in mixed areas (N = 1,618)				(c) Victimisation in transit in urban areas (N = 318)			
	Exp(B)	95% C.I.for EXP(B)		Sig	Exp(B)	95% C.I.for EXP(B)		Sig	Exp(B)	95% C.I.for EXP(B)		Sig
		Lower	Upper			Lower	Upper			Lower	Upper	
Woman	0.932	0.497	1.745	0.825	**1.687**	**1.188**	**2.395**	**0.003**	1.489	0.704	3.152	0.298
LGTBQI + /Non-binary/Other	0.000	0.000		0.998	**2.955**	**1.590**	**5.491**	**0.001**	0.485	0.055	4.305	0.516
Young (18 or under)	**4.721**	**1.655**	**13.463**	**0.004**	1.158	0.655	2.047	0.613	1.476	0.511	4.270	0.472
Reduced mobility	1.662	0.496	5.572	0.411	**2.322**	**1.363**	**3.956**	**0.002**	0.393	0.048	3.246	0.386
University education	**0.583**	**0.314**	**1.083**	**0.088**	0.938	0.682	1.289	0.692	**0.343**	**0.162**	**0.726**	**0.005**
Travel only at nighttime	0.577	0.214	2.144	0.507	1.319	0.812	2.142	0.253	1.537	0.559	4.231	0.405
Restaurant/café	1.319	0.561	3.101	0.526	0.818	0.581	1.154	0.253	1.370	0.164	11.458	0.771
Tunnel	0.667	0.280	1.586	0.359	**2.141**	**1.393**	**3.291**	**0.001**	1.669	0.196	14.208	0.639
Lack of staff	**1.942**	**1.063**	**3.548**	**0.031**	1.228	0.908	1.662	0.182	0.715	0.346	1.478	0.365
Poor maintenance	1.201	0.618	2.335	0.588	1.017	0.718	1.439	0.924	1.738	0.782	3.859	0.175
Drug use/selling at station	**2.046**	**1.111**	**3.767**	**0.022**	**1.501**	**1.102**	**2.045**	**0.010**	1.517	0.715	3.219	0.278
(a) Nagelkerke R Square	0.151											
(b) Nagelkerke R Square	0.062											
(c) Nagelkerke R Square	0.109											

Table 7.3 Determinants of safety perceptions across the rural-urban continuum

	(a) Fear in transit in rural areas (N = 362)				(b) Fear in transit in mixed areas (N = 1,649)				(c) Fear in transit in urban areas (N = 330)			
	Exp(B)	95% C.I.for EXP(B)		Sig	Exp(B)	95% C.I.for EXP(B)		Sig	Exp(B)	95% C.I.for EXP(B)		Sig
		Lower	Upper			Lower	Upper			Lower	Upper	
Woman	**3.119**	**1.750**	**5.561**	**0.000**	**1.942**	**1.506**	**2.503**	**0.000**	**1.911**	**1.047**	**3.488**	**0.035**
LGTBQI_Non-binary_Other	1.295	0.328	5.120	0.712	**3.110**	**1.829**	**5.289**	**0.000**	**3.685**	**1.095**	**12.401**	**0.035**
Young (18 or under)	1.062	0.349	3.237	0.915	0.874	0.554	1.378	0.562	1.207	0.443	3.292	0.713
University education	**0.587**	**0.347**	**0.994**	**0.047**	**0.673**	**0.528**	**0.858**	**0.001**	1.043	0.559	1.944	0.895
Travel only at nighttime	**3.359**	**1.309**	**8.618**	**0.012**	**1.641**	**1.091**	**2.469**	**0.017**	**2.675**	**1.177**	**6.080**	**0.019**
Victim to crime	**4.988**	**2.480**	**10.033**	**0.000**	**4.530**	**3.243**	**6.328**	**0.000**	**7.823**	**3.494**	**17.517**	**0.000**
Restaurant/Cafe	0.737	0.370	1.468	0.385	**0.518**	**0.395**	**0.678**	**0.000**	**3.997**	**0.800**	**19.967**	**0.091**
Tunnel	1.072	0.521	2.208	0.850	**1.950**	**1.461**	**2.603**	**0.000**	**11.166**	**2.195**	**56.806**	**0.004**
Lack of staff	**2.712**	**1.646**	**4.468**	**0.000**	**2.333**	**1.858**	**2.931**	**0.000**	**4.089**	**2.285**	**7.318**	**0.000**
Poor maintenance	**2.028**	**1.122**	**3.665**	**0.019**	**1.481**	**1.138**	**1.928**	**0.004**	1.132	0.564	2.274	0.727
Drug use/selling at station	**2.896**	**1.702**	**4.929**	**0.000**	**1.880**	**1.483**	**2.382**	**0.000**	1.048	0.576	1.905	0.878
(a) Nagelkerke R Square	0.340											
(b) Nagelkerke R Square	0.256											
(c) Nagelkerke R Square	0.320											

regression models underlied the methodology of this previous study (Ceccato et al. 2024b) that show that up to 48% of girls under 18 years old express fear of victimisation while using rail-bound transportation, compared to 35% of women over 18 years old (Chi-square = 8.5, df = 1, p = 0.004). However, they report lower levels of fear regarding property crimes (Chi-square = 20.5, df = 1, $p < 0.001$) and violence (Chi-square = 8.5, df = 1, p = 0.004). In contrast, they exhibit greater fear of experiencing sexual harassment (Chi-square = 24.6, df = 1, $p < 0.001$) and stalking (Chi-square = 14.6, df = 1, $p < 0.001$). The findings of this study (Ceccato et al. 2024b) call for gender-age-sensitive mobility policies that consider the intersectionality of transit safety and reflect women's and girls' safety needs in these particular rural contexts. Improvements in the environment to which women are exposed along their trips are necessary to make women safer, especially those who have longer trips from their homes to the stations. An important area for future research is identifying settings and places that should be improved to decrease victimisation and maximise safety perceptions for women and girls in areas on the rural-urban continuum.

7.5.5 Safety Precautions

Related to the higher fear levels among individuals from rural stations, rural travellers tend to take more precautions than the urban ones, especially during the day. For instance, 47% of rural individuals report taking safety measures during the day, compared to 39% of urban travellers, a statistically significant difference (p = 0.013). This suggests that travellers in rural areas may generally be more cautious when navigating their environment.

In addition, rural travellers are more likely to travel with someone during the day, with 10% of rural respondents reporting this compared to only 6% of urban travellers (Chi-square = 4.05, df = 1, p = 0.044). This could indicate a heightened perception of risk in rural areas, where travelling alone may make one feel more vulnerable. As many as 16% of rural individuals also avoid carrying a handbag compared to 11% of those living in urban areas (Chi-square = 4.94, df = 1, p = 0.026), further suggesting that rural travellers may feel more exposed to the risk of theft or assault in the evening hours.

One of the most significant differences is the precaution of carrying some form of defence weapon (i.e. pepper spray). Individuals living in rural areas are significantly more likely to report this, with 11% reporting this precaution in the evening compared to just 4% of urban travellers (Chi-square = 15, df = 1, p = 0.000). This difference is also notable during the day, where 4% of rural travellers carry a defence weapon compared to only 1% of urban travellers (Chi-square = 8.15, df = 1, p = 0.004). This suggests that those living in more rural areas may experience a greater need for self-defence due to their perception of being at greater risk of victimisation.

Rural respondents also report trying to look confident as a precautionary measure to a higher degree, with 35% declaring taking this precaution in the evening compared to 26% of urban respondents (Chi-square = 8.18, df = 1, p = 0.004). Furthermore, a higher proportion of rural individuals (15%) avoid travelling at certain times in the evening, compared to 9% of urban respondents (Chi-square = 9.13, df = 1, p = 0.003). This indicates that rural individuals may be more mindful of when they choose to travel and may fear being left alone at the station during the evening when fewer passengers are present, especially in rural settings as opposed to urban areas.

Overall, rural travellers tend to take more precautionary measures and are more likely to engage in behaviours such as travelling with someone, carrying some defence weapon, or avoiding certain times for travel. The difference is particularly significant for daytime precautions, where almost half of the rural travellers take some precaution compared to 39% of urban travellers (Chi-square = 6.19, df = 1, p = 0.013).

7.6 Conclusions

This study underscores the significant challenges posed by victimisation and safety perceptions across the rural-urban continuum, particularly for women, individuals identifying as LGBTQI+ , and those with reduced mobility. Rural transit environments exhibit unique safety challenges imposed by travellers' higher victimisation rates and, heightened levels of fear when compared with those travellers in urban areas. Fear of crime, particularly at stations, remains a pervasive issue, with rural travellers reporting significantly higher levels of fear even during the daytime. Poorer safety perceptions during the day than evening hours is not a typical result found in the international literature, given that higher levels of fear are often found during the evening hours. One explanation could be that daytime hours reveal more clearly the physical and social disorder at stations, such as panhandling, which may shape perceptions of insecurity. Another possibility is that travellers' expectations of safety are higher during the day. Regardless of the possible explanations, these findings point to the interplay of social, environmental, and situational factors influencing both victimisation and perceptions of safety, highlighting the urgent need for targeted interventions that consider the specific vulnerabilities of rural transit users. Urban areas, while benefiting from better connectivity, still face challenges related to nighttime travel and the safety of vulnerable groups. Enhancing lighting and ensuring staff availability during late hours could help address these concerns. The protective role of travellers' education observed across all contexts suggests that information campaigns and educational programmes about personal safety and available resources could empower travellers to navigate transit environments more confidently.

A recommendation is to improve staffing and maintenance at rural transit stations, as the lack of personnel and poorly maintained facilities strongly contribute to both victimisation and fear. An increased staff presence, better lighting, and enhanced surveillance could foster a greater sense of safety for travellers. Moreover, incorporating inclusive and accessible design for individuals with reduced mobility is essential in transit environments in rural and mixed areas. Policymakers should prioritise funding for these infrastructure improvements as part of broader long term efforts to ensure equitable and sustainable transit systems that meet the needs of vulnerable groups.

The study reveals distinct patterns of victimisation and fear in mixed areas, with gender and LGBTQI+ identities serving as significant predictors. Therefore, transit systems should implement diversity-sensitive policies, including anti-discrimination training for staff and campaigns to foster a welcoming and inclusive environment. In addition, addressing environmental factors such as tunnels and areas with drug-related activities is critical and urgent. Adopting shared goals, local governments and transit authorities could work collaboratively to redesign these environments to mitigate risks and promote safety.

References

Abraham J, Ceccato V (2022) Crime and safety in rural areas: a systematic review of the English-language literature 1980–2020. J Rural Stud 94:250–273. https://doi.org/10.1016/j.jrurstud.2022.05.010

Aransiola TJ, Ceccato V (2020) The role of modern technology in rural situational crime prevention: a review of the literature. Rural Crime Prevention-Theory, Tactics and Techniques, pp 58–72

Barclay E, Donnermeyer JF, Jobes PC (2004) The dark side of gemeinschaft: criminality within rural communities. Crime Prevent Community Saf 6(3):7–22. https://doi.org/10.1057/palgrave.cpcs.8140191

Ceccato V (2015) Rural crime and community safety. J Rural Stud 39:157–159. https://doi.org/10.1016/j.jrurstud.2015.04.001

Ceccato V, Abraham J (2022) Crime and safety in the rural: lessons from research. Springer Nature

Ceccato, V., Gliori, G., and Sundling, C (2024a) LGBTQI+ safety perceptions in transit environments. Journal of Urban Mobility, 6, 100087. https://doi.org/10.1016/j.urbmob.2024.100087

Ceccato V, Sundling C, Gliori G (2024b) Transit safety of women in rural–urban contexts. Fem Criminol 19(4):372–394

Ceccato V, Sundling C, Gliori G (2024c) What makes a railway station safe and for whom? The impact of transit environments on passengers' perceptions of victimisation and safety. Eur Transport Res Rev

Clarke RVG (1992) Situational crime prevention: successful case studies

Cohen LE, Felson M (1979) Social change and crime rate trends: a routine activity approach. Am Sociol Rev 588–608

DeKeseredy WS (2020) Woman abuse in rural places. Routledge

Donnermeyer JF (2016) Introduction to the international handbook of rural criminology. In: The Routledge international handbook of rural criminology. Routledge, pp. 1–10
Gray D, Farrington J, Shaw J, Martin S, Roberts D (2001) Car dependence in rural Scotland: transport policy, devolution and the impact of the fuel duty escalator. J Rural Stud 17(1):113–125
Hale R, Harkness A (2022) Contemporary dimensions of rural victimisation. Rural Victims of Crime: Representations, Realities and Responses
Harkness A (2020) Rural crime prevention: theory, tactics and techniques. Routledge
Harkness A, White R (2021) Crossroads of rural crime: representations and realities of transgression in the Australian countryside. Emerald Publishing Limited
Hogg, R., and Carrington, K (2006) Policing the rural crisis. Federation Press.
Kenyon S, Lyons G, Rafferty J (2002) Transport and social exclusion: investigating the possibility of promoting inclusion through virtual mobility. J Transp Geogr 10(3):207–219
Mawby RI (2015) Exploring the relationship between crime and place in the countryside. J Rural Stud 39:262–270. https://doi.org/10.1016/j.jrurstud.2014.12.003
McGuire R, Longo A, Sherry E (2022) Tackling poverty and social isolation using a smart rural development initiative. J Rural Stud 89:161–170
Sheller M (2018) Theorising mobility justice. Tempo Social 30(2):17–34
Statistics Sweden (2020) Folkmängden per region efter ålder och kön. Retrieved from: https://www.statistikdatabasen.scb.se/pxweb/sv/ssd/START__BE__BE0101__BE0101Y/FolkmDesoAldKonN/
Sundling C, Ceccato V (2022) The impact of rail-based stations on passengers' safety perceptions. A systematic review of international evidence. Transport Res f: Traffic Psychol Behav 86:99–120
Swedish Transport Administration (2009) Number of passengers, Annual Average Daily Traffic (AADT) 2009
Tillväxtverket (2021) Städer och landsbygder. Tillväxtverket. https://tillvaxtverket.se/tillvaxtverket/statistikochanalys/statistikomregionalutveckling/regionalaindelningar/staderochlandsbygder.1844.html
UN, United Nations (2015) Transforming our world: the 2030 Agenda for Sustainable Development. United Nations: New York, NY, USA
Velaga NR, Beecroft M, Nelson JD, Corsar D, Edwards P (2012) Transport poverty meets the digital divide: accessibility and connectivity in rural communities. J Transport Geogr 21:102–112. https://doi.org/10.1016/j.jtrangeo.2011.12.005
Weisheit RA, Falcone DN, Wells LE (2005) Crime and policing in rural and small-town America. Waveland Press
World Bank (2017) Sustainable mobility for all. Global Mobility Report 2017: Tracking Sector Performance. https://documents1.worldbank.org/curated/en/920101508269072500/pdf/120500-REPL-PUBLIC-GM-Report-2017-Online-04-06-18.pdf
Yarwood R (2001) Crime and policing in the British countryside: some agendas for contemporary geographical research. Sociol Ruralis 41(2):201–219. https://doi.org/10.1111/1467-9523.00178

Safety and Station Management: Perspectives from Sweden, the Netherlands, and Denmark

8

Abstract

This chapter examines the historical development, organisational structures, and management of safety issues within European railway systems, focusing on Sweden, the Netherlands, and Denmark. It draws on 14 semi-structured interviews with safety experts, gathering insights into their knowledge and perspectives on the organisation of railway systems and station preparedness for safety-related concerns. Key challenges identified include the ongoing evolution of threats and the critical role of human presence in ensuring passenger safety. Interviewees also emphasised technological and organisational obstacles, highlighting the need to address different user groups' diverse safety requirements when planning future stations' safety conditions.

Keywords

Safety · Crime · Public transportation · Stations · Physical and social environment · Rail-bound

8.1 Introduction

The physical and social characteristics of a station's environment impact crime levels and travellers' perceptions of safety (Loukaitou-Sideris et al. 2002). However, they are not the only factors affecting safety conditions. Research has shown that the quality of the way a place is managed is essential in influencing what happens within these environments (Clarke and Bichler-Robertson 1998; Eck et al. 2023; Madensen and Eck 2012). As suggested by Eck (2003), station managers—those who control, supervise, or own spaces—are essential in detecting and deterring potential criminal behaviours. For example, transit stations and the surrounding public areas are controlled by the people who own and/or manage them. They may have a single owner or multiple private and public actors as owners, such as transit operators, municipalities, real estate companies, retailers, etc.

In some countries, station managers possess a comprehensive understanding of how transportation hubs function and, depending on national guidelines, have the autonomy to make safety-related decisions that impact thousands of daily passengers. In other countries, the actions of station managers (if they exist) are limited by job descriptions and institutional mandates. These variations reflect decades of organisational reform in Europe. Since the late 1980s, many European countries have restructured their national railway systems, introducing vertical separation

V. Ceccato et al., *Transit Safety for a Sustainable City*, Sustainable Development Goals Series,
https://doi.org/10.1007/978-3-031-97427-4_8

of infrastructure from operations and deregulation (Alexandersson and Hultén 2008). We wonder how these organisational structures influence passenger safety in Sweden, the Netherlands, and Denmark.

This study aims to examine the historical evolution of European railway systems, focusing on Sweden, the Netherlands, and Denmark. It investigates how national organisational frameworks shape station management, particularly service provision and safety priorities. Through interviews with transport operators and stakeholders, the study highlights challenges and opportunities in maintaining a secure transit environment. Additionally, it explores interviewees' expectations regarding station design, technology, and security to address evolving passenger needs and improve their travel experience. The characterisation of railway systems presented in this chapter is based mainly on the information (often uneven) obtained from interviews with professionals in each country. We do not claim to provide a comprehensive account of all the facts required to classify these systems, but instead, we have aimed to describe their main features as far as possible with the material available. This chapter should therefore be read as indicative rather than exhaustive, calling for future research to establish an in-depth research agenda in this field.

8.2 Research Design

Sweden was selected as a reference because this study was carried out as part of the research project "The effect of the station environment for crime and passenger safety", funded by the Swedish Transport Administration (Trafikverket) from 2021 to 2023. In Sweden, the Swedish National Council for Crime Prevention (BRÅ) is a government agency that supports crime prevention efforts through research, statistics, and development initiatives. It enhances knowledge about crime and safety, guides local and regional actors, and evaluates the effectiveness of crime prevention measures. At the regional level in Sweden, county administrative boards support and coordinate crime prevention efforts under this regulation. Their role is to reduce crime and enhance safety within the region by supporting knowledge- and problem-based local and regional crime prevention initiatives, fostering collaboration and knowledge exchange among relevant actors at local and regional levels, promoting competence development and cross-sectoral training in crime prevention, and initiating and supporting crime prevention measures that span multiple municipalities (Swedish government 2016). Their work is based on evidence and proven practices and is aligned with the activities of national authorities. Each year, county administrative boards must report their regional and local crime prevention efforts to the Swedish National Council for Crime Prevention (BRÅ), following BRÅ's guidelines. Municipalities are expected to implement knowledge-based crime prevention initiatives tailored to local needs; this has become more tangible with a recent new regulation (Swedish government 2021). The law requires municipal executive committees to take ultimate responsibility, appoint a safety coordinator, and establish collaborative councils with other actors to address ongoing crime prevention efforts. This involves fostering collaboration among various local actors, including transit operators and station managers, other actors that represent local schools, social services, and community organisations, to create a cohesive approach to crime prevention.

A semi-structured interview template was developed to explore how countries organise their efforts to address transit safety challenges. The template aimed at helping identify similarities and differences in countries approaches, trace developments over recent decades, identify examples of good practices, and gather perspectives on how an ideal stations and other transport nodes should be in the future. The interviews focused mainly on the role of the station environment in crime prevention and passenger safety, as well as on broader expectations regarding user safety and the specific needs of various passenger groups. The interviews were carried out digitally with key experts from different stakeholders in Sweden (9 people), the Netherlands (2), and Denmark (2). An expert from IUC (International Union of Railways, a worldwide railway organisation) was also interviewed using an adapted version of the interview template. Besides the questions in the template, follow-up questions were asked, depending on the expert answers. The authors conducted all the interviews over a period of six months in 2022–2023. In the following sections, we discuss the national contexts of the railways and how interviewed experts see the work devoted to safety issues in their contexts.

8.3 The Railway Systems in Sweden, the Netherlands, and Denmark

Since the late 1980s, several European countries have gradually reformed their national railway monopolies by implementing vertical separation of infrastructure from operations, introducing intra-modal competition through various procedures and contracts, and, in some cases, partially privatising the railway industry (Alexandersson and Hultén 2008). According to these authors, EU railway deregulation has followed four main approaches: the British model, which rapidly privatised and fragmented the market; the Swedish incremental approach, focused on gradual changes; the German and Dutch model, aligned with EU legislation; and the French late compliance model, which later embraced competition. These approaches reflect differing national priorities, such as market liberalisation, efficiency, and compliance with EU mandates, each impacting the pace and nature of reforms in the railway sector.

Rail-bound traffic is predominantly privately owned or operated, and there is a lack of entities responsible for systematically overseeing critical infrastructures and the various actors delivering services essential to society (Boin and Smith 2006). This imposes challenges for governments and experts in dealing with security problems and crises (Olsen et al. 2007). As suggested by Strandh (2017), there is no one-size-fits-all solution to the challenge of managing security and safety issues affecting rail-bound traffic. Ensuring safety throughout an entire journey requires the collaborative effort of multiple stakeholders. There are also expectations that delivering transit safety across railway systems demands coordinated cooperation among transportation service providers and other relevant actors (police, private companies, local crime prevention councils, and NGOs), yet they may not look the same across different countries. Below, we present the railway infrastructure and country contexts in some detail. Our description of railway systems is limited to insights from the interviewed professionals in each country. We do not claim to offer a complete classification, but rather a brief mommentary characterisation based on the available information from various sources and interviews.

8.3.1 Sweden

The Swedish Transport Administration (Trafikverket) is responsible for the platforms, tracks, and signals, including their construction, operation, and maintenance. They oversee and ensure the functionality and safety of railway infrastructure across the country. State-owned SJ (formerly Statens Järnvägar) is one of several train-operating companies that provides passenger train services. The distinction between Trafikverket and SJ allows SJ and other train operators to focus on running train services while Trafikverket ensures the infrastructure is safe and operational. At the same time, the operation and maintenance of public transportation systems are either the responsibility of public companies or entrusted to private contractors. For example, Hong Kong-based MTR is one of the companies currently operating regional transport in Sweden through its subsidiaries. By the late 2010s, Sweden had 17 passenger train operators and 21 freight operators (Strandh 2017). This constellation of actors results from the deregulation of the national transport system, which has been a gradual process unfolding over several decades. This transformation shifted the system from a centralised to a more decentralised one, aiming to foster competition and enhance service quality. Service provision in rail-bound traffic has transitioned to an inter-organisational approach, focusing on cross-sectoral collaboration management. However, according to Strandh (2017), there are differing views on how collaboration should be implemented. This issue seems to appear in Sweden between actors and across levels, from national to local levels. Currently, an apparent lack of coordination also hinders effective (crisis) management and communication between actors.

8.3.2 The Netherlands

In the Netherlands, ProRail manages national railway infrastructure, including maintenance, expansion, and traffic control, while Dutch Railways (NS) operates passenger services. ProRail is responsible for the public areas of stations, such as platforms and bridges, while NS stations handle the commercial aspects. Each station has a station manager overseeing its operations, with larger stations having multiple managers and smaller stations typically having one manager responsible for several locations. Security managers collaborate with station managers to monitor and address safety issues, focusing on integrating all stakeholders to ensure safety and security. This collaboration includes keeping the station "clean, whole, and safe", meaning, for example, ensuring that the lift works. According to the interviewed experts, there is a general agreement that having everything functioning in the stations is important for the feeling of safety and for *being* safe. Therefore, it is also deemed important to work with the municipalities regarding the environment around the station. A recurrent issue has been the role of CCTVs in railways in the Netherlands. Dutch Railways has CCTV surveillance on trains and stations and manages gate access to stations. Their security staff works to maintain safety onboard trains and stations and can be directed to a particular place in need with the highest priority.

8.3.3 Denmark

In Denmark, the state-owned DSB (formerly Danske Statsbaner) acts as the station manager, but ownership of stations is divided between Banedanmark, which is state-owned, and DSB. The division of ownership complicates coordination. There is ongoing discussion about whether DSB should take over more responsibilities to streamline operations and create a unified customer experience. Customers should experience one station, not a station with different owners. The Danish infrastructure plan allocates funds for station modernisation, including safety improvements and accessibility for people with reduced mobility. DSB is responsible for recommending how these funds should be used and ensuring their proper implementation. Banedanmark manages railway infrastructure but does not have a dedicated safety team. Safety regulations are strictly enforced by national authorities, with guidelines from UIC influencing practices. UIC promotes cooperation between stakeholders, emphasising the importance of joint exercises with authorities to improve safety and security across the rail network.

8.4 Station Safety and Management

8.4.1 Sweden

In Sweden, several entities manage the railway system as an organisation. The Swedish Transport Administration (Trafikverket) is responsible for planning, constructing, and maintaining railway infrastructure. This national organ also coordinates safety measures across the network, which focuses on preventing accidents much more than crime or other undesirable incidents, such as suicide or acts of terrorism (Thodelius 2018; Harms-Ringdahl 2004). The Swedish Transport Agency (Transportstyrelsen) regulates and supervises the safety of railway operations, issuing safety certificates to ensure compliance with national and international safety standards. Railway operators are responsible for safely running trains and adhering to the regulations of the Swedish Transport Administration and the Swedish Transport Agency. Incidents occurring along the trip and in transit environments may be recorded by different companies responsible for railway services. Travellers themselves contact the company to create the record via phone or apps. This means that a national integrated incident reporting system is lacking for issues of crime and safety happening in railway stations and related transit environments.

8.4.2 The Netherlands

In the Netherlands, Dutch Railways has a control room available 24/7 to assist passengers and staff, ensuring rapid response to incidents. According to our respondents, the security staff is deployed based on priority, focusing on train safety, though incidents often extend to the stations. Incidents are registered through a separate system, and information is shared with stakeholders such as the police, operators, and ministries to ensure coordinated responses. While communication is generally efficient, there are ongoing discussions about improving safety responsibilities, especially at stations where third-party employees may feel unsafe. Security guards are deployed regionally and work in collaboration with the police, although police availability can sometimes be limited due to staff shortages. In the Netherlands, ticket gates have been chosen to keep unauthorised people out and ensure that only users will enter the station. However, for people who are travelling and thus authorised to be there, a strategy is to increase safety by creating more activity with commercial facilities within the station area to make the place more attractive.

8.4.3 Denmark

In Denmark, the surveillance office located at the headquarters in Copenhagen oversees CCTV monitoring and receives reports from staff in the event of an incident. According to our respondents, passengers are encouraged to contact the police directly if they witness an emergency on board, with clear signs on the train to guide them. While regional and long-distance trains typically have staff on board to assist in such situations, S-trains (local trains) in Copenhagen do not have staff, which may limit immediate support for passengers in these trains. The system relies heavily on passengers being aware of how to respond and the availability of staff on specific routes to manage incidents effectively. In Denmark, the ambition is for stations to be integrated into the surrounding environment to create more life. This is done by attracting shops, bakeries, and restaurants to attract people, even those who are not going to travel. As more activity increases perceived safety (Lorenc et al. 2013), the goal is to create attractive places at the station, as part of the city, where people want to sit at a restaurant or café. Another measure to increase safety is improved wayfinding, for example, by establishing new wayfinding signs, but also to increase visibility, for example, by moving trees that stand in the way.

8.5 Current Challenges in Delivering Safety Services

In this section, we combine answers from all the interviewed experts in this study, which means that some of these reflections discussed below may reflect country cases, while others may reflect the railway sector at the European level.

UIC is a worldwide professional association representing the railway sector and promoting rail transport. UIC members are railway stakeholders such as passenger operators, infrastructure managers, freight operators, and, in other words, rail service providers. In addition to its members, UIC also has invited partners such as authorities and associations.

According to an advisor from UIC, the railways' first challenge is establishing a comprehensive security strategy that must rely on adequate information about where and when most problems occur. However, the expert highlights that there must be a balance between ensuring safety and keeping stations open as public places.

> "The focus must be on identifying the most critical parts of the network and stations. This requires conducting a risk assessment to determine which areas should be prioritised for protection. The next challenge is selecting the most effective and proportionate measures and balancing financial constraints, ethical considerations, and operational impacts. Additionally, there is the ongoing challenge of maintaining an open railway system, avoiding the more restrictive security measures often seen in airports."

Thus, the main challenge, according to UIC, is risk management, assessing the risk, and choosing the best way to deliver protection to travellers. This is challenging as the threat is constantly evolving. The protective measures need to evolve and be adapted to the whole picture. Another challenge as seen by the UIC expert is the cost of security interventions. "Our key business is to run trains, not prevent trains from running, so you must do it most smoothly". An example addressed by UIC is the problem of graffiti in the railway system, which causes a lot of damage when a carriage is found to be sprayed all over and must be replaced. Finding another carriage creates delays, and the cleaning is expensive. Thus, even though graffiti is a seemingly petty crime, it can create much damage, is costly to clean and, at the same time, may affect travellers feelings of safety. Moreover, part of this problem is, according to UIC expert, caused by organised crime producing graffiti all over Europe and therefore points out the need for the national police to recognise that this is a problem that needs to be dealt with at the European level.

UIC expert mentioned three kinds of measures to deal with crime and safety issues at the European level: *human-oriented measures, technology measures, and organisational procedures. Human-oriented measures* refer to variables such as human presence in the stations. In common with questionnaire responses from users (see, for example, Chapters 5, 7, 9, 10, 12, 13) and interviews with key experts in Sweden, Denmark, and the Netherlands, UIC confirms the importance of human presence for the feeling of safety—in fact, the most important point, according to the UIC representative. In this field, measures are aimed at the public but also towards the staff, including security staff—for example, how to behave in "abnormal" situations, and how to report. Awareness campaigns regarding security are organised by the railways. Included in the human measures are training of security staff, training together with the police, and coordination of interventions.

For examples of *technology measures*, UIC expert mentions guidance on the security of video protection at stations. Even though staff and surveillance are seen as essential for safety both by key persons in all countries and by the users, the presence of staff is decreasing, and digitalised solutions are taking over more and more in stations as in many other areas in society. In many station areas, this problem is dealt with by using CCTV and guards as potential substitutes. However, many of the interviewees emphasise the importance of human contact and available staff, as many stations are unmanned. Experts from the countries we interviewed about the need of technical solutions to contact staff even if they are not physically present. The users need someone to turn to. Also, they agree that stakeholders need to know what happens at the stations. Even smaller incidents that are not reported to the police need to be known if they are to be remediated. Those incidents, although "small", can be the reason for lower safety levels and have an impact on travel, for example by causing users to avoid certain stations or times of the day. As long as these problems are not known, the mapping and creation of a shared overview are prevented. Here, users can be the "eyes and ears" that complement that of the staff. A standardised incident-reporting system and a central location for synthesis were emphasised in the Swedish interviews to handle safety problems. However, when inadequacies are analysed and rectified to increase safety, they also need to be followed up to be evaluated in terms of effectiveness. These systems are organised differently in the different countries. There are several ways of reporting incidents in the Netherlands, such as via a "WhatsApp" number. In Denmark, there has been a focus on increasing staff. In Sweden, the reporting system for travellers is fragmented. Thus, there is no national unified system for reporting incidents in Sweden's rail system. However, some operators have their reporting systems, sometimes for staff only and sometimes also for travellers.

The third aspect concerns the *organisational procedures* as mentioned by UIC expert. This typically starts with a topic, organises a workshop around it, gathers all the stakeholders, and organises presentations to share experiences. Common exercises are arranged in which stakeholders cooperate in simulated scenarios to get to know each other and to clarify contact routes in case something happens. A concern acording to UIC expert is to make clear 'who is responsible for what' in specific threatening situations, from terrorism to petty crime, and, depending on the situation, the responsibility often differs. For example, certain technological measures can only be used by the police, such as in a situation with a bomb threat, when the police need to take the lead.

Moreover, when several stakeholders are involved, cooperation between the different actors is important for solving station safety problems. However, interviewed experts in Sweden point to a lack of national guidelines and a need for changes to the law to provide actors with mandates and clear responsibilities for coordination. On a more preventive international level, for standards and regulations to be developed, it must be clear what the most critical parts of a network or stations are and what should be protected. A long term plan with risk assessment therefore needs to be in place and include information on how the best measures, in a proportionate way, should be selected.

According to UIC expert, one challenge is that security (e.g., regarding crime) and safety considerations (involving accidents or in case of fire, for example) can sometimes conflict. While fewer entry and exit points improve security by making the monitoring of who enters and exits easier, multiple exits are needed for enhanced safety during evacuations. Additionally, older stations pose complications due to hidden areas within their design. Another concern relates to the stations as hyperspaces, and in particular how technology, such as CCTV systems, should be implemented to avoid creating potential vulnerabilities, like entry points for cyberattacks. The environment surrounding the station also presents safety challenges, such as the need to prevent vehicles from being driven into the station area in cases of terrorist attacks.

Interviews with key Swedish experts highlight the importance of detecting and understanding the diverse needs of different user groups. Safety concerns vary for groups such as the elderly, women, or individuals with disabilities, and this diversity calls for tailored solutions. While there are regulations in place to support individuals with specific disabilities, such as reduced mobility or visual impairments, other groups may lack adequate adaptations, and support for tailored safety interventions.

8.6 Expectations for Stations in the Future

8.6.1 Sweden

In Sweden, there are, as in all countries, financial restraints that make strategic and operative work more difficult; "the price tag is the big challenge every time" as reminded an interviewed expert. However, experts also pointed out that minor modest solutions can often be found across the country that may solve some of the safety challenges in railway systems.

The physical environment and the design of stations are seen as being fundamental, a prerequisite for further work to improve safety. Experts highlight the importance of particular features of the stations as key for ensuring safety in transit environments. Examples are open environments, good materials, clear signage, wide platforms, and no hidden corners are examples of this, as well as a lack of crowdedness or spaces where people can hide. Lighting will continuously need to be improved. Lifts, stairs, and escalators must work and be maintained in a way that they do not become littered, dirty, and covered with graffiti. It is important to create safe environments which people enjoy being in, and a blend of people who remain in the environment will make it feel safe. There are currently projects where women and children are counted in a station over time to see if the number of women increases after an intervention. This is viewed as an equality issue as it is primarily women who feel unsafe and mostly women who use public transport. However, it must be decided which threats the design should be based on. For example, design based on the threat of terrorism was not relevant some years ago, but it is necessary today and needs interventions other than, for example, those to deal with pickpocketing.

There are expectations that cooperation between stakeholders will need to continue to be improved. For example, municipalities are responsible for the area around the station and the way to and from the station. There must thus be cooperation regarding the physical environment to make it feel safe. Bushes must be kept low to give unrestricted sightlines. Good management is essential in transit environments. Maintenance must be followed up continuously, if not, it will still be unsafe.

There is also an agreement that cooperation is beneficial only if it is filled with content and followed up to evaluate what is functioning and what is not. Cooperation is needed with the police when it comes to drug selling, which moves between different places. Here, it is a priority to continue developing analyses and situational action. However, for cooperation to work smoothly, responsibility needs to be more evident in specific fields, according to interviews in Sweden. Someone must be designated to have coordination responsibility, so all parties know who is responsible for what. A newly changed law in Sweden gives more responsibility to the municipalities for safety, which may be one step towards making responsibility borders more explicit between stakeholders.

There are expectations among those interviewed of a continued focus on surveillance, such as security guards and CCTV. However, according to the Swedish interviews, there is room for better coordination as each stakeholder procures guards for their part of the station area, although someone would need to be assigned to connect these pieces to a whole system. Another possible change is although security guards and camera surveillance systems are in place, there is still a lack of 24/7 operative coordination that can react in real time. This creates "great potential" if it becomes possible to enter the operational mode when there is an indication that something is happening on a platform. For example, a clear governmental mission is needed for this development. In the Swedish interviews, coordination responsibility is a central theme as no one has the overarching picture. When no one has the overarching picture, it is difficult to prove that there is a problem, as one of the Swedish experts notes: "What happened at Uppsala station in 2020? No one has the whole picture".

Intervieweed experts also suggest that stakeholders want to listen more to users, both in their positive and in their negative feedback. Improved incident reporting would increase knowledge of "where the problem lies or where we should invest more resources", as suggested by one expert. However, there must be a receiver, which is not the case for all actors today. With that in place, security campaigns could target travellers and ask them to pay attention to certain things and report when they see something. It would also provide the opportunity, as mentioned above, to activate camera surveillance to see what is happening and to send security guards who are nearby. There is a great desire for more human contact, but many stations are small, and staff is not a realistic solution, especially not around the clock. However, new technology will facilitate and can be used to monitor and improve access to help when someone has concerns. AI and other tools will be further developed and can help even without human contact. Also, continuing to work from a "door-to-door" perspective, that is, safety throughout a journey, is highlighted as an important perspective on safety work. The station only constitutes a small part of the trip, and, especially at night, the walk home can be desolate and associated with a lack of safety.

8.6.2 The Netherlands

The expectations regarding future of stations in the Netherlands focus on improving safety, comfort, and accessibility. Platforms, being the most dangerous areas due to crowding, will continue to have staff to manage passenger flow. Increased visibility of employees, mainly security personnel, has been prioritised in order to make passengers feel safer, with flexible deployment of security teams based on incidents and major events. Station designs will prioritise passenger needs, using open layouts with glass, plants, and greenery to create welcoming, comfortable spaces. According to the interviwed expert, there is a vision of making stations places where passengers want to remain, not just pass through, while ongoing learning and improvements will shape future station designs, funded when possible. See Chapter 16 for examples of future stations.

8.6.3 Denmark

According to our respondents, future stations will prioritise accessibility, better infrastructure, modernised technology, and improved passenger experiences with real-time surveillance and increased staff presence. The combination of upgraded surveillance cameras, real-time surveillance, and increasing the number of staff means that they can be more proactive around the clock. There will be a stronger focus on sustainability, with greener stations and eco-friendly designs that promote safety, comfort, and environmental responsibility. Expectations are that stations should be designed to create a welcoming environment with passenger comfort and safety in mind, featuring glass walls, greenery, and visible employees. Enhanced physical and digital connectivity will ensure that stations are more efficient hubs, supporting transport, community engagement, and local business growth. Ultimately, stations are expected to evolve into dynamic, secure, and user-friendly environments that meet the diverse needs of passengers and staff alike. Tunnels should be redesigned with better lighting to enhance visibility and safety. Additionally, incorporating EU guidelines for responsible station management could improve safety and user experience.

8.7 Conclusions

We finish this chapter with more questions than answers. We have exploratorily investigated how the organisation of railway systems in a selection of European Countries is structured from the national to the local level, focusing on the role of those managing the station's environments. Although Sweden, Denmark, and the Netherlands are organised differently, they share many perspectives both in terms of challenges and solutions to safety problems. For example, about the importance of the physical environment in enhancing safety, the use of new technology to compensate for the absence of staff, and a strong reliance on improved surveillance and social control in transit environments.

A key challenge mentioned by interviewed experts is that, even with the best intentions, safety interventions in railway systems can conflict with one another or their impact may be unevenly distributed across the system, leading to other problems. This issue is not unique to transit operators and agencies as recognised by the interviewees; it is also prevalent in other sectors of society. In systems thinking, such conflicts among supposed partners are referred to as "accidental adversaries". This occurs when two or more actors who should be working towards a shared goal inadvertently undermine each other due to misaligned incentives or conflicting actions. In the Swedish case, this is most likely the result of poor coordination, which leads to a lack of a shared vision or an incomplete understanding of the broader problem. Although Sweden places a greater emphasis on stakeholder cooperation than many other countries, experts highlight the need for better coordination at both local and international levels, with clear accountability at every stage. To enhance this coordination, it is crucial to understand individuals' safety needs and preferences and establish national guidelines that provide clear direction for safety management and interventions across the railway system.

This analysis is exploratory and is based on insights from a limited number of experts from each country. As a result, the perspectives presented may not fully capture the entire scope of safety issues across the railway systems in each transit agency or even less over a particular nation. The differences in expertise of each interviewed expert in this study made it challenging to obtain a common base for comparison across countries. Despite these limitations, this chapter provides insights and raises important questions for further research. For example, it remains unclear how national regulations differ in their approach to safety management at the local level and how these variations impact the overall safety problems and operational efficiency of railway systems. Another key question, particularly for Sweden, is how improved coordination among stakeholders from local to national levels can be better implemented to enhance the safety conditions of travellers across railway systems. Regardless of the answer, such coordination at least in the Swedish case would demand the existence of a national integrated "incident-reporting system" to enable interventions to be tailored to specific issues, locations, and times, supporting evidence-based planning. Moreover, it would provide a long-term perspective for both interventions and assessments, ensuring sustained improvements in safety at stations, and throughout the journey, ultimately making public transportation more attractive and accessible.

References

Alexandersson G, Hultén S (2008) The Swedish Railway deregulation path. Rev Netw Econ 7(1):18–36. https://doi.org/10.2202/1446-9022.1136

Boin A, Smith D (2006) Terrorism and critical infrastructures: implications for public–private crisis management. Public Money Manag 26(5):295–304

Clarke RV, Bichler-Robertson G (1998) Place managers, slumlords and crime in low rent apartment buildings. Secur J 11(1):11–19

Eck J (2003) Police problems: the complexity of problem theory, research and evaluation. Crime Prevent Stud 15:79–114

Eck, J. E., Linning, S. J., and Herold, T. D (2023) Place management. In Place Management and Crime (pp. 23–32). Springer.

Harms-Ringdahl L (2004) Relationships between accident investigations, risk analysis, and safety management. J Hazard Mater 111(1–3):13–19

Lorenc T, Petticrew M, Whitehead M, Neary D, Clayton S, Wright K, Renton A (2013) Fear of crime and the environment: systematic review of UK qualitative evidence. BMC Public Health 13:1–8

Loukaitou-Sideris A, Liggett R, Iseki H (2002) The geography of transit crime: documentation and evaluation of crime incidence on and around the green line stations in Los Angeles. J Plan Educ Res 22(2):135–151

Madensen TD, Eck JE (2012) Crime places and place management. In: Cullen FT, Wilcox P (eds) The Oxford handbook of criminological theory. Oxford Handbooks

Olsen OE, Kruke BI, Hovden J (2007) Societal safety: concept, borders and dilemmas. J Contingen Crisis Manag 15(2):69–79

Strandh V (2017) Exploring vulnerabilities in preparedness – rail bound traffic and terrorist attacks. J Transp Secur 10(3):45–62. https://doi.org/10.1007/s12198-017-0178-5

Swedish government (2016) Förordning (2016:1258) om regional samordning inom det brottsförebyggande området, https://www.riksdagen.se/sv/dokument-och-lagar/dokument/svensk-forfattningssamling/forordning-20161258-om-regional-samordning-inom_sfs-2016-1258/

Swedish Government Official Report (2021) Kommuner mot brott: Betänkande av Utredningen om kommunernas brottsförebyggande ansvar (SOU 2021:49). Swedish Government. https://www.regeringen.se/rattsliga-dokument/statens-offentliga-utredningar/2021/06/kommuner-mot-brott-sou-202149/

Thodelius, C. (2018). *Rethinking Injury Events: Explorations in Spatial Aspects and Situational Prevention Strategies.* Tekniska Högskola (Sweden).

9 Safety Around the Clock: Temporal Variations of Victimisation and Safety Perceptions in Transit Environments

Abstract

This chapter explores the temporal dimension of victimisation and safety perceptions in transit environments, analysing key factors across different timeframes: daily variations (day and night), differences between weekends and weekdays, and seasonal patterns. Drawing on police statistics and survey data from 4,893 participants across 28 municipalities, the analysis examines how different fears, times of day, and types of victimisation interact to shape safety perceptions. Findings show that most crimes occur in the evening, with offence types varying according to the time of day. Crime rates are higher at weekends, while seasonal patterns show only minor differences overall but some variation is found when analysed according to crime type. Safety perceptions also shift significantly, with the lack of safety peaking at night. Previous victimisation is the strongest predictor of daytime fear. Understanding these temporal dynamics is essential for addressing the safety concerns of transit users both at stations and along their routes, making these places more accessible and sustainable.

Keywords

Crime · Safety · Day · Night · Weekends · Seasonal variations · Police statistics · Survey

9.1 Introduction

The city moves according to its rhythm, shaped by the daily routines of its inhabitants. From the quiet early hours to the bustling midday rush and the slower, dimly lit evenings, urban life unfolds in predictable yet dynamic patterns. As Kevin Lynch (1972) explores in "What Time is This Place?", time in the city is experienced through movement, transitions, and rhythms that give structure to daily life. The early morning calm gives way to a rush hour peak in the morning, afternoon, and evening, shifting in intensity and location as the city's pulse changes. While most activities and social interactions—such as meeting a friend, going to school, or commuting to work—are positive and routine, others, like accidents or becoming a victim of crime, are less frequent, less desirable but often preventable.

Such temporal dynamics highlight the importance of considering time when analysing urban safety, as crime does not merely occur in space but within the lived rhythms of the city. Felson (2006) examines how the daily routines of both potential offenders and victims—their sleeping patterns, walking, working, and eating—affect the occurrence of crime. He emphasises the importance of studying these rhythms to understand criminal behaviour.

V. Ceccato et al., *Transit Safety for a Sustainable City*, Sustainable Development Goals Series,
https://doi.org/10.1007/978-3-031-97427-4_9

> "The daily life of a city provides the targets for crime and removes them. The sleeping, walking, working, and eating patterns of offenders affect the metabolism of crime. We must study these rhythms of life if we wish to understand crime" (Felson 2006, pp. 6–7)

These rhythms also shape crime opportunities, which, as Ratcliffe (2010) argues, are neither random nor uniform, but closely tied to human activity patterns. For example, areas with high pedestrian traffic during rush hours may experience crimes differently from those with more sporadic activity as guardianship and crime opportunities change. These relationships are influenced by the diverse urban environments, particularly along routes to the station and within the station itself. To make cities safer, it is necessary to understand the temporal and spatial rhythms of human behaviour, which is essential to identifying when and where crime is more likely to occur, allowing for more effective crime prevention strategies.

This chapter examines the temporal dimension of victimisation and safety perceptions in transit environments (within 100 m of the stations), analysing key factors across different timeframes: daily variations (day and night), differences between weekends and weekdays, and seasonal patterns. It analyses the temporal safety perceptions and patterns of actual crime occurrence at rail-bound stations with global relevance. It offers insights into urban planning, public transport policies, and crime prevention strategies. Ensuring accessible travel and public transport is a key principle of Agenda 2030 and the global goals for sustainable development. This chapter's findings align with several of these objectives, and we discuss our contribution at the end of this chapter.

9.2 Temporal Variation of Victimisation and Safety Perceptions in Transit Environments

In Sweden, the winters are long and dark, with dusk already at 3 PM, so going to work in the winter often means travelling in the dark both to and from work. On the other hand, summers come with daylight until late evening, or in the north of the country, throughout the night. A place visited in broad daylight is perceived differently at night when visibility, an important factor for safety, is lower. Thus, travellers generally feel less safe in the dark (Börjesson 2012; Ceccato et al. 2021). Therefore, lighting is repeatedly found important for perceived and actual safety at nighttime (e.g. Sundling 2020). The elderly are more at risk of falling accidents (van Hoven and Meijering 2019) in obscure environments. Women and older people, especially as in Deniz (2019), find the time of day important for safety; women's fear of crime in urban public environments is especially apparent at night (Sundling and Jakobsson 2023). Being able to see others—where there are no hiding places for perpetrators—and being seen by others if there is a threat appears to increase safety (Lorenc et al. 2013; Loukaitou-Sideris 2014). At night, departures are less frequent, which can also prolong waiting at the station, increasing the lack of safety (Chowdhury and van Wee 2020).

Feeling safe is crucial to psychological well-being in public spaces, particularly in transport environments. Conversely, feelings of poor safety can create significant barriers to using these spaces freely, often prompting individuals to take precautions or avoid them altogether. Fear can lead to limited mobility when the feeling of poor safety prevents people from moving in the external environment, for example, at night (Newton et al. 2020). Fear of crime is believed to lower the quality of life and well-being in other ways. It justifies costly precautions, such as long detours, damages social trust and relationships between groups, and reduces social control (Hale 1996).

Crime is more anticipated at certain times, and Deniz (2019) found that more than 70% preferred to use public transport when they did not expect criminal or anti-social behaviour and that it was more important than the station's location and surroundings. Different weekdays can be perceived differently, for example, travelling late on a Tuesday night as opposed to a Saturday night. In the Paris region, young women were found to feel most unsafe at weekends (Vanier and de Jubainville 2017). Peak hours and off-peak hours also attract different crimes. While pickpocketing is more common in crowded environments, empty places can increase the risk of robbery and other sorts of crime that demand anonymity (Zahnow and Corcoran 2019; Ceccato et al. 2015). Thus, crime and perceived safety studies should address temporal patterns to provide a nuanced picture.

9.3 The Swedish Case

This chapter investigates temporal differences in victimisation and safety perceptions in Swedish rail-bound travel based on police data and answers from a safety survey. We present police data for daily, weekly, and seasonal variations in crime rates, and the survey results include daily variations in perceived safety.

For the police recorded data, a database was created for 28 municipalities with a population of 1.78 million and standardised using Swedish criminal codes. A Geographic Information System (GIS) was used to visualise and analyse spatial criminal patterns in and around station areas and over time. A selection of crimes was conducted within 100 m of the stations. Moreover, the survey data used in this study is based on responses from travellers and non-travellers in the same 28 municipalities. The data includes 4,893 participants, of which 63% are women, 34% are men, and three per cent identify themselves as LGBTQI+ (lesbian, gay, bisexual, transgender, queer, or intersex) and two per cent as non-binary or other. Women are, thus, overrepresented in our sample. For more details about the datasets used in this analysis, see Chapter 4 of this book.

9.3.1 Daily Variations of Crime around the Stations

9.3.1.1 Crime and Victimisation

According to Swedish police data (Fig. 9.1), most crimes within 100 m of the stations occur during the evenings, even though it varies depending on the crime type. In total, 30% of all crimes around the stations occurred during the last four hours of the day (20:00–00:00). For vandalism, this is especially obvious, where the equivalent number is 45%. Incidents of violence, as well as weapon- and drug-related crimes, also show a clear pattern of an increase during the evening.

The variations in crime levels between different times of the day were also assessed with the use of an ANOVA post-hoc Scheffe test, where a 24-h day was divided into four categories: (1) Morning = 06:00–12:00, (2) Afternoon = 12:00–18:00, (3) Evening = 18:00–00:00, and (4) Night = 00:00–06:00. To conduct the test, all crime incidents between 2019 and 2021 within 100 m of the stations were distributed according to the 24-h clock based on their time stamp, which gave a mean value per minute for all time categories and each crime type. The results show that the mean number of crimes per hour is highest in the evening for all crime types except theft (vandalism: M = 1.91; weapons & drugs: M = 1.38; violence: M = 0.49; theft: M = 1.03). Theft shows a different variation in time from that of the others, as it is more evenly distributed throughout the morning, afternoon, and evening. The highest level of theft occurs during the hour of 16:00–17:00 (M = 1.38). Peak-hour crowdedness facilitates crimes such as pickpocketing (Ceccato et al. 2022). It could also be a "lagged effect" from reports of people returning home from school or work and noticing a burglary or a bicycle theft that may have happened during the day. This could also be the case for the spikes in theft in the morning at 07:00–08:00 and 10:00–11:00. These are also peak hours in the public transport system when most people are travelling.

Reported incidents of violence are very low in the morning (M = 0.12), but seem to increase from the afternoon until they peak in the early evening (M = 0.49). For weapon- and drug-related crimes, there is a similar pattern, where there is a continuous increase from the morning onwards until it peaks in the evening. According to Harries (1997), a "lag effect" may occur in people's manifestation of stress. Their stress builds up during the day and then "blows up" later, for example, when they travel somewhere else after work. If this is correct, it suggests that crime (violence, in this case) may happen where there is possibility of externalising it by changing places, such as from work to a bar (Ceccato 2010). Also, serious violence is more

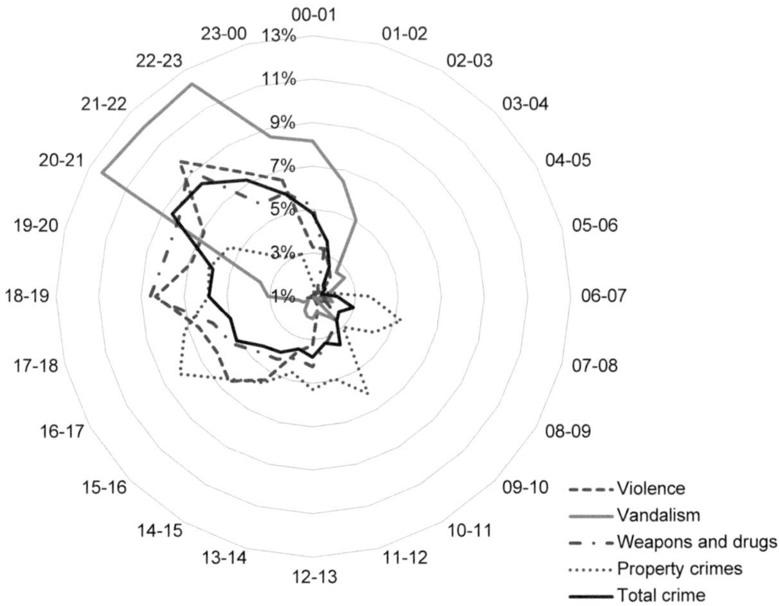

Fig. 9.1 Temporal variation between crime types within 100 m of the stations according to hours of the day 2019–2021 (N = 4,172). (*Data source* Swedish Police Authority)

frequent at times when most people have time off, such as in the evenings (Ceccato 2005).

For violence, both the afternoon and the evening have a significantly higher mean value than the morning and the night, according to the ANOVA test. Vandalism mainly occurs late in the day, evening and night. The mean value during the evening is significantly higher than all other periods of the day, and the night has a significantly higher mean value than the morning and afternoon. As noted for weapon- and drug-related crimes, they increase throughout the day, which is also evident in the table as the afternoon is significantly higher than the morning, and the evening is significantly higher than the afternoon. Perhaps it is a result of people finishing school or work and consuming drugs to relax after having built up stress levels during the day. Morning and nighttime are low points for these crimes. Theft crimes are relatively consistent from the morning until the evening but low at night-time. However, presumably, many thefts, such as burglaries, take place during the night, but are not reported until the morning when they are noticed. Thus, the differences in crime incidents between high- (afternoon/evening) and low-risk hours (morning/night) are statistically significant (95% level) for all types of crimes. Looking at the total, the evening registers significantly higher levels of crime incidents than all other times of the day.

9.3.1.2 Perceived Safety

There are significant differences in perceived safety during the day compared to at night. The survey shows that while 30% (N = 1,023) always or sometimes feel unsafe at the station in the daytime, it increases to twice as much, 63% (N = 2,025), at nighttime. Similarly, on the way to or from the station, 29% (N = 990) feel unsafe during the day and 63% (N = 2,031) at night. On board the train, people feel safer; 16% (N = 574) feel unsafe during the day and 39% (N = 1,264) at night.

Almost all of those who feel unsafe during the day, 99%, also feel unsafe during nighttime (N = 1,271). Inversely, of those who are afraid at night (N = 2,341), more than half, 53%, also feel unsafe during the daytime. Fear of crime is much higher among those who feel unsafe during the day (67%) compared to those who feel unsafe only at night (28%) (Chi-square = 350.37, df = 1, $p < 0.001$). Those who feel unsafe during the day have higher levels of fear across all different crime types compared to those who feel unsafe only at night, but both groups are most fearful of violence and theft (Fig. 9.2). Among those who

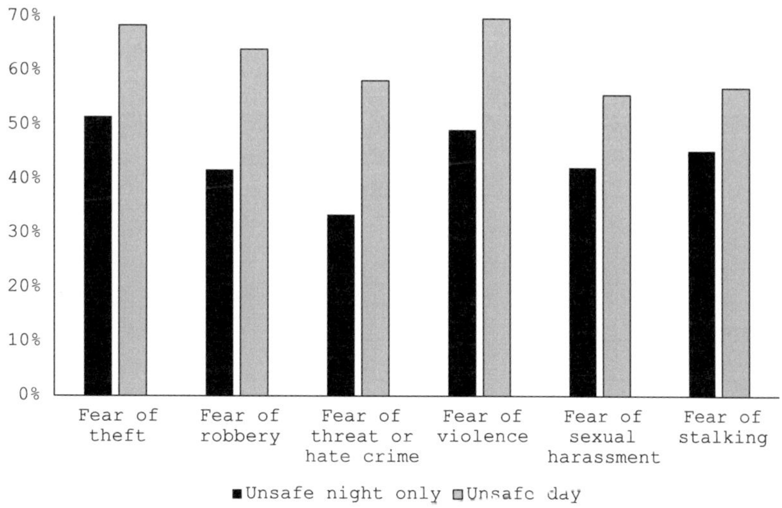

Fig. 9.2 Fear of different types of crime among those who experience fear of crime. $N_{unsafe\ night\ only} = 287$, $N_{unsafe\ day} = 849$. (*Data source* Survey)

are fearful of crime, fear is significantly higher at the stations (77% vs. 63%; Chi-square = 23.80, df = 1, $p < 0.001$) and on trains (49% vs. 28%; Chi-square = 38.08, df = 1, $p < 0.001$) among those who feel unsafe during the day, compared to those who feel unsafe only at night. There is, however, no significant difference between the groups for fear on the way to the station (74% vs. 71%). For all analyses, the significance level is set to 5% ($\alpha = 0.05$).

Safe or unsafe during the day. A logistic regression model (Table 9.1a) was estimated to assess more about those who are afraid of using the transport system during daytime and what individual and place characteristics are most important to explain their safety perceptions. The result shows that, compared to those who feel safe during the day, travellers who feel *unsafe* are more likely to have previously been a victim of crime. People are almost five times as likely to be afraid if they have been victimised. Being unsafe in one's neighbourhood makes reporting fear of crime during daytime more than three times as likely. Lack of staff at the station and drug use/selling at the station both imply a double risk. A protective factor is if there is a restaurant or café in the station, meaning that people feel safer in those environments.

Table 9.2 shows the results of the logistic regression analysis of fear of various crime types; the strongest predictor of feeling unsafe both day and night is the experience of threat or hate crime. Other significant predictors of daytime fear include concerns about robbery, violence, and stalking, while fear of theft and sexual harassment shows no significant difference (Table 9.2c).

Safe or unsafe during the night. When comparing those who feel unsafe to those who feel safe at *nighttime* (Table 9.1b), there are several statistical differences. Even here, having been a victim of crime is the most influential factor, making it more than three times as likely that people will feel unsafe. Being a woman or a lack of staff both make it twice as likely that people will feel unsafe. Suppose there is drug selling spot at the station, identifying as LGBTQI+, non-binary, or other, having to use a tunnel—all these factors, together with poor illumination, all significantly increase the likelihood of feeling unsafe when travelling at nighttime. Having a university education and there being a café or restaurant at the place are protective factors, increasing the likelihood of feeling safer.

Thus, both during daytime and nighttime travels, earlier victimisation, feeling unsafe in one's neighbourhood, lack of staff, and drug selling or use all decrease feelings of safety. However, other variables often found in safety research, such as gender differences, "disappear" when analysed separately between day and night. Thus, being a woman is not a variable that can predict feeling safe or unsafe during the day,

Table 9.1 Logistic regression for factors describing respondents who (a) Y = 1 = always feel unsafe during daytime, 0 = Safe, (b) Y = 1 = always feel unsafe during nighttime, 0 = Safe, (c) Y = 1 = always or sometimes feel unsafe during daytime, 0 = unsafe during nighttime

	(a) Unsafe during daytime vs. Safe (N = 112)				(b) Unsafe during nighttime vs. Safe (N = 458)				(c) Unsafe during daytime vs. Unsafe during nighttime (N = 816)			
	Exp(B)	95% C.I.for EXP(B)		Sig	Exp(B)	95% C.I.for EXP(B)		Sig	Exp(B)	95% C.I.for EXP(B)		Sig
		Lower	Upper			Lower	Upper			Lower	Upper	
Woman	0.936	0.595	1.473	0.775	**2.568**	**1.934**	**3.410**	**0.000**	**1.406**	**1.085**	**1.822**	**0.010**
LGTBQI +, non-binary or other	1.610	0.689	3.763	0.271	**1.811**	**1.004**	**3.264**	**0.048**	**2.487**	**1.376**	**4.493**	**0.003**
Young (18 years of age or under)	0.776	0.342	1.759	0.544	0.967	0.614	1.523	0.886	**1.724**	**1.098**	**2.706**	**0.018**
University education	0.711	0.462	1.093	0.120	**0.688**	**0.535**	**0.885**	**0.004**	0.814	0.639	1.038	0.097
Reduced mobility	1.692	0.829	3.453	0.149	**1.589**	**0.995**	**2.537**	**0.053**	1.129	0.883	1.443	0.333
Frequent traveller (4–7 days/w)	1.097	0.704	1.711	0.683	1.111	0.859	1.439	0.422	1.242	0.758	2.036	0.389
Victim of crime	**4.794**	**3.158**	**7.277**	**0.000**	**3.351**	**2.532**	**4.433**	**0.000**	**2.920**	**2.132**	**3.998**	**0.000**
Unsafe in neighbourhood	**3.275**	**2.052**	**5.225**	**0.000**	**3.131**	**2.253**	**4.351**	**0.000**	**3.723**	**2.511**	**5.520**	**0.000**
Poor illumination	1.015	0.646	1.593	0.949	**1.309**	**1.017**	**1.684**	**0.036**	0.782	0.611	1.002	0.052
Drug use/ selling at station	**1.978**	**1.274**	**3.070**	**0.002**	**1.835**	**1.435**	**2.346**	**0.000**	**1.338**	**1.046**	**1.713**	**0.021**
Tunnel	1.427	0.863	2.359	0.166	**1.726**	**1.307**	**2.279**	**0.000**	0.923	0.729	1.168	0.503
Isolated station	1.067	0.679	1.677	0.778	1.185	0.918	1.530	0.193	0.869	0.683	1.106	0.254
Lack of staff	**2.358**	**1.493**	**3.722**	**0.000**	**2.432**	**1.905**	**3.106**	**0.000**	**1.450**	**1.158**	**1.816**	**0.001**
Poor maintenance	1.012	0.636	1.610	0.961	1.205	0.922	1.574	0.172	**1.365**	**1.052**	**1.771**	**0.019**
Restaurant/café	**0.586**	**0.379**	**0.908**	**0.017**	**0.718**	**0.554**	**0.929**	**0.012**	**1.826**	**1.445**	**2.306**	**0.000**

(a) Nagelkerke R Square = 0.202.
(b) Nagelkerke R Square = 0.253.
(c) Nagelkerke R Square = 0.172

and neither can identification as LGBTQI+ or non-binary, although both variables could predict feeling unsafe at night. Likewise, having to use a tunnel and, not surprisingly, poor illumination only predict a lack of safety at night. It is also clear that both at nighttime and daytime, bringing activity to a station area, for example, in the form of a café or a restaurant, could increase safety, according to the results (e.g. Chapter 8). The independent variables in the logistic regression models that did not predict poor safety, either in nighttime or daytime travel, are age (under 18 years of age) and reduced mobility, although the following are close to having a significant effect on the 5% level: being a frequent traveller (4–7 days a week), the station being an isolated one, and poor maintenance.

Unsafe day or night. A comparison between those who feel unsafe during the day and those who feel unsafe only at night (Table 9.1c) shows that the strongest factor is feeling unsafe in one's neighbourhood, significantly increasing the likelihood of daytime fear. Previous victimisation and identifying as LGBTQI+ , non-binary, or other also contribute to higher levels of daytime fear.

Table 9.2 Logistic regression describing fear for different crime types and locations for respondents who (a) Y = 1 = always feel unsafe during daytime, 0 = safe, (b) Y = 1 = always feel unsafe during nighttime, 0 = safe, (c) Y = 1 = always or sometimes feel unsafe during daytime, 0 = unsafe during nighttime

	(a) Unsafe during daytime vs. Safe (N = 159)				(b) Unsafe during nighttime vs. Safe (N = 629)				(c) Unsafe during daytime vs. Unsafe during nighttime (N = 1166)			
	Exp(B)	95% C.I.for EXP(B)		Sig	Exp(B)	95% C.I.for EXP(B)		Sig	Exp(B)	95% C.I.for EXP(B)		Sig
		Lower	Upper			Lower	Upper			Lower	Upper	
Fear of crime (way to station)	1.106	0.691	1.771	0.676	**1.771**	**1.325**	**2.368**	**0.000**	0.963	0.701	1.322	0.814
Fear of crime (at station)	**1.971**	**1.189**	**3.267**	**0.009**	**1.609**	**1.215**	**2.131**	**0.001**	**1.546**	**1.162**	**2.059**	**0.003**
Fear of crime (onboard train)	**1.759**	**1.181**	**2.618**	**0.005**	1.092	0.835	1.426	0.521	**2.367**	**1.736**	**3.226**	**0.000**
Fear of theft	0.911	0.533	1.557	0.733	1.329	0.964	1.833	0.082	1.314	0.945	1.828	0.105
Fear of robbery	**2.441**	**1.362**	**4.375**	**0.003**	**1.461**	**1.038**	**2.058**	**0.030**	**1.600**	**1.110**	**2.305**	**0.012**
Fear of threat or hate crime	**1.666**	**1.034**	**2.684**	**0.036**	**1.681**	**1.255**	**2.253**	**0.001**	**1.725**	**1.243**	**2.394**	**0.001**
Fear or violence	1.384	0.798	2.402	0.248	**1.800**	**1.312**	**2.470**	**0.000**	**1.544**	**1.108**	**2.150**	**0.010**
Fear of sexual harassment	0.989	0.613	1.595	0.963	**1.620**	**1.195**	**2.195**	**0.002**	0.989	0.708	1.383	0.951
Fear of stalking	1.545	0.955	2.499	0.076	1.105	0.874	1.607	0.275	**1.435**	**1.028**	**2.002**	**0.034**

(a) Nagelkerke R Square = 0.217.
(b) Nagelkerke R Square = 0.337.
(c) Nagelkerke R Square = 0.274.

Additional factors include the presence of drug activity, being under 18, the lack of staff, being a woman, poor maintenance, and the need to use a tunnel. While no variables strongly predicted increased safety during the day, having a university education or a restaurant/café nearby showed some association with reduced fear. Factors such as frequent travel, reduced mobility, poor illumination, and station isolation did not significantly predict daytime poor safety.

Similar to the Chi-square analysis, logistic regressions indicate that those who feel unsafe during the day are less likely to fear the journey to or from the station than those who feel unsafe only at night (Table 9.2c). However, they are twice as likely to feel afraid on board the train, and their fear of crime at stations is also significantly higher.

9.3.2 Weekly Variations of Crime around the Stations

According to police data, there are variations in the number of crime incidents that occur on weekdays compared to weekends (Fig. 9.3). On Fridays and Saturdays, there is an increase in reported crime. This can be explained by human activity patterns, as activities during the week generally follow a structured pattern, like going to work or school,

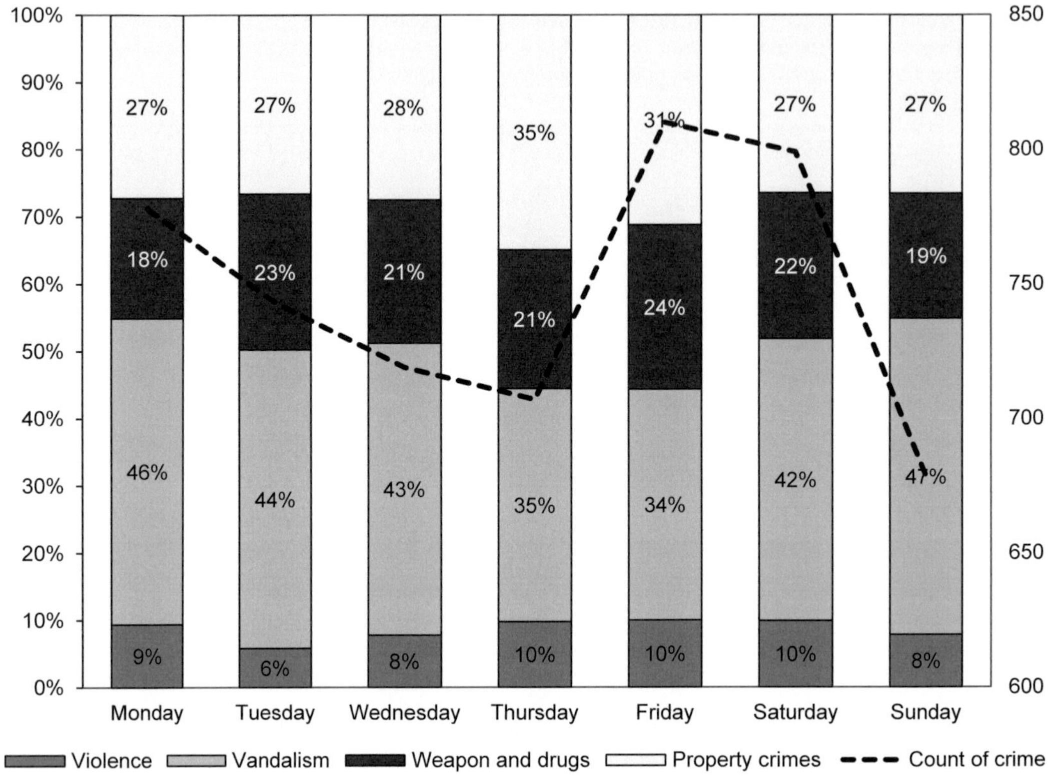

Fig. 9.3 Weekly variation between crime types within 100 m of the stations, 2019–2021 (N = 5,231). (*Data source*: Swedish Police Authority)

buying food, or working out at the gym. The weekends, on the other hand, tend to be more "unstructured" with more spontaneous activities. This spontaneity is more likely to attract criminogenic activities such as drinking and partying (Ceccato and Attig 2020). In the graph above (Fig. 9.3), it is noticeable that Tuesday, Wednesday, and Thursday have a lower crime activity, with a sharp increase on Friday and Saturday. Sunday has the lowest crime level of all days, and Mondays are relatively high compared to the other weekdays. This is believed to be a lagged effect in the dataset, with the implication that incidents that occurred during the weekend are being noticed and/or reported on Monday.

The ANOVA test with post-hoc Scheffe was used to test the variations for each crime category between the days of the week. However, it produced only one significant result at the 5% significance level (vandalism reports were significantly higher on Mondays than on Thursdays). To make a further assessment, the variations between weekdays (Monday, Tuesday, Wednesday, and Thursday) and weekends (Friday and Saturday) were tested by conducting a t-test with independent samples. Crime incidents based on over three years of crime data were distributed weekly, giving a mean value per day. Noticeably, this mean value is higher for the weekends in all crime categories and the total, except for incidents of vandalism, where the mean value remains relatively unchanged, 1.99 on weekdays and 1.96 on weekends. Vandalism reports did not produce any significant variations in this test. However, for property crimes, the mean value on weekdays is 1.32, which increases to 1.49 on weekends; for weapons and drugs, a mean value of 0.94 on weekdays is 1.19 on weekends, and the mean value of 0.38 for daily violent incidents rises to 0.51 at weekends. For these three crime categories, property crimes ($p = 0.002$), violence ($p = < 0.001$), and weapons- and drug-related crimes ($p = < 0.001$), the differences (higher mean values for incidents per day at weekends) are all statistically significant.

9.3.3 Seasonal Variations of Crime around the Stations

The seasonality of crime is influenced by several factors, such as variations in temperature and/or changes in human activity patterns (e.g. Anderson et al. 2000; McDowell et al. 2012). During the summer months, higher temperatures often lead to more people spending time outdoors in public spaces, which can increase opportunities for street crimes such as assaults or thefts. In contrast, colder winter months reduce outdoor activities, which may lower these types of crimes but increase indoor offences. Figure 9.4 shows the seasonal variations of different kinds of crime around the stations according to police data. The variations are then measured and analysed by conducting a one-way ANOVA test with the post-hoc Scheffe method, using data over three years distributed over one calendar year. The year was divided into four seasonal categories: (1) Winter = December-February, (2) Spring = March–May, (3) Summer = June- August, and (4) Autumn = September-November.

The differences in the levels of crime records between seasons are only minor, but when breaking this down according to each crime category, the variations in mean value per day become more evident. Incidents of violence within 100 m of the stations are at the highest level in the autumn, with a daily mean value of 0.49, and in summer, 0.43, which, at least for the summer, may be a consequence of people being on holiday, having more spare time, with an increase in alcohol consumption. Vandalism within 100 m of the stations is at its lowest level in the winter with a mean value per day of 1.78 when people have fewer outdoor activities and generally spend less time outside in climates typical for Sweden (Tucker and Gilliland 2007). In the categories of property crimes and weapon- and drug-related crimes, a shift can be seen between them in the summer, as property crimes increase when weapon- and drug-related incidents decrease. The winter has the highest number of police recorded drug-related crimes with a mean value of 1.23 cases per day, which falls to 0.74 per day in the summer. On the contrary, property crimes crimes are at their highest during the summer months, averaging 1.64 per day, compared to the lowest level in the winter of 1.20 property crimes crimes per day.

Two of the crimes, weapons- and drug-related crimes and property crimes within 100 m of the stations, had statistically significant seasonal variations in ANOVA analyses. The ANOVA test with the post-hoc Scheffe shows that weapon- and drug-related crimes are significantly lower during the summer compared to the other seasons (summer and winter: $p = < 0.001$; summer and autumn: $p = 0.021$; summer and spring: $p = 0.045$). The reason for the lower frequency of these kinds of crime is not clear, as drug-related crimes are habitual and persistent (McDowall et al. 2012) and could, therefore, have been thought to have only minor fluctuations over the years. It could be that fewer police resources would be available during summer leading to fewer recorded crime.

The number of incidents of property crimes within 100 m of the stations is significantly higher in summer than winter and spring ($p = < 0.001$). This aligns with previous research that shows that property offences, such as property crimes, increase in the summer. During that period, people leave for leisure activities or longer holiday travels, leaving many homes with open windows unguarded for longer periods, increasing the risk of burglary. Motor vehicle property crimes, larceny (Hipp et al. 2004), and bicycle property crimes (Shiode et al. 2023) are also more common during the summer. This follows the element in routine activity theory addressing the absence of a guardian (Cohen and Felson 1979). In urban areas, larger numbers of tourists also attract crime, such as pickpockets, which can be another reason for higher summer frequency at certain places (Marteache and

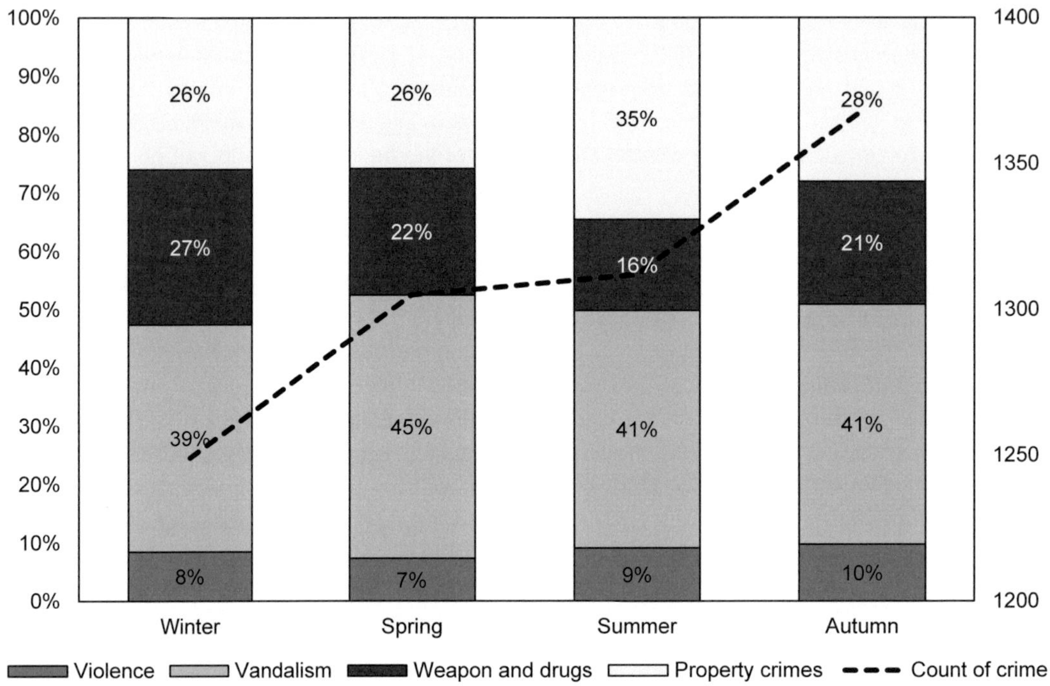

Fig. 9.4 Seasonal variation between crime types within 100 m of the stations, 2019–2021 (N = 5,234). (*Data source*: Swedish Police Authority)

Trinidad 2024). The role of temperature variations in crime has been suggested as a possible reason for violent crimes due to increased alcohol-related consumption and consequently aggression. Hipp et al. (2004) also compared the routine activity theory with the temperature aggression theory for property crimes. The latter was, however, not found to explain seasonal variations for this type of crime. Note that although some crimes vary seasonally, as shown in this section and in the literature, our study cannot establish a causal relationship between seasonality and these crime patterns around the stations.

9.4 Discussion of Results

Considering temporal patterns of crime is essential for making transit environments safer and more sustainable. Day or night, weekday or weekend, and summer or winter imply different travel experiences, and the risk of different crimes will vary. Therefore, it is important to separate different crimes by type and see that they can vary over time. Additionally, the temporality of crime and safety is dependent on context. For example, in Sweden and other Nordic countries, seasonal differences are more pronounced with cold and dark winters and long bright days in the summer, compared to many other regions in the world, which can affect travellers patterns of victimisation and safety in transit environments. It is also important to ask different groups, as the experience may differ. Many of the results in this chapter can probably be generalised to other public places. For example, feeling unsafe at night is not confined to stations only. To increase safety in public places across the city and in other areas, we need to know what leads to a lack of safety and for whom. To reach such a goal, it is essential to account for temporal factors of transit safety, which is what we have attempted to do in this chapter.

Findings from this chapter can be linked to several of the global Goals for Sustainable Development. Fear of victimisation and actual unsafe conditions, especially at night, can deter people from engaging in physical activities such as walking or cycling to transit stations, impacting their physical health. Safe and accessible transportation systems connect people to job opportunities and other activities. Moreover, exposure to unsafe environments can have psychological effects, contributing to stress and anxiety. This links directly to Goal 3, good health and well-being, as avoiding walking or cycling to transit stations due to safety concerns can reduce physical activity. Goal 11 is relevant as it emphasises the need for safe, inclusive, and sustainable urban environments at all times of day, during the week and throughout the year. Lastly, this knowledge base links to Sustainable Development Goal 13, climate action. If people are prevented from using public transport and walking at certain times because of a perceived lack of safety, it can hinder environmentally friendly development, such as reduced car use. Thus, encouraging public transportation through improved safety can reduce reliance on private vehicles, contributing to lower greenhouse gas emissions

9.5 Conclusions and Recommendations

Our temporal analysis of the police records around the station reveals that most crimes are recorded late at night, with vandalism against the station being the most common recorded offence. Consistent with international research, these findings show that travellers feel significantly less safe at night. Those who already feel unsafe during the day experience even greater fear at night, with heightened anxiety about crimes, particularly violence and property crimes. These findings have a clear impact on policy and practice. Below, we discuss a few ways ahead.

Urban planners, safety experts and transport operators can benefit from this type of evidence to detect patterns and better allocate resources. This helps identify when and where crimes are more likely to happen, highlighting the most vulnerable times (whether certain hours of the day, days of the week, or even specific seasons) as well as those most at risk. Our results show that previous victimisation is the strongest predictor of fear, especially during daytime. During the week, crime is most frequent on Fridays and Saturdays. Seen over a year, the seasonal variation depends on the crime type. While property crimes are higher in summer than in other seasons, crime levels are lowest for weapons and drugs. One recommendation is to deploy additional security personnel during late-night hours and weekends at and around the stations when crime rates are elevated to deter potential offenders and reassure passengers that they have someone to turn to in case anything happens.

Another recommendation is to encourage community involvement in safety initiatives, such as neighbourhood watch schemes and volunteer escorts during late hours, or to encourage bus drivers to stop any time during their route after dark. One example implemented in Kalmar municipality in Sweden is called "Nattstopp", a service offered by Kalmar regional public transportation system managed by a county to enhance passenger safety during late-night travel. After 9 PM, individuals travelling alone can request to disembark between regular bus stops along the standard route, allowing them to alight at a location where they feel most secure. Another example is the "Guarded Bus Stop" in São Paulo, Brazil, where the bus stops "talk" and keep company until the bus arrives. This has become an especially requested service by young women in the evenings (Sérvio 2023). Finally, having well-maintained environments at and around the stations can reduce crime opportunities and improve safety perceptions at particular times. With long dark winters, the maintenance of lights along the paths towards public transportation is fundamental for the safety of travellers.

As we are going to discuss in the following chapters, certain groups of travellers—including women, individuals with disabilities, and LGBTQI+ individuals—tend to experience higher levels of fear in transit environments, especially during particular "time windows".

By aligning individuals' safety needs with the city's natural rhythms, planners, public transportation and safety experts can improve people's accessibility to public transport throughout the year. For this to happen, we need to detect when people experience victimisation or feel most unsafe to be able to shape the future of mobility and encourage more sustainable urban living.

References

Anderson CA, Anderson KB, Dorr N, DeNeve KM, Flanagan M (2000) Temperature and aggression. Adv Exp Soc Psychol 32:63–133

Börjesson M (2012) Valuing perceived insecurity associated with use of and access to public transport. Transp Policy 22:1–10

Ceccato V (2005) Homicide in Sao Paulo, Brazil: assessing spatial-temporal and weather variations. J Environ Psychol 25(3):307–321

Ceccato V (2010) Safety in underground stations: a research project. Seminar in the Stock- holm city technical office, City of Stockholm

Ceccato V, Attig S (2020) Brott på tunnelbana 2009–2019. KTH Royal Institute of Technology

Ceccato V, Cats O, Wang Q (2015) The geography of pickpocketing at bus stops: an analysis of grid cells. In: Ceccato V, Newton A (eds) Safety and security in transit environments. Palgrave Macmillan UK, London

Ceccato V, Langeförs L, Näsman P (2021) Young people's victimization and safety perceptions along the trip. Nordic J Criminol 22(1):106–125

Ceccato V, Gaudelet N, Graf G (2022) Crime and safety in transit environments: a systematic review of the English and the French literature, 1970–2020. Public Transport 14(1):105–153

Chowdhury S, van Wee B (2020) Examining women's perception of safety during waiting times at public transport terminals. Transp Policy 94:102–108

Cohen LE, Felson M (1979) Social change and crime rate trends: a routine activity approach. Am Sociol Rev 44:588–607

Deníz D (2019) Improving perceived safety in public transportation through design. WIT Trans Built Environ 182:199–208

Felson M (2006) Crime and nature. Sage, Thousand Oaks

Hale C (1996) Fear of crime: a review of the literature. Int Rev Victimol 4(2):79–150

Harries K (1997) Serious violence: patterns of homicide and assault in America. Springfield, Ill: Charles Thomas

Hipp JR, Curran PJ, Bollen KA, Bauer DJ (2004) Crimes of opportunity or crimes of emotion? Testing two explanations of seasonal change in crime. Soc Forces 82(4):1333–1372

Lorenc T, Petticrew M, Whitehead M, Neary D, Clayton S, Wright K, Renton A (2013) Fear of crime and the environment: a systematic review of UK qualitative evidence. BMC Public Health 13:1–8

Loukaitou-Sideris A (2014) Fear and safety in transit environments from the women's perspective. Secur J 27:242–256

Lynch K (1972) What time is this place? MIT Press

Marteache N, Trinidad A (2024) Property crimes victimization in urban settings: comparison of tourist and resident case profiles. Crime Delinq 70(11):3102–3126

McDowall D, Loftin C, Pate M (2012) Seasonal cycles in crime, and their variability. J Quant Criminol 28:389–410

Newton A, Loukaitou-Sideris A, de Jubainville HDA, Huang D, Norgaard JK, Solymosi R (2020) Precautions and responses. Transit crime and sexual violence in cities. Routledge, pp 276–294

Ratcliffe JH (2010) Crime mapping: Spatial and temporal challenges. In: Piquero AR, Weisburd D (eds) Handbook of quantitative criminology. Springer

Sérvio, G (2023) Novo ponto de ônibus que 'conversa' e faz companhia chega a SP.

Shiode N, Shiode S, Nishi H, Hino K (2023) Seasonal characteristics of crime: an empirical investigation of the temporal fluctuation of the different types of crime in London. Comput Urban Sci 3(1):19

Sundling C (2020) Huddinge, Sweden. Transit crime and sexual violence in cities. Routledge, pp 167–175

Sundling C, Jakobsson M (2023) How do urban walking environments impact pedestrians' experience and psychological health? A systematic review. Sustainability 15(14):10817

Tucker P, Gilliland J (2007) The effect of season and weather on physical activity: a systematic review. Public Health 121(12):909–922

van Hoven B, Meijering L (2019) Mundane mobilities in later life—exploring experiences of everyday trip-making by older adults in a Dutch urban neighbourhood. Res Transp Bus Manag 30:100375

Vanier C, de Jubainville H (2017) Feeling unsafe in public transportation: a profile analysis of female users in the Parisian region. Crime Prevent Community Saf 19(3):251–263

Zahnow R, Corcoran J (2019) Crime and bus stops: an examination using transit smart card and crime data. Environ Planning B: Urban Anal City Sci 48(4):706–723

Part III
Stations and People

10 A Place of One's Own: Women's Safety in Spaces Between Home and Stations

Abstract

Growing concerns about women's safety in public spaces, particularly due to the heightened risk of sexual violence in these places, have sparked significant academic interest. Both direct and indirect experiences of victimisation shape women's fear in public environments. However, most transport policies remain gender-neutral, overlooking the specific vulnerabilities women face in their daily lives. Based on a case study of municipalities in southern Sweden, this chapter examines women's transit safety, focusing on their experiences between home and transit stations. It explores how individual and environmental factors influence women's vulnerability and safety perceptions of public spaces. Using a systems-thinking approach, we highlight interconnected elements that must be addressed to improve women's safety. We emphasise that effective strategies require coordinated efforts by policymakers, urban planners, and transportation authorities across multiple planning dimensions.

Keywords

Gendered safety · Crime · Safety perceptions · Queer · Intersectionality

10.1 Introduction

Virginia Woolf's "A Room of One's Own", published in 1929, addressed the status of women, especially how women and men differed in their experiences and how materially their worlds regulated their life opportunities. Woolf raised her voice to draw attention to the chronic underinvestment in society in women's education, the persistent disregard for women's voices, and the urgent need for women to assert their financial and intellectual independence. Nearly a century later, we are still searching for reasons for why little has changed—or, in some cases, why development has moved unexpectedly in the wrong direction. In 2025, in Afghanistan, women are denied education, a voice, and a place of their own, both privately and particularly in public spaces. Needless to say, we can see that their situation is a complete violation of their rights as we learn that:

> "…women are not allowed in parks, gyms, or beauty salons…. They are forbidden from going outside for a walk or exercise, speaking or showing any part of their face or body outside the home… or even singing or reading from within their homes if strangers can hear them". (BBCNews 2023)

Unfortunately, this is not only the case in Afghanistan; in many countries, women still lack guaranteed full access to public spaces and the freedom of mobility despite having access to education and employment. Sexual harassment

V. Ceccato et al., *Transit Safety for a Sustainable City*, Sustainable Development Goals Series,
https://doi.org/10.1007/978-3-031-97427-4_10

and other forms of sexual violence in public spaces—whether urban or rural—are pervasive, daily experiences for women and girls worldwide (UN-Women 2024). Women's fear is most intense during the evening hours (Ceccato et al. 2022). Previous research indicates that while men are statistically more likely to be victims of crime on public transport, women report higher levels of fear than men (Ceccato 2017). Women and gender-diverse individuals face not only sexual violence and harassment, but also bear the burden of "safety work" (Kelly 2012) to protect themselves and feel secure while using public transport or being in public places. It is not uncommon for women to be forced to adopt various strategies to avoid victimisation or to steer clear of areas deemed unsafe (Whitzman 2013). For women living in rural areas, the challenges are even more significant due to the scarcity of resources in these remote regions (Ceccato et al. 2024a). However, identifying these places and understanding their characteristics is crucial to ensuring safety and free mobility for all.

Even though the 67th session of the United Nations Commission on the Status of Women emphasised achieving gender equality and the adopted conclusions highlight the importance of schools, jobs, and healthcare for women's empowerment (Commission on the Status of Women Sixty-Seventh Session 2023), it has overlooked the crucial need for safe, reliable, and affordable transportation, which is key to their economic and social autonomy. Despite its importance for women's opportunities, this issue remains underprioritised in international policy. The #MeToo movement has helped to reveal that a particular hindrance to women's mobility is their fear of sexual harassment while travelling, especially when using public transport. Still, research has been lacking on transit safety among women of all ages.

This chapter addresses women's safety in the city, focusing on public spaces between places, particularly on the way to stations or in transit to other public spaces. Through research findings and personal narratives, we discuss examples from women who participated in a case study in 28 municipalities in Sweden. We discuss the problems they encounter during their daily routine, whether they are victimised and their feelings of safety in these public spaces. We conclude by recommending practices that create welcoming and affirming spaces for all passengers, regardless of gender identity or sexual orientation. By promoting safer and more inclusive transit spaces, the chapter supports SDG 5 (Gender Equality) by addressing the specific safety needs of women. It contributes to SDG 10 (Reduced Inequalities) by fostering welcoming environments for all, regardless of gender or sexual orientation. Moreover, it aids SDG 11 (Sustainable Cities and Communities) by advocating public spaces that are accessible, safe, and equitable for all citizens.

We adopt a systems-thinking perspective to better understand the dynamics of victimisation of women during their trip and on the way to the station (see Chapters 1–3 for in-depth reference) as well as making suggestions for change; for instance, improving women's safety in transit requires understanding how factors such as urban design, transportation infrastructure, management, and public policies interact. This can involve redesigning public spaces, enhancing safety measures, improving public transport accessibility, and promoting awareness around gender inclusivity (Ceccato and Loukaitou-Sideris 2020a; Gekoski et al. 2017; Gekoski et al.). By considering these interconnections, systems-thinking enables a comprehensive strategy that addresses multiple layers of challenges in urban environments, advancing SDG 11 substantially.

10.2 Women's Transit Safety: Victimisation and Fear

Criminological theories provide the theoretical background that can help us understand why crime occurs in transit environments, which settings and which individuals may be more prone to victimisation, and how elements of the physical and social environment may reduce fear or increase the risk of victimisation. Two sets of theories focus on the opportunities of offenders, the vulnerability of targets, and inadequate guardianship, explaining women's transit risks, and the other theories emphasise environmental cues, prior experiences, and gendered social norms shaping women's transit safety concerns.

These theories highlight how transit environments can unintentionally foster victimisation when physical and social safeguards are insufficient at that particular place and time. The Routine Activity approach emphasises the convergence of a motivated offender, a suitable target, and the absence of capable guardians (Cohen and Felson 1979), and it is often used to explain the occurrence of criminal activity in poorly monitored transit spaces. In this case, isolated or poorly lit areas increase women's vulnerability, as they create opportunities for offenders to act without being observed. In addition, Social Disorganisation Theory (Shaw and McKay 1942) associates victimisation risks with structural inequalities, such as high poverty levels and weak community control in urban transit hubs, making perhaps women more vulnerable in areas of high housing mobility andsegregation.

There are micro-situational circumstances that also affect crime occurrence. An empty train station may provide anonymity for verbal sexual abuse or rape, while a crowded bus increases the risk of physical sexual abuse, such as groping. The behaviour of bystanders and the characteristics of these micro-contexts are important factors in determining whether or not sexual harassment occurs (Ceccato and Loukaitou-Sideris 2020b). In a case study in India, Tripathi et al. (2017) found that sexual harassment often went unnoticed by other passengers, possibly because such incidents were so common that they became normalised as part of everyday life (Mellgren et al. 2018).

Women often experience heightened anxiety while waiting at transit stops or stations, with feelings of apprehension particularly pronounced during extended waiting times. This sense of vulnerability is even greater for women of colour, who report feeling unsafe both during the day and at night. To cope with these challenges, many female travellers rely on their mobile phones, using them as a tool to stay connected and create a sense of security in these public spaces (Chowdhury and van Wee 2020). Chowdhury et al. (2023) show that women also employ adaptive strategies, such as pre-planning trips, using phones for maintaining connection with others, and appearing disinterested, to manage fear and prevent victimisation. All these activities are included in what we call women's daily "safety work" (Vera-Gray and Kelly 2020). These behaviours highlight their agency but also reveal the persistent fear they face, often intensified by awareness of their ethnicity and appearance.

Previous research shows that several factors influence victimisation in transit environments. Formal/informal surveillance reduces the risk; emptiness increases the risk of crime due to isolation; crowdedness can reduce risk through social guardianship but may also increase the risk of crime, such as pickpocketing, if it leads to anonymity; CCTV reduces risk through monitoring; while territoriality and control of access both reduce the risk of crime by enhancing perceived ownership and restricting entry; underpasses increase the risk of crime due to isolation and entrapment; open environments can increase or decrease risk depending on visibility and accessibility; maintenance reduces the risk of crime by promoting a sense of order; lighting can reduce risk if effective, but increase it if poorly implemented; emergency alarms reduce the risk of crime by providing a means of calling for help; ICT improves safety through better communication while body-worn cameras reduce risk by increasing accountability (Ceccato et al. 2022; Sundling and Ceccato 2022).

Fear of crime theories address the psychological and environmental factors that shape women's perceptions of safety in transit. Various factors, including socio-psychological influences, shape feelings of fear. Intersectionality theory (Crenshaw 2013) highlights how overlapping vulnerabilities—such as being young, from an ethnical minority, or poor—magnify women's fear. People often form fears as a result of past experiences, media, and accounts from others (Koskela and Pain 2000). Gender is an important factor that affects both mobility and safety (Ceccato and Moreira 2021; Loukaitou-Sideris and Fink 2009).

Previous victimisation tends to increase fear (Ferraro 1995), and fear can also be heightened by parental warnings and police advice (Loukaitou-Sideris 2012). This is particularly evident for women, who are often exposed to warnings about "stranger danger" from a young age through both family and societal messages (Pain 2001).

Transit operators and urban planners must better understand the interplay between individual experiences, physical infrastructure, and broader social dynamics. Each element of a journey across public spaces—such as poorly lit pathways, isolated transit stops, or societal norms—can shape women's safety perceptions. Factors influencing safety perceptions include open environments, which can both increase safety through visibility and decrease it if overly exposed. Illumination, the presence of staff, cafés/kiosks, CCTV, and security guards all enhance safety perceptions by improving visibility, providing assistance, and offering natural surveillance. Police presence may either increase or decrease safety perceptions, depending on the context. Isolated places reduce safety due to feelings of vulnerability, while basic amenities, maintenance, and mobile phone coverage enhance safety by promoting comfort, signalling care, and ensuring connectivity. Lastly, trust in staff and others boosts safety perceptions through a sense of community support accountability (Ceccato et al. 2022; Sundling and Ceccato 2022).

Based on previous knowledge and in order to investigate women's safety through a systems-thinking perspective, we examine the following research questions:

- What are the patterns of women's victimisation?
- Using a whole journey approach to safety, how do women express their safety perceptions during the trip, at the station, and on the way there?
- What precautions do women take to enhance their sense of safety?
- Finally, from a systems-thinking perspective, what recommendations can be drawn from these findings to inform gender-sensitive urban planning and proactive safety interventions?

This chapter explores women's transit safety, drawing from a case study in 28 municipalities in Sweden. For more details, see Chapter 4 in this book. In the next section, we discuss the study, starting with the study area.

10.3 The Swedish Case Study

As previously mentioned in other chapters, the dataset covers 28 municipalities in this study, comprising 3,191 respondents, 61% identifying themselves as women, 34% as men, and 5% as LGBTQI+ or non-binary. The study area comprises 47 rail-bound stations distributed over three railway lines stretching through central and southern Sweden; for more information, see, for example, Chapter 4 of this book. Regarding age, with 2,589 participants, 30% are under 30% are between 30 and 59, and 14% are 60 or older. Ethnicity data gathered from 3,151 respondents shows that 88% were born in Sweden, while 12% were born abroad. This diverse representation offers valuable insights into the demographic distribution of the study population, including an emphasis on LGBTQI+ inclusivity and generational differences. The dataset further describes participants based on education, disabilities, and station size. Regarding schooling, 59% of the respondents have completed higher education or university studies, 32% have a secondary school ("gymnasium") level of education, and 9% have primary school as their highest level of education. Of 3,092 respondents, 91% reported no disabilities, 8% indicated minor disabilities, and 1% reported major disabilities. This demographic and contextual information provides essential insights for further analysis.

Fig. 10.1 Women worldwide tend to engage in activities such as caregiving and are therefore more often transit captives than men, facing unique safety threats while in transit. Yet, gender-neutral public transportation policies prevail. (*Source* Nils Lukas Harder 2024)

10.3.1 Women's Victimisation in Public Spaces Between Places

Women are more likely than men to rely on public transport, and they often engage in "trip chaining", combining multiple trips for activities such as shopping or caregiving (Fig. 10.1), also known as "mobility of care" (Levy 2019; Sánchez-de Madariaga and Zucchini 2020). The need for multiple stops increases women's exposure in public spaces, which may increase their vulnerability to street harassment. Getting around by train should be a convenient and safe experience, but for many, it can come with troubling experiences. To better understand the reality of the transit environments, respondents were asked about their experiences of crime over the past five years, whether as victims or witnesses. More specifically, the question posed to respondents was: "In the past 5 years, have you experienced or witnessed any of the following on the train, at the station, or on your way to/from the station?".

A total of 14% of the 3,749 respondents reported being directly victimised by a crime. Although the most commonly reported incident was aggressive begging, observed by 23% of women, they are also exposed to other crimes, including stalking (6%), sexual harassment (5%), threats and hate crimes (4%), theft (4%), violence (3%), and robbery (1%). Among those who were victimised, stalking stands out as the most common form of victimisation, with 46% of women reporting being affected (Fig. 10.2). This is followed closely by sexual harassment, which impacts 45% of women, emphasising the pervasive nature of these offences in Sweden. Property crime affects 26% of women who were victimised, while violence is reported by 12%, reflecting the serious physical and emotional toll these crimes take. Additionally, threats and hate crimes account for 20% of the crimes women were victims of, underlining the widespread fear of targeting these acts.

It is noteworthy that young women (up to 29 years old) were the most common victims of crime in these 28 Swedish municipalities (23% reported victimisation compared to 11% of those 30 years or older) (Chi-Square = 41.249, df = 1, $p < 0.001$). The differences were especially notable in crimes of a sexual nature, such as sexual harassment (Chi-Square = 97.837, df = 1, $p < 0.001$) and stalking (Chi Square = 44.427, df = 1, $p < 0.001$). One reason for these results could be that young women are more likely to be in public spaces, such as transit environments, during times when the risk of victimisation is higher (evening and rush hours, having an obvious effect on environmental factors, for example, empty stations) and are also more vulnerable to crimes such as sexual harassment and stalking due to their age (Loukaitou-Sideris and Ceccato 2020) (Fig. 10.2).

Table 10.1 unpacks these statistics in more detail and allows comparisons between men, women and LGBTQI+ individuals. Women are particularly vulnerable to sexual harassment and stalking, both reported at 7%, significantly higher than men (1% and 2%, respectively). Within

Fig. 10.2 Women's victimisation by type and safety perceptions among those who declared being victimised (N = 297) and declared being fearful (N = 795) 2023

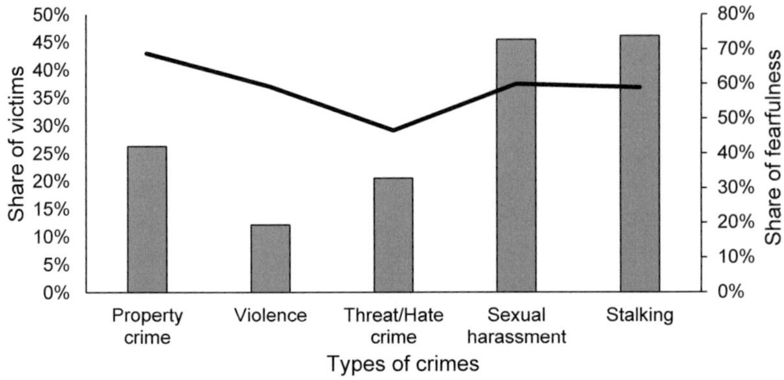

Table 10.1 Fear of crime among women as a percentage of those who declare being fearful

	Fear < 30 years old (N = 217)	Fear > = 30 years old (N = 851)
Property crime	59.0%	75.8%
Violence	45.6%	67.2%
Threat/Hate crime	44.7%	50.8%
Sexual harassment	79.3%	41.7%
Stalking	68.7%	47.1%

the LGBTQI+ group, there are some differences, although not statistically significant, that women who identify as LGTBQI+ are more exposed to stalking and property crimes. In contrast, the male counterpart reported more threats and hate crimes as well as aggressive panhandling, but the difference is not significantly different from women who identify themselves as LGBTQI+ .

Victimisation on the way to the station (10%), at the station (9%), and on the train (8%) also affects women more than men, though less than LGBTQI+ individuals. Aggressive panhandling impacts 14% of women, less than men (17%), but this is still notable. While women report lower rates of hate crimes (3%) compared to LGBTQI+ individuals (13%), they experience distinct safety concerns, particularly related to harassment and stalking. These findings underscore the need to address gendered victimisation in public places, which will be discussed in more detail in the next chapter. The patterns of victimisation across different locations (on the way to the station, at the station, or on the train) show no significant gender differences, suggesting that the discrepancies in victimisation are primarily linked to the types of crimes experienced rather than the types of environments and perhaps other situational circumstances. However, tunnels are often perceived as unsafe spaces for passengers, especially women, with the risk of crime heightened due to their enclosed, isolated nature and limited visibility (Buckley 1996). These features can create an environment where women feel particularly vulnerable, especially when travelling alone or during off-peak hours. Research also highlights the fact that poor visibility, restricted sight lines, and isolated spaces heighten the fear of crime, as visibility and the ability to see and be seen boost people's confidence and sense of safety (Umaña-Barrios and Gil 2017). The lack of proper lighting, clear sightlines, and the absence of other people contribute to a heightened perception of danger, making tunnels a critical focus for safety improvements in transit planning (Fig. 10.3).

More interestingly, there is a significant gap between victimisation and fear among women, especially for crimes that threaten personal safety (Fig. 10.1). While victimisation rates for property crime (26% of victims) and violence (12% of victims) are lower, fear levels are much higher among women, at 69% and 59% respectively of those who declare being fearful. Sexual harassment (45% of victims) and stalking (46% of victims) have high victimisation rates, but fear remains similarly elevated, reflecting their emotional and psychological impact. For threats

Fig. 10.3 Safety in (a) tunnels, (b) risk of crime on the way to the station and (c) older adults using the stairs to exit the station. (*Source* Authors)

and hate crimes, 20% of those who have been victimised report experiencing these offences, while 46% of those who declare being fearful identify threats and hate crimes as significant concerns. These statistics reveal that women are disproportionately affected by fear, which threatens their well-being in everyday activities. This underscores the need for targeted interventions, enhanced support systems, and stronger societal efforts to prevent victimisation and reduce fear.

10.3.2 Women's Safety Perception

Fear of crime shows significant gender differences. Women are more likely to express fear, with 39% reporting concerns, compared to 24% of men. However, the group most affected by fear is LGBTQI+ individuals, with nearly half declaring fear of crime (Chi-square = 84.053, df = 2, $p < 0.001$). Within this group, women experience the lowest perception of safety, with 59% reporting fear of crime, compared to 36% of their male counterparts. Women are more fearful of all types of crimes, but especially more fearful of sexual harassment (24% compared to 5% of men) and stalking (23% compared to 6% of men), whereas the difference is smaller for property crimes (27% compared to 20% of men) and violence (23% compared to 17% of men).

There are significant differences between young travellers (under 30) and older travellers (over 30) regarding their fear of various crimes. The analysis shows that young travellers report higher levels of fear regarding certain crimes than older travellers. Younger women (under 30) tend to be even more fearful, with nearly half of them declaring fear of crime (Chi-square = 33.811, df = 1, $p < 0.001$). In this group, sexual harassment and stalking are particularly prevalent, with 79% and 69% of those who declared being fearful reporting fear of these crimes, respectively. These concerns reflect the reality that women are disproportionately affected by such offences, which often take place in public spaces or during vulnerable times, such as at night. In contrast, while older women (30 years of age and above) show a lower level of fear regarding sexual harassment and stalking, they are more likely to fear property crime, violence and threats, or hate crimes.

The data highlights a gendered pattern where women, especially younger women, are more vulnerable to crimes that invade personal space and safety, contributing to their overall poor safety perceptions. The disparities between age groups in terms of fear also point to how different life stages and experiences shape one's perception of safety.

This highlights the need for targeted interventions to address the specific fears and risks women face in both public and private spaces, issues that we will discuss in the following sections.

As night falls, the overall experience of travelling by public transport undergoes a noticeable change, with a significant decline in the sense of safety. Women who responded to our survey consistently report that the period spent going to or from the station is perceived as the most dangerous part of their journey, where 28% of women reported fear of crime. Additionally, waiting at the station itself is also commonly perceived as unsafe with 27% expressing fear of crime, particularly when compared to the relative safety of being onboard the train, where only 16% of women reported experiencing fear of crime.

During the day, a majority of women feel relatively safe across all aspects of their journey whether on the train, waiting at the station, or commuting to and from the station. However, this perception shifts significantly at night: 44% of women report sometimes or always feeling unsafe on the train during nighttime, more than double the percentage of those who feel unsafe during the day. The most significant shift, however, occurs in the station areas and when going to and from the station. While 33% of women report sometimes or always feeling unsafe during the day at the station, 73% express fear at night. Similarly, 33% of women sometimes or always feel unsafe on the way to or from the station during the daytime, but this number surges to 74% after dark.

Women who answered the survey identified key measures to enhance station safety, prioritising improved social control, such as more staff, CCTV, police/security patrols, and emergency help buttons at stations and on trains. Good-quality lighting ranked fifth in importance. Around one-third suggested better station maintenance, followed by recommendations for increased train frequency, fewer stops, more information, and new technologies like apps or hotlines for reporting issues during trips (Ceccato et al. 2024c). In the next section, we present our recommendations for addressing women's safety in public spaces using a systems-thinking approach—the "Tree Model".

10.4 Improving Women's Safety

The "Tree Model" places women's safety at its core, highlighting the interdependence of various domains essential for addressing victimisation and fear in public spaces. A critical component of this system is knowledge, which plays a pivotal role in identifying the problem and developing solutions. The model illustrates that no single intervention can address women's safety in transit alone. Instead, these factors depend on each other; for instance, improving only transportation infrastructure is not enough without supportive urban design or public policies (Fig. 10.4).

It is essential to recognise the proactive measures women take as part of their everyday routines to mitigate transit-related risks (women's safety work, Vera-Gray and Kelly (2020)), such as pre-planning trips or using their phones for safety. While these actions reflect individual agency, they highlight the need for systemic responses that ease the burden of "safety work" on women, such as incident-reporting systems, improved lighting, CCTV, and emergency alarms (where it is possible to call for help as it happens)—and create safer, more equitable transit environments.

Policymakers and urban planners should be encouraged to integrate women's lived experiences (including their lifestyles and caregiving activities) into their initiatives by designing gender-sensitive safety strategies tailored to their needs. Policymakers can work to design safety interventions at the national, regional, and local levels. At the same time, urban planners and transportation operators and authorities can support the process by integrating safety measures into urban design and routine planning operations in places around public transportation nodes. The study by Gekoski et al. (2015) highlights the importance of collaboration between urban planners, transportation operators, and authorities in enhancing safety around public transportation nodes. The report emphasises that integrating safety measures—such as improved lighting, surveillance, and design-related features—into urban planning and routine operations can significantly reduce incidents of sexual harassment and other offences. The "tree model" suggests that a strategy requires

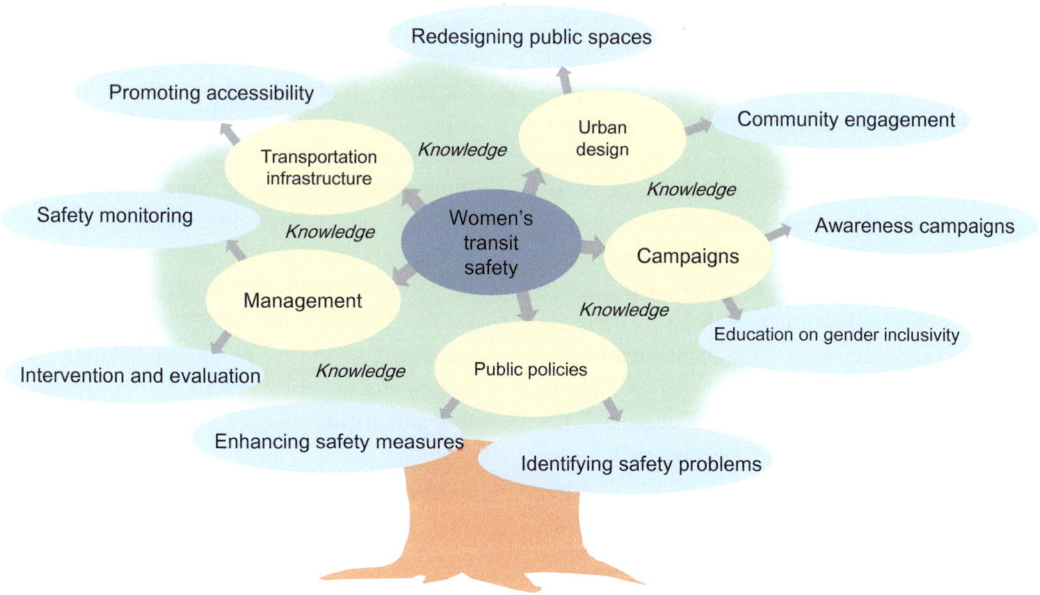

Fig. 10.4 Women's transit safety tree: A multilevel systems-based approach to tackle crime and safety problems. (*Source* Authors)

addressing multiple dimensions simultaneously, but not necessarily immediately. The first step is identifying the system's most critical problems or "leverage points" (Ceccato and Newton 2024). Having an incident reporting system at the national level across the railway system is a basic condition for identifying where most problems are.

Worldwide, women engage in different daily activities (e.g. caregiving), and women are more often "transit captives" than men and face distinct safety threats while in transit. However, gender-neutral policies still dominate the field. This demands that local voices of women and girls should be encouraged to advocate gender-sensitive interventions that are particularly important for their locality. Many calls have already been put forward asking transport operators to take the lead and establish systems that inform passengers that they can report incidents and ensure support is accessible (Ceccato et al. 2019, 2024b; Sundling and Ceccato 2022; Sundling et al. 2024). The absence of a dedicated reporting centre for travellers in Sweden today means that tailored interventions are not feasible for the broader population or, even more critically, for specific groups, such as women. If station personnel are unavailable, alternative reporting measures such as hotlines, help buttons, or apps must be in place. Meaningful change is unlikely without a system that allows individuals to report safety incidents directly and ensures their concerns are addressed (Ariel et al. 2024).

It is important to examine how intersections between gender identity and gender status interact with other factors, such as age, disability, and the specific characteristics of transit settings along the journey. Interventions must be designed with sensitivity to these potential differences, considering factors such as transit use frequency and temporal patterns of victimisation. Finally, contextual elements—such as variations in management practices across companies, municipalities, and counties—should not be overlooked.

10.5 Conclusions

We started this chapter by referring to Virginia Woolf's iconic assertion that women need a "room of one's own," which echoes as strongly today as it did nearly a century ago. As this chapter has shown, the persistent lack of safe, inclusive public spaces continues to undermine the rights and freedoms of women and gender-diverse individuals worldwide. Whether it is the extreme restrictions faced by women in Afghanistan or the pervasive fear and harassment in public transport systems globally (Ceccato and Loukaitou-Sideris 2020a), these challenges highlight the urgent need for interventions that align with local contexts and are connected to a long-term sustainable perspective. A society's commitment to equality can be measured by its ability to safeguard the rights of all individuals to occupy and move freely through public spaces. Just as Woolf envisioned a world where women could assert their intellectual and financial independence, we must envision a world where women can feel safe—not as a privilege but as a fundamental right. We recommend a systems-thinking perspective to tackle women's safety (the "tree model"), highlighting the interdependence of various domains essential for addressing victimisation and fear in public spaces. A critical component of this system is knowledge, which plays a pivotal role in identifying safety problems and developing solutions.

Improving the environments women encounter during their trips is essential to enhance their safety, particularly for those with long journeys to stations. Future research should identify areas needing improvement to reduce victimisation and boost safety perceptions. Transport hubs showing poor social control, inadequate lighting, signs of disorder, and desolation are often cited as problematic. At the local level, planners and transport operators can collaborate, as illustrated in the three examples below:

- Identify the specific contexts where issues occur and prioritise stations and routes with the highest victimisation and safety concerns.
- Monitor user incident reports regularly and conduct field inspections by transport planners and safety experts to tailor solutions to local conditions.
- Implement infrastructure for reporting incidents, such as hotlines, help buttons, or apps, to ensure passengers can report safety concerns, especially in areas without on-site staff.
- Carefully plan the location of new stations, especially in rural areas. Avoid desolate or poorly supervised locations and consider visibility and safety during Scandinavia's nighttime or long winter dark months. However, seasonal variations of light in other parts of the globe should be considered.
- Design stations and surrounding areas with increased visibility, eliminating barriers like fences and ensuring proper lighting to reduce fear and crime.

A holistic approach to safety in transit is challenging due to the need for better coordination among transport operators, police, municipal administrations, and other public entities, and not least, travellers. Interventions should be supported by security technology, media campaigns, and policies to reduce crime and fear. The "whole journey approach" aligns with the UN's 2030 Agenda for sustainable development, emphasising the need to understand women's safety patterns across rural and urban areas. Research must focus on the varying safety risks throughout the journey, particularly for rural women, those with disabilities, and LGBQTI+ individuals, as their vulnerabilities to crime and fear require further investigation. See an in-depth discussion of the recommendations in Ceccato et al. (2024a). In one of the next chapters, we focus on the challenges faced by LGBTQI+ and gender nonconforming individuals in transit environments and what needs to be done to better support this group.

References

Ariel B, Ceccato V, McDonnell A, Webster K (2024) Experiences and reporting of unwanted sexual behaviors on Great Britain's Rail Network: a survey of victims and witnesses with an embedded randomized vignette experiment on callback effects. Victims Offend. https://doi.org/10.1080/15564886.2024.2330094

BBCNews (2023) Taliban order Afghanistan's hair and beauty salons to shut. BBC News. https://www.bbc.com/news/world-asia-66094490

Buckley JB (1996) Public transit and crime: a routine activities/ecological approach

Ceccato V (2017) Women's victimisation and safety in transit environments. Crime Prevent Community Saf 19(3):163–167

Ceccato V, Gaudelet N, Graf G (2022) Crime and safety in transit environments: a systematic review of the English and the French literature, 1970–2020. Public Transport 14(1):105–153

Ceccato V, Loukaitou-Sideris A (2020a) Transit crime and sexual violence in cities: International evidence and prevention. Routledge

Ceccato V, Loukaitou-Sideris A (2020b) Why sexual crimes and fear of crime happen in transit environments: a review of theories. In: Transit Crime and Sexual Violence in Cities, pp 24–31

Ceccato V, Moreira G (2021) The dynamics of thefts and robberies in São Paulo's Metro, Brazil. Eur J Crim Policy Res 27(3):353–373. https://doi.org/10.1007/s10610-020-09462-3

Ceccato V, Newton A (2024) Systems thinking for sustainable crime prevention in risky places. Routledge

Ceccato V, Näsman P, Catherine S, Langefors L (2019) Trygghet i kollektivtrafiken i Stockholm i ett internationellt perspektiv: En handlingsplan mot sexuella trakasserier och brott i transitmiljöer. In: Kungliga tekniska högskolan

Ceccato V, Sundling C, Gliori G (2024a) Transit safety of women in rural–urban contexts. Fem Criminol 19(4):372–394

Ceccato V, Sundling C, Gliori G (2024b) Vad gör en järnvägsstation säker och för vem? Effekten av stationsmiljön för brott och resenärers trygghet. https://www.researchgate.net/publication/378653634_Vad_gor_en_jarnvagsstation_saker_och_for_vem_Effekten_av_stationsmiljon_for_brott_och_resenarers_trygghet_Slutlig_rapport?channel=doi&linkId=65e22d4ec3b52a11700306c9&showFulltext=true

Ceccato V, Sundling C, Gliori G (2024c) What makes a railway station safe and for whom? The impact of transit environments on passengers' victimisation and safety perceptions. Eur Transp Res Rev 16(1):21

Chowdhury S, Patel P, Giridharan V, Ceccato V (2023) Formation of fear and adaptive behavior in young ethnic minority women riding public transport. Transp Res Rec 2678(3):687–697. https://doi.org/10.1177/03611981231182712

Chowdhury S, van Wee B (2020) Examining women's perception of safety during waiting times at public transport terminals. Transp Policy 94:102–108. https://doi.org/10.1016/j.tranpol.2020.05.009

Cohen LE, Felson M (1979) Social change and crime rate trends: a routine activity approach. Am Sociol Rev 588–608

Crenshaw KW (2013) Mapping the margins: intersectionality, identity politics, and violence against women of color. In: The public nature of private violence. Routledge, pp 93–118

Ferraro KF (1995) Fear of crime: Interpreting victimisation risk. SUNY press

Gekoski A, Gray JM, Adler JR, Horvath MA (2017) The prevalence and nature of sexual harassment and assault against women and girls on public transport: an international review. J Criminol Res Policy Pract 3(1):3–16

Gekoski A, Gray JM, Horvath MA, Edwards S, Emirali A, Adler JR (2015) What works in reducing sexual harassment and sexual offences on public transport nationally and internationally: a rapid evidence assessment. British Transport Police and Department for Transport

Kelly L (2012) Standing the test of time? Reflections on the concept of the continuum of sexual violence. Handbook on sexual violence, xvii–xxvi

Koskela H, Pain R (2000) Revisiting fear and place: women's fear of attack and the built environment. Geoforum 31(2):269–280. https://doi.org/10.1016/S0016-7185(99)00033-0

Levy C (2019) Travel choice reframed: "deep distribution" and gender in urban transport. In: Scholten CL, Joelsson T (eds) Integrating gender into transport planning: from one to many tracks. Springer International Publishing, pp 43–65

Loukaitou-Sideris A (2012) Safe on the move: the importance of the built environment. In: The urban fabric of crime and fear, pp. 85–110. https://doi.org/10.1007/978-94-007-4210-9_4

Loukaitou-Sideris A, Ceccato V (2020) Sexual harassment on transit: evidence from the literature. In: Transit Crime and Sexual Violence in Cities. Routledge, pp 12–23

Loukaitou-Sideris A, Fink C (2009) Addressing women's fear of victimisation in transportation settings: a survey of US transit agencies. Urban Affairs Rev 44(4):554–587

Mellgren C, Andersson M, Ivert A-K (2018) "It happens all the time": women's experiences and normalization of sexual harassment in public space. Women Crim Just 28(4):262–281

Pain R (2001) Gender, race, age and fear in the city. Urban Stud 38(5–6):899–913

Sánchez-de Madariaga I, Zucchini E (2020) "Movilidad del cuidado" en Madrid: nuevos criterios para las políticas de transporte. Ciudad y Territorio Estudios Territoriales 52(203):89–102

Shaw CR, McKay HD (1942) Juvenile delinquency and urban areas

Sundling C, Ceccato V (2022) The impact of rail-based stations on passengers' safety perceptions. A systematic review of international evidence. Transport Res F: Traffic Psychol Behav 86:99–120

Sundling C, Ceccato V, Gliori G (2024) Disability, victimisation, and safety in train travel. Transp Res Interdiscip Perspect 25:101131

Tripathi K, Borrion H, Belur J (2017) Sexual harassment of students on public transport: an exploratory study in Lucknow, India. Crime Prevent Community Saf 19:240–250

Umaña-Barrios N, Gil A (2017) How can spatial design promote inclusivity, gender equality and overall sustainability in Costa Rica's urban mobility system? Procedia Eng 198:1018–1035. https://doi.org/10.1016/j.proeng.2017.07.147

United Nations Commission on the Status of Women (2023) Innovation and technological change, and education in the digital age for achieving gender equality and the empowerment of all women and girls. Agreed Conclusions (E/CN.6/2023/L.3). https://www.unwomen.org/en/csw/csw67-2023/session-outcomes

UN-Women (2024) Creating safe and empowering public spaces with women and girls. United Nations. https://www.unwomen.org/en/what-we-do/ending-violence-against-women/creating-safe-public-spaces

Vera-Gray F, Kelly L (2020) Contested gendered space: public sexual harassment and women's safety work. In: Crime and Fear in Public Places, Routledge, pp 217–231

Whitzman C (2013) Women's safety and everyday mobility. In: Building inclusive cities. Routledge, pp 35–52

11 When Travel Makes You Sick! Women's Asthma and Allergy Experiences on Trains

Abstract

Few studies have addressed the issue of allergies when travelling on trains from the user's perspective. This chapter helps address this gap by presenting the findings of a small sample study on the perceived safety of asthmatic and allergic individuals. A questionnaire was sent to members of the Swedish Asthma and Allergy Association in the spring of 2022 (N = 66). Although most respondents felt safe during their trip, 88% declared feeling afraid of allergens onboard the train (e.g. nuts and peanuts, perfumes, and fur-bearing animals) when they were asked more specifically. Allergens were identified as the most frightening factor, imposing changes in individuals' travel behaviour. The most common precaution to counteract poor safety was avoiding contact with certain people. Clearly designated areas without animals and perfumes, staff encouraged not to wear perfume, and accessible staff are commonly requested improvements for more accessible travelling.

Keywords

Safety · Allergen · Nut · Perfume · Animal · Precautionary measure

11.1 Introduction

Sustainable transport is part of the United Nations' 2030 global goals for sustainable development and facilitates convenient access to public transport (UN 2022). The Swedish Government also has a goal of accessible transport for all, although this goal has been repeatedly moved forward, partly due to a lack of knowledge of vulnerable travellers' needs (Swedish Parliament 2013). As much transport research focuses on "the general traveller", the needs of subgroups may be overlooked. Therefore, a focus on specific vulnerable groups in the travel environment can reveal needs that would otherwise remain hidden. Furthermore, to improve environmental and social sustainability, public transport must be made attractive and accessible for as many potential travellers as possible, so the needs of different subgroups must be researched.

Internationally, between 10 and 30% of the population have allergies or asthma, according to the World Allergy Organization, WAO (Pawankar et al. 2013). The prevalence of allergic diseases is rapidly increasing around the world, in both countries of Global North and Global South. Allergies are the most common immunologic diseases (for a review, see De Martinis et al. 2020). The mechanisms behind this remain

V. Ceccato et al., *Transit Safety for a Sustainable City*, Sustainable Development Goals Series,
https://doi.org/10.1007/978-3-031-97427-4_11

largely unclear although there are both individual and environmental factors. The WAO (Pawankar et al. 2013), concludes that exposure to allergens is one of the environmental risk factors for developing different kinds of allergic diseases. However, knowledge of major allergens, both indoors and outdoors, that cause allergies, is lacking. They should therefore be mapped and quantified to enable implementation of preventive measures (ibid), particularly because certain allergic diseases, such as food allergies, are potentially fatal. Moreover, allergens from foods such as nuts and peanuts have been believed to be airborne, meaning that, even without direct contact, they can still provoke symptoms just by being inhaled (Ramirez and Bahna 2009). However, recent research has had conflicting results, for example, it has in some cases not been possible to induce allergic reactions through the air (Lovén, Björkman et al. 2021).

The global cost of allergies is high because of their high prevalence among the population. For example, in Sweden, the total annual cost to society of allergic rhinitis was estimated at EUR 1.3 billion for a population of 9.5 million (Cardell et al. 2016) because of absence from work, reduced working capacity at work, healthcare contacts, and medications. Moreover, there is poor awareness overall of asthma and allergies and their consequences in terms of morbidity and potential mortality in societies (Pawankar et al. 2013), which leads to major long-term implications for those most affected by asthma and allergies.

Public transportation, such as buses and trains, is one of many public environments where people with allergies might be at risk of allergen exposure. Uncertainty as to whether this risk will be encountered or not during a trip potentially affects passengers' safety perceptions in these environments.

In this chapter, we investigate the safety perceptions of passengers with asthma and allergies while travelling by train and what, in their opinion, would make travelling safer for them. The study also focuses on which situational factors of the train trip affect their safety. Finally, the chapter finishes by discussing the implications of these results to passengers and society in general.

The decision to carry out an in-depth analysis focusing on travellers with asthma and allergy was born when respondents with allergy diseases in a larger study carried out by the authors reported feeling more unsafe than travellers without disability. With the intention of knowing more about the nature of transit safety among those travelling with asthma and allergies, we analysed responses to a questionnaire by a sample of members of the Swedish Asthma and Allergy Association collected in the spring of 2022 ($N = 66$). Thus, this chapter is therefore part of a larger project which aims to understand the safety perceptions of different types of users in the railway environment and is funded by the Swedish Transport Administration. The analysis is based in Sweden and constitutes a relevant case study as a third of the population has asthma, allergies, or other kinds of hypersensitivity (Swedish Asthma and Allergy Association 2020), which is a similar percentage to in the USA (Morgan Griffin 2021), but higher than in other countries in the European Union (EAACI 2016).

This chapter is composed of six sections of which this introduction is the first. Next, in section two, it presents the theoretical background on safety perceptions while travelling, including previous research on asthma and allergy sufferers. In section three, the sample and the process of data collection and analysis are described. Section four presents the results followed by a discussion in section five. The paper ends with conclusions and recommendations for research and practice.

11.2 Theoretical Background

Asthma, allergies, and other kinds of hypersensitivity are the most common chronic diseases in Europe. Up to 20% of patients with allergies struggle daily with the fear of a possible asthma attack, anaphylactic shock, or even death from an allergic reaction (EAACI 2016). Public transport environments, such as trains, can become fearful places for those with an allergic condition. Not being able to distinguish which situation is safe and which is dangerous can be especially stressful (Rosen et al. 2020). In this section, we discuss the concept of fear with a particular focus on transit environments.

11.2.1 Safety During the Trip: Situational Health-Related Risks and Safety Perceptions

Fear is "an emotion, a feeling of alarm or dread caused by awareness of expectation of danger" (Warr 2000, p. 453). Individuals may feel fear for a variety of reasons, some of which may relate to the likelihood of being a victim of accidents, crime, or, in this case, being exposed to allergens.

"Unsafety" according to the Collins English Dictionary (2023) means "unsafe state or condition; exposure to danger or risk; insecurity" and can be viewed as a combination of cognitions of risk and affective reactions. According to Chua et al. (2021), health risks can be perceived through three dimensions: susceptibility, severity, and psychological risk. Susceptibility to illness is the probability of being negatively impacted if encountering a health-related event (Witte 1996). If the perceived severity of the feared event is high, a person's concern for their health-related risk can be high (Liu et al. 2016). The psychological risk, according to Kurtz and Clow (1997), is when an event triggers a reaction of anxiety, worry, or other kinds of psychological discomfort. Together the three dimensions can be used to model the psychological mechanism for health-related risk perception in travelling. The perceived health risk can also in turn impact mental well-being and perceived uncertainty (Chua et al. 2021).

Much of the research on safety perceptions while travelling by train is focused on fear of crime and other "deviant" behaviour, or on accidents. However, poor safety can be experienced for many different reasons in public transport. For example, during the recent COVID-19 pandemic, poor safety was also related to the risk of contagion (Dong et al. 2021). Or, for a traveller with restricted movement, a lack of predictability (e.g. if the elevator is suddenly out of order) may be a source of poor safety (Sundling 2015). Or, for people with asthma or allergies, allergens constitute a constant risk of encountering situations when travelling that may impact their health.

Poor safety is not perceived equally by individuals; it is a function of the settings an individual is taking part in at a particular moment, at a particular setting, as previously discussed, but is also a result of what constitutes the individual in terms of gender, age, disability, ethnic background, or socioeconomic status or a combination and intersection of these characteristics. By adopting an intersectional lens (Crenshaw 1989, 1991), we suggest that the experiences of a "transit captive" with extreme allergy conditions are different from someone with extreme allergy conditions who can move around by car or by other means. Each of these identities (e.g. disability, age, gender) brings unique challenges and dynamics when travelling on public transportation that cannot be understood solely through a single identity category. Adopting an intersectional lens is important because it allows us to understand and address the complexities of human experiences in transit environments. In contrast to being a secondary, auxiliary category after gender and race (Watermeyer and Swartz 2023; Goldberg 2016), the need to specifically consider disability (such as having an allergy) as a starting point for analysis of transit safety is also worthy of recognition.

While understanding the factors that influence an individual's fear is important, it is also important to consider how fear impacts an individual's everyday life. Fear affects mobility and travel patterns (Ceccato and Loukaitou-Sideris 2022), for example, by avoiding travel situations or places that are considered especially risky or even by avoiding public transport completely. Jackson and Gray (2010) distinguish between "functional" and "dysfunctional" fear. On the one hand, functional fear leads to precautionary actions that may reduce both fear and the risk of something bad happening, sometimes even prompting individuals to support activities that make it more difficult for the event to occur, such as participating in night patrols or neighbourhood watch schemes (Gray et al. 2011). On the other hand, dysfunctional fear can paralyse individuals, leading to overly negative impacts such as constrained mobility and avoidance of public spaces, which in turn reduce life opportunities.

11.2.2 Previous Studies on Public Transportation, Health-Related Risks, and Perceived Safety

Travel needs differ depending on the situation, whether one is commuting to work, travelling with heavy luggage or children, or whether one is young, old, etc. In that sense, all travellers are "vulnerable travellers" at different times in their lives. Thus, public transport that is considered safe and accessible must be flexible to balance the needs both within and between travellers. As for allergies, the needs of pet owners must be balanced with the needs of allergy sufferers to travel without being exposed to pet allergens. In Sweden, as in many countries, there are usually designated pet areas on trains. However, other kinds of allergens are generally not restricted to certain areas. Another example of conflicting interests is scents that have been used purposely in transport to attract travellers and to enhance perceived travel quality. For example, a cleaning product scent has been used in Dutch trains to decrease littering behaviour (de Lange et al. 2012). In Singaporean buses, pleasant scents have been used to induce people to use public transport. In aeroplanes, scents have been discussed in relation to decreasing the anxiety of passengers with a fear of flying or decreasing motion sickness (for an overview, see Spence 2021). However, perfumes may induce symptoms in individuals with allergies or asthma (Millqvist et al. 1999). Thus, what is perceived as a facilitator for some may be a barrier for others.

There are few studies addressing allergies in public transport from the users' perspective; this group's experiences are rarely represented in transportation literature. More generally, evidence-based information is scarce regarding allergens, both indoors and outdoors (Pawankar et al. 2013). However, there are studies that indicate that many individuals with allergic diseases feel restricted in their use of public transport (Partti-Pellinen et al. 2000; DunnGalvin et al. 2020). In a Finnish study, 53% of adults with different kinds of allergies or asthma had experienced symptoms in public transport (Partti-Pellinen, et al. 2000). Also, the extent of direct exposure from pets was uncertain, as only 0.13% of all passengers travelled with a pet. Instead, a major cause of exposure may be the clothing of travellers and staff and/or covered seats where allergens can remain (Custovic et al. 1994). Food allergies often threaten to produce sudden allergic reactions, including life-threatening anaphylaxis, which may restrict mobility (Lange 2014). In a European survey addressing individuals with peanut allergies, most (84–93%) reported lifestyle restrictions regarding food and 53% felt restricted when using public transport (Dunn Galvin et al. 2020).

As shown above, perceived safety implies cognitions of risk and affective reactions but also leads to behavioural consequences such as avoidance of certain situations. Vulnerable groups adapt their behaviour to a greater extent (Newton et al. 2020). Thus, in the present study, we wanted to acquire more knowledge from asthma and allergy sufferers to understand how they perceive the safety of the train transport environment. What do they perceive as barriers, what do they avoid, and what improvements would they consider effective?

11.3 The Case Study

The questionnaire examines the following research questions: How safe are those individuals with asthma and allergies while travelling by train? What is the individual profile of those individuals with asthma and allergies who travel by train? Which situational factors of the train trip affect their safety? How do poor safety perceptions impact their journey? And what are the recommendations of individuals with asthma and allergies who travel for improving safety during the train trip?

Of Sweden's 10.5 million inhabitants (Statistics Sweden 2023), more than three million individuals have asthma, allergies, or other kinds of hypersensitivity (Swedish Asthma and Allergy Association 2020), which is about 30% of the country's population. Swedes frequently use trains, which is the means of public transportation with which passengers travel farthest; in 2021, the average train trip was 28 kms (Transport Analysis 2022). The state-owned train operator, SJ AB, provides (long-distance) train services along all major railway lines in Sweden as well as to neighbouring countries. In the 1980s, the system was opened up to private train operators (Spaven 1993; Alexandersson and Hultén 1999), and nowadays, rail traffic is composed of local and regional trains, which are the responsibility of the regional public transport authorities (Swedish Transport Agency 2019). The pandemic has negatively affected the number of passenger kilometres travelled since 2020. However, in June 2022, the numbers started approaching pre-pandemic levels (Swedish Public Transport Association 2022).

In Sweden, pets are usually allowed on trains in certain designated areas (e.g. SJ 2022), but rules may differ depending on the operator. Smoking is prohibited by law on board trains and in stations, as well as in all outdoor passenger areas, such as platforms (Swedish Parliament 2018).

11.3.1 Data and Methods

The present study is a small sample study that aimed to obtain in-depth knowledge of a particular group of users who declared themselves as feeling unsafe during their trip. A questionnaire was advertised digitally to Swedish Asthma and Allergy Association members during the spring of 2022 via social media. The Association has around 15,000 members (Swedish Asthma and Allergy Association 2022). The questionnaire contains 21 questions, divided into five main parts: control questions such as frequency of use of public transport; questions on perceived safety; questions on the impact of safety perceptions on behaviour and attitudes; questions regarding recommendations for improvements; and background questions (e.g. age, gender, and potential disability). Most questions have fixed answer alternatives. A few questions have open-ended answers to allow respondents to formulate their own answers instead of choosing from a predetermined list of alternatives. Two questions with open-ended answers concern the safety and attractiveness of train travelling: "Are there aspects other than those mentioned above that would make you feel safer when you travel by train?" and "Are there aspects other than those mentioned above that would make train travelling *more attractive for you?"*.

A total of 66 questionnaires were completed. Most identified themselves as women—61 or 92%. The mean age was 54 years old (range 24–76). The Association's membership also skews towards women (67%), and of those who are 30 years of age and older, 71% are women; internationally, allergies are more prevalent among women (Jensen-Jarolim 2017). According to the Association, women are more active in the Association apart from being more numerous (Enberg, pers. comm., September 5 2022).

Of the 66 respondents, 26 (39%) lived with a spouse or partner, 19 (29%) with a spouse/partner and a child/children, while the remaining respondents lived alone (11 respondents; 17%), with a child/children (7; 11%), or with parents/siblings/other adults (3; 5%). A total of 76% had a university degree. Nearly half of the respondents reported they had a minor disability (31; 47%), while 18 (27%) reported a major disability; also, 15 (23%) reported that they did not have a disability and two (3%) did not want to answer. Apart from asthma/allergy (58 of 66 respondents; 89%), the most common disabilities were reduced mobility (22; 34%) and bowel or urinary problems (12; 18%).

The two open-ended questions were collapsed. A total of 44 open-ended answers were collected; several of the answers were detailed and extensive. Inductive thematic coding (Braun and Clarke 2006) was undertaken as allergen unsafety became apparent from the fixed answers. Combining closed-ended and open-ended questions is a way of capturing both breadth and depth (Hansen and Swiderska 2023). The analysis of the data is, therefore, both quantitative and qualitative.

Because of the limited degree of generalisation, we have analysed the questionnaire answers in an exploratory manner. In practice, this means that our findings are not applicable to the general population, even within the Swedish Asthma and Allergy Association. Despite the small sample, the opportunities that this exploratory research provides far outweighed the sample limitations.

11.4 Results

11.4.1 The Profile of Passengers with Asthma and Allergies

Most respondents travelled by train only occasionally—59% (39 individuals out of 66) travelled less than once per month—and 11% (7 respondents) never travelled by train. When asked about general safety when travelling by train (N = 59 of 66), a majority declared feeling safe during their trip, especially during daytime; 66% (39 of 59 respondents) always felt safe onboard the train, 59% (35 of 59) on their way to or from the train, and 51% (30 of 59) at the station. During nighttime, they felt less safe; 34% (20 of 59) always felt safe onboard the train, 36% (21 of 59) on their way to or from the station, and 29% (17 of 59) at the station. In the station area, the respondents felt most safe on the platform (78%; 46 of 59) and least safe in an underpass or on a pedestrian bridge (20%; 12 of 59). For 20 respondents (34%), underpasses or bridges were not encountered at their station.

11.4.2 Situational Factors that Affect Safety

When respondents were asked what factors resulted in the highest levels of poor safety during an entire trip, "allergens" was the most common answer regardless of the part of the travel chain (Table 11.1). It is, however, onboard the train (88%) and inside the station (76%) that respondents were especially afraid of allergens. Only seven of the 59 respondents (12%) who answered this question declared they were not afraid of exposure to allergens during a trip. The second most impactful factor was a "disturbing smell", especially onboard the train (73%), followed by "COVID-19 contagion" onboard the train (56%).

Given the profile of the sample of respondents, it is not surprising that when respondents answered open-ended questions, allergens were, by far, the most noted problem for travel safety and attractiveness of public transportation; from a total of 44 answers, 35 addressed problems with allergens. Below we describe in more detail what the respondents (R) highlighted. They mentioned especially perfumes, animals, smoking, and food, such as peanuts, as the main source of concern:

> "I have asthma and allergies, and I risk asthma attacks when people have fur-bearing animals in the wrong carriages or smell strongly of perfumes and smoke. I also have a peanut allergy and also have airborne reactions." R48

When asked about the consequences of poor safety perceptions, at nighttime, 83% (43 out of 52 respondents) said they avoid certain individuals or groups. In daytime, it was 63%. The increased fear at nighttime, in combination with what is said in the quotes below, suggests a combination of perceived risk of crime victimisation and fear of allergens. The respondents were extremely alert (81%) and chose to place themselves where they could be seen by others (58%). All alternatives have lower rates in the daytime. However, more importantly, in the open-ended answers, again, most of these relate to allergens. The respondents describe how they had adapted their behaviour to be able to travel by train.

> "As an asthmatic, I often have to change seats, sometimes several times, because people are wearing perfume. It is a major accessibility problem." R19
>
> "In the subway or connecting tram, it happened that I had to rush out coughing at the wrong station, but I have no solution to the problem." R48
>
> "Fur-bearing animals are allowed to an increasing extent and if you have serious allergies in the family, it becomes increasingly difficult to travel." R52

Some did not find it possible to travel by train at all:

> "I have a nut allergy and chemical intolerance/sensory hyperreactivity. It would be wonderful to be able to go by train." R21

Table 11.1 Answers to the question "When you are travelling by train, are you afraid of being exposed to any of the following?" (N = 59 respondents answered this question)

	Yes, on the way to or from the station N (%)	Yes, at the station N (%)	Yes, onboard the train N (%)	No N (%)
Allergens	22 (37)	45 (76)	52 (88)	7 (12)
COVID-19 contagion	10 (17)	16 (27)	33 (56)	26 (44)
Disturbing smell	19 (32)	38 (64)	43 (73)	9 (15)
Disruptive behaviour	19 (32)	29 (49)	20 (34)	24 (41)
Theft	11 (19)	13 (22)	10 (17)	35 (59)
Robbery	15 (25)	12 (20)	1 (2)	41 (69)
Violence	12 (20)	11 (19)	4 (7)	43 (73)
Threat or hate crime	12 (20)	11 (19)	4 (7)	43 (73)
Sexual harassment	9 (15)	10 (17)	8 (14)	45 (76)
Stalking	10 (17)	6 (10)	3 (5)	48 (81)
Aggressive begging	11 (19)	13 (22)	7 (12)	40 (68)
Train accident/risk of being hit	4 (7)	4 (7)	5 (8)	50 (85)
Terror attack	5 (8)	8 (14)	7 (12)	49 (83)
Total	159	205	197	460

"The most important thing for me is to travel when there are few passengers; in that way, I can sometimes avoid getting sick from smells. Unfortunately, most people smell of perfume, for example, from fabric softeners for laundry. I got really sick when I used public transport, now I don't dare to go by train anymore. I am not afraid of anything else, just smells." R36

11.4.3 Recommendations by Respondents to Improve Safety During the Train Trip

When asked what safety improvements would be most effective (fixed-alternatives question), an "accessible train attendant" onboard the train was the most desired improvement (66%; 38 of 58 respondents). "More staff at the station" was the second most popular alternative (64%) followed by "a 'help button' to get help on the train" (62%). In other words, all three of the most chosen alternatives relate to more accessible staff.

Many of the open-ended answers also address suggestions for improvements. Several participants wished to have special carriages or areas designated for individuals with allergies in general, and not only a ban on pets in these areas:

"Separate compartment for people with fragrance hypersensitivity, allergy to nuts/peanuts and allergy to fur-bearing animals." R25

"That there were perfume-free and fragrance-free compartments so that I would not risk shortness of breath, coughing, and acute fatigue. Therefore, I almost never use public transport." R65

"That the smoking ban is complied with, and that attention is also paid to "perfume-free" areas." R39

There is a desire to have the designated areas clearly defined and delimited and to have staff available:

"That there is ALWAYS direct contact with the train host or staff present. If the entire train carriage cannot be made available to people with respiratory allergies and asthmatics, then a plexiglass area with doors from ceiling to floor." R61

"Greater knowledge of allergies and special places where dog owners/cat owners must stand on indoor platforms so that allergy sufferers can avoid close contact! Dogs have seats at the back of trains or buses and go in the back and don't pass everyone! Carriages for fur-bearing animals only on all trains!" R24

Some respondents stressed that they think staff should not use perfume or smell of smoke and that perfume-free products should be used on premises:

> "That staff are not perfumed. Allergic to perfume. That it is only cleaned with unscented agents, as these are also more environmentally friendly. That there were unperfumed zones." R49

Some respondents also highlighted the fact that allergens can stay on surfaces, for example, from a dog even if the dog itself is gone. They wished for actions that can reduce the problem:

> "Ban on having fur-bearing animals on seats! Animal epithelium sticks on carpets and upholstered seats causing illness." R24
>
> "Better cleaning in the passenger compartment due to allergens, especially seat cushions and tables." R52

Some respondents also specifically addressed the problem of food allergens and the changes that would be welcome:

> "Nut-free seats." R14
>
> "Not selling nuts on the trains due to allergy/anaphylaxis (allergy shock); the same as on flights. Clear marking of ingredients in the restaurant. Food and coffee for everyone, i.e. alternatives for allergy sufferers. Allergy seats/train carriages where you may not have pets, nuts, or other airborne allergens." R33

However, one respondent said that nut sales had been discontinued by a certain operator:

> "I was very pleased when SJ (Swedish State Railway) turned out to have stopped selling peanuts. In the subway or connecting tram, it happened that I had to rush out coughing at the wrong station." R48

11.5 Discussion

The aim of this study was to understand how individuals with allergies and asthma perceive the safety of train travel, what factors impact their safety, the consequences of their feeling unsafe, and what would make travelling safer according to them. A large majority (88%) found allergens to be the major problem. Also, most open-ended answers (80%) concerned allergens, especially perfumes and other chemicals, fur-bearing animals, and foods such as nuts and peanuts. The findings also show that many respondents recommended creating allergen-free zones that include animals, perfumes, nuts, and peanuts. The respondents also proposed that nuts and peanuts should not be sold in transport environments.

Previous research has shown that much of the allergens from cats and dogs in public transport come from people's clothes, and from covered seats and floors, not from pets that are present (Partti-Pellinen, et al. 2000). However, the same study also showed that allergen levels were lower in dust from uncovered seats and after cleaning (also in pet-free carriages). Similarly, a study in Britain showed that upholstered seats in public transport functioned as an allergen reservoir (Custovic et al. 1994). Thus, clear boundaries between spaces for pets and not bringing pets through "allergen-free" zones, together with thorough cleaning, might decrease the health-related risks for passengers with allergies. Also, the respondents' proposal to refrain from selling nuts or peanuts onboard has been put into practice in some transport areas. For example, several airlines, such as British Airways, have stopped selling peanuts (The Scotsman 2019). It is, however, now being debated whether the presence of nuts can actually cause allergic reactions through the air (Lovén Björkman et al. 2021), so research in this area still seems to need more study.

When asked a general question about safety during train travel in the present study, 66% of the respondents declared feeling safe in the daytime. This is in contrast with 88% fearing being exposed to allergens on the train. A possible interpretation could be that most respondents generally feel safe, despite the risk of allergens. Another possible reason could be that, when answering the general question, the respondents had been primed to think about other kinds of poor safety, such as "fear of crime", accidents or contagion. In any case, allergens constituted a major problem when travelling by train for the respondents.

When the respondents declared that they avoid certain people because of feelings of poor safety in public places where people are in transit, this could be because of a fear of crime or other threats. However, according to the written answers, it is apparent that there are other reasons why one can be fearful of other people in the travel environment. For example, because of the use of perfume, travellers with allergies can feel forced to adapt their behaviour to avoid health problems, for example, changing seats, leaving the train at the next stop, or avoiding travelling by train. Thus, in order to know what "fear" means more specifically, it is important to understand the context.

Most of the respondents travelled by train only occasionally. According to the written answers, many respondents will not risk using public transport because they are afraid of becoming ill. Thus, allergens could be a reason for not travelling by train more often, although it is not possible to determine the extent in the present paper. This is, however, in line with previous research showing that many individuals with allergies feel restricted in their use of public transport (Partti-Pellinen et al. 2000; Dunn Galvin et al. 2020).

The present study was part of a larger project aiming at understanding the causes of safety for different types of users. Because of the relatively high number of earlier respondents with asthma or allergy feeling unsafe in the travel environment, we decided to further investigate this phenomenon via the present study by asking specifically about allergens. More research on allergy sufferers' perceptions of travelling by public transport (including metro, bus, train, etc.) is needed and should be included in future studies.

While this study is of limited generalisability due to its small sample size (number of respondents and limited group of study), it could serve as the basis for future large-scale surveys focusing on allergies in transit environments in Sweden and abroad. It could be argued that the respondents of this questionnaire might present more severe symptoms than many other individuals with allergies as they are members of the asthma/allergy association. As with other disabilities, mild symptoms might not pose serious problems in travel environments, while more severe symptoms do. However, it is important to focus on the most affected groups of potential travellers in order to find out what barriers they experience and how accessibility could be improved for them. In the case of allergy sufferers, this would probably reduce barriers for those with milder symptoms as well. Future research should seek to untangle the nature of poor safety among allergy sufferers adopting an intersectional lens, putting disability in first place, and not as a secondary category of analysis as previous research has suggested. This approach has the potential to ensure tailored interventions towards the specific safety needs of particular groups while in transit, helping therefore to address the multifaceted nature of individual experiences which is the basis for a more equitable and inclusive society.

11.6 Conclusion and Recommendations

"Allergens" was the most feared factor when taking a train trip, especially onboard the train, for those who suffer from allergies. Allergens from animals, perfumes, and foods were considered a health-related risk, and the respondents had adapted their behaviours in different ways to deal with this risk; for example, changing seats or travelling when there are few passengers. As suggested by the respondents, improvements could include transit environments that clearly designate areas that do not allow allergens, encouraging staff not to wear perfume, and finally, making staff more accessible for travellers.

Researchers can also contribute by considering the needs of this group in their transport- and safety-related studies. Research on allergens in transit environments also requires detailed data collection during the trip as well as testing designated areas in carriages for individuals suffering from a high degree of asthma or allergies. Further research is needed in each context to understand where the problem lies and how to deal with it. In cases of health-related risk caused by allergens, for example, there may be variations in transportation modes as well as temporal variations such as differences between weekdays and weekends and between seasons. Policymakers should be made aware and experience first-hand the conditions and situation of public transit environments regarding allergens and the threats they impose on those who suffer from asthma and/or allergies. It would be beneficial to conduct regular evaluations of train stations/transit routes in consultation with actual users who are sufferers of asthma and/or allergies to reduce the threats imposed by allergens and to enhance awareness among the population of the problem.

Creating educational campaigns and policy frameworks is also essential. Public transport authorities could launch awareness campaigns for the general public about allergens, sharing strategies and getting institutions, organisations, and the public to work together to reduce these threats. The societal costs of asthma and allergies are high because of their high prevalence. Thus, less allergen exposure in public places such as trains would be beneficial. Considering that approximately one-third of the Swedish population suffers from asthma or allergies and up to 30% internationally, addressing the needs of this group of potential passengers so as to increase their use of trains or public transport, in general, could also positively impact environmental and social sustainability. This issue is present but is currently invisible on the part of the authorities. Policy interventions should create clear and easy protocols for reporting cases where problems with allergens appear in transit environments. As such, it is urgent that transit operators prioritise the needs of this group of passengers.

References

Alexandersson G, Hultén S (1999) Competitive tendering of railway services in Sweden: extent and effects 1989–1999. [Conference paper]. Thredbo Conference Series. https://thredbo-conference-series.org/downloads/thredbo6_papers/thredbo6-theme1-Alexandersson-Hulten.pdf

Braun V, Clarke V (2006) Using thematic analysis in psychology. Qual Res Psychol 3(2):77–101

Cardell LO, Olsson P, Andersson M, Welin KO, Svensson J, Tennvall GR, Hellgren J (2016) TOTALL: high cost of allergic rhinitis—a national Swedish population-based questionnaire study. NPJ Primary Care Respir Med 26(1):1–5

Ceccato V, Loukaitou-Sideris A (2022) Fear of sexual harassment and its impact on safety perceptions in transit environments: a global perspective. Violence against Women 28(1):26–48

Chua BL, Al-Ansi A, Lee MJ, Han H (2021) Impact of health risk perception on avoidance of international travel in the wake of a pandemic. Curr Issue Tour 24(7):985–1002

Collins English Dictionary (2023) Unsafety. In Collins English Dictionary. https://www.collinsdictionary.com/dictionary/english/unsafety

Crenshaw K (1989) Demarginalizing the intersection of race and sex: a black feminist critique of antidiscrimination doctrine, feminist theory and anti-racist politics. Univ Chicago Legal Forum 140:139

Crenshaw K (1991) Women of color at the center: Selections from the Third National Conference on Women of Color and the Law: Mapping the margins: Intersectionality, identity politics, and violence against women of color. Stanford Law Rev 43(6):1241–1279

Custovic A, Taggart SCO, Woodcock A (1994) House dust mite and cat allergen in different indoor environments. Clin Exp Allergy 24(12):1164–1168

de Lange MA, Debets LW, Ruitenburg K, Holland RW (2012) Making less of a mess: scent exposure as a tool for behavioral change. Soc Influ 7(2):90–97

De Martinis M, Sirufo MM, Suppa M, Di Silvestre D, Ginaldi L (2020) Sex and gender aspects for patient stratification in allergy prevention and treatment. Int J Mol Sci 21(4):1535

Dong H, Ma S, Jia N, Tian J (2021) Understanding public transport satisfaction in post COVID-19 pandemic. Transp Policy 101:81–88

Dunn GA, Blumchen K, Timmermans F, Regent L, Schnadt S, Podestà M, Fisher HR (2020) APPEAL-1: a multiple-country European survey assessing the psychosocial impact of peanut allergy. Allergy 75(11):2899–2908

EAACI, European Academy of Allergy and Clinical Immunology (2016) Statistics and figures. Allergy UK. https://www.allergyuk.org/about-allergy/statistics-and-figures/ (Retrieved 14 Dec 2022)

Enberg, S (2022) Mail correspondence.

Goldberg C (2016) Is intersectionality a disabled framework? Presenting PWIVID: In: visibility and variability as intracategorical interventions. Critical Disability Discourses

Gray E, Jackson J, Farrall S (2011) Feelings and functions in the fear of crime: applying a new approach to victimisation insecurity. Br J Criminol 51(1):75–94

Hansen K, Świderska A (2023) Integrating open- and closed-ended questions on attitudes towards outgroups with different methods of text analysis. Behav Res Methods 1–21

Jackson J, Gray E (2010) Functional fear and public insecurities about crime. Br J Criminol 50(1):1–22

Jensen-Jarolim E (2017) Gender effects in allergology—secondary publications and update. World Allergy Organ J 10(1):1–3

Kurtz DL, Clow KE (1997) Services marketing. John Wiley & Sons

Lange L (2014) Quality of life in the setting of anaphylaxis and food allergy. Allergo J Int 23(7):252–260

Liu B, Pennington-Gray L, Krieger J (2016) Tourism crisis management: can the extended parallel process model be used to understand crisis responses in the cruise industry? Tour Manag 55:310–321

Lovén Björkman S, Sederholm U, Ballardini N, Beck O, Lundahl J, Nopp A, Nilsson C (2021) Peanuts in the air: clinical and experimental studies. Clin Exp Allergy 51(4):585–593

Millqvist E, Bengtsson U, Löwhagen O (1999) Provocations with perfume in the eyes induce airway symptoms in patients with sensory hyperreactivity. Allergy 54(5):495–499

Morgan Griffin R (2021) Allergy statistics and facts. Medically reviewed by Nayana Ambardekar, MD on December 12, 2021. Retrieved from https://www.webmd.com/allergies/allergy-statistics

Newton A, Loukaitou-Sideris A, de Jubainville HDA, Huang D, Norgaard JK, Solymosi R (2020) Precautions and responses. In: Transit crime and sexual violence in cities. Routledge, pp. 276–294

Partti-Pellinen K, Marttila O, Mäkinen-Kiljunen S, Haahtela T (2000) Occurrence of dog, cat, and mite allergens in public transport vehicles. Allergy 55(1):65–68

Pawankar R, Canonica GW, Holgate ST, Lockey RF, Blaiss M (2013) WAO white book on allergy: Update 2013. World Allergy Organization

Ramirez DA, Bahna SL (2009) Food hypersensitivity by inhalation. Clin Mol Allergy 7(1):1–6

Rosen CS, Glassman LH, Morland LA (2020) Telepsychotherapy during a pandemic: a traumatic stress perspective. J Psychother Integr 30(2):174

SJ AB, Swedish State Railways (2022) Sällskapsdjur är också välkomna ombord. Retrieved December 14, 2022, from https://www.sj.se/sv/reseinfo/sallskapsdjur.html

Slovic P (1987) Perception of risk. Science 236(4799):280–285

Spaven M (1993) Rail privatisation: the Swedish experiment. Retrieved from. https://doi.org/10.3366/scot.1993.0073?journalCode=scot

Spence C (2021) Scent in motion: on the multiple uses of ambient scent in the context of passenger transport. Front Psychol 12

Statistics Sweden (2023) Befolkningsstatistik. Retrieved February 22, 2023, from https://www.scb.se/hitta-statistik/statistik-efter-amne/befolkning/befolkningens-sammansattning/befolkningsstatistik/

Sundling C (2015) Travel behavior change in older travelers: understanding critical reactions to incidents encountered in public transport. Int J Environ Res Public Health 12(11):14741–14763

Swedish Asthma and Allergy Association (Astma- och allergiförbundet) (2021) Medlemsantalet för 2021 fastställt. Retrieved November 2, 2022, from https://astmaoallergiforbundet.se/medlemsantalet-for-2021-faststallt/

Swedish Asthma and Allergy Association (Astma och allergiförbundet) (2020) Astma- och allergirapporten. Retrieved from https://astmaoallergiforbundet.se/content/uploads/2020/11/Astma-och-Allergirapporten_e-rapport.pdf

Swedish Parliament (2013) The entire journey all year round! A follow-up of the accessibility of the transport system for people with functional disabilities (2013/14:RFR5). Stockholm, Sweden: Committee on Transport and Communications, The Riksdag Printing Office

Swedish Parliament (2018) Lag (2018:2088) om tobak och liknande produkter. Retrieved from https://www.riksdagen.se/sv/dokument-lagar/dokument/svensk-forfattningssamling/lag-20182088-om-tobak-och-liknande-produkter_sfs-2018-2088#K6

Swedish Public Transport Association (Svensk Kollektivtrafik) (2022) Halvårsrapport Kollektivtrafikbarometern 2022. Retrieved from https://www.svenskkollektivtrafik.se/globalassets/svenskkollektivtrafik/dokument/verktyg-och-system/kollektivtrafikbarometern/kollektivtrafikbarometern-halvarssammanstallning-2022.pdf

Swedish Transport Agency (Transportstyrelsen) (2019) Resandeflöden på Sveriges järnvägsnät—analys av utbud och efterfrågan på tågresor. Retrieved from https://www.transportstyrelsen.se/globalassets/global/publikationer-och-rapporter/marknadsovervakning/resandefloden-pa-sveriges-jarnvagsnat20190411.pdf

The Scotsman (2019) Can I eat nuts on the flight? Regulations from EasyJet, Ryanair, British Airways. Retrieved from https://www.scotsman.com/lifestyle/travel/can-i-eat-nuts-on-a-flight-regulations-from-easyjet-ryanair-british-airways-1419013

Transport Analysis (Trafikanalys) (2022) Regional linjetrafik. Retrieved from https://www.trafa.se/kollektivtrafik/kollektivtrafik/

United Nations (UN) (2022) Goal 11: Sustainable cities and communities. Retrieved September 22, 2022, from https://unstats.un.org/sdgs/report/2022/goal-11/

Warr M (2000) Fear of crime in the United States: avenues for research and policy. Crim Justice Behav 4:451–489

Watermeyer B, Swartz L (2023) Disability and the problem of lazy intersectionality. Disabil Soc 38(2):362–366

Witte K (1996) Predicting risk behaviors: development and validation of a diagnostic scale. J Health Commun 1(4):317–342

12 Inclusive Transit: Accommodating Disabilities in Train Travel

Abstract

A vulnerable but also heterogeneous group in public transport is made up of travellers with disabilities. Based on a case study of municipalities in southern Sweden, this chapter characterises the victimisation and perceived safety among train travellers with disabilities, namely asthma/allergies, depression/anxiety, and/or reduced mobility (N = 3,374). Results show that people with these disabilities are more likely to be victimised in transit than those with no disability and that they feel more unsafe, especially if they had been previously victimised. According to participants with disabilities, key improvements would include more staff, greater accessibility of staff, a more substantial police presence, and enhanced CCTV coverage. Including disability groups in planning is essential to make public transportation safer and more inclusive.

Keywords

Asthma or allergy · Depression or anxiety · Reduced mobility · Railway · Crime

12.1 Introduction

According to the World Health Organization (WHO), one in six people, or 1.3 billion people, experience a disability (2023). In their definition, a disability emerges in the interaction between personal factors (such as specific health conditions) and the environment. Therefore, if an environment is adapted to better match the needs of the users, ability can increase and disability decrease (Sundling et al. 2015). Several United Nations' 2030 global Sustainable Development Goals (UN 2024) are relevant for travellers with disabilities. Goal 11 concerns making cities and human settlements more inclusive, safe, resilient, and sustainable. This includes the facilitation of transport. We know that people with disabilities are more at risk of exclusion from public transport than others (WHO 2023). Goal number 7, "sustainable energy for all", is relevant as increased safety in public transport will lead to decreased car dependence. Goal 10, "reduced inequality", directly includes people with disabilities. Finally, Goal 5, "gender equality", is relevant because we know that women feel less safe in the transport environment (Ceccato et al. 2024) and that women with disabilities are especially vulnerable (Brownridge 2006). However, the knowledge of the safety needs of "people with disabilities" in train travel is relatively scarce in the international literature (see, e.g. Sundling et al. 2022).

This chapter enhances the understanding of victimisation and perceived safety among

V. Ceccato et al., *Transit Safety for a Sustainable City*, Sustainable Development Goals Series,
https://doi.org/10.1007/978-3-031-97427-4_12

individuals with disabilities, specifically those with asthma/allergies, depression/anxiety, and/or reduced mobility. It draws on data from a survey (N = 3374) conducted in municipalities in southern Sweden, providing valuable insights into their experiences within this context. This chapter builds on a previous study carried out by Sundling et al. (2024).

12.2 Theory of Transit Safety and Disability

People with disabilities constitute a vulnerable group of travellers (Yavuz and Welch 2010). They are excluded to a higher extent from public transport than those without disabilities. The reasons can be poor accessibility or unaffordability (WHO 2023). For example, travellers with restricted mobility may still encounter several barriers, such as lifts that are not working, gaps between the train and platform, or steps to vehicles (Bezyac et al. 2017). Travellers with disabilities are also found to be more victimised than others (Iudici et al. 2017), especially women, who are more vulnerable to harassment and sexual violence (Hughes et al. 2012). In a study of women in Europe, Australia, and North America, more than 50% of women with disabilities had been subjected to physical violence compared to one-third of women without disabilities (Brownridge 2006). Travellers with disabilities also feel more unsafe (Yavuz and Welch 2010) than others because of, for example, discrimination, vulnerability, and feeling exposed (Wayland et al. 2022). Other passengers' attitudes and the fear of being harassed can constitute travel barriers for travellers with disabilities (Sundling et al. 2014). Therefore, people with disabilities can also be more afraid to go out after dark (Ceccato 2015; Ceccato and Bamzar 2016).

However, perceived safety and victimisation for people with disabilities is an under-researched area. Only one of 52 included publications focused on disability in a recent systematic literature review on perceived safety in train travel (Sundling and Ceccato 2022). Therefore, more research is needed that addresses people with disabilities. Also, depending on the kind of disability, needs may differ; for example, allergens may be a barrier for travellers with allergies and asthma (see Chapter 11), while this is not the case for others. This study addresses the following research questions:

- Are respondents with asthma/allergy, depression/anxiety, or reduced mobility more victimised than others when travelling by train? Do they feel more unsafe?
- What individual and place factors could predict unsafety among people with disabilities?
- What improvements would make travelling safer, according to the respondents?

Building on the discussions in earlier chapters, a systems thinking approach is adopted here to guide a perspective that aim to create transport environments that are both safe and inclusive. This study illustrates, on a small scale, how such an approach can help address the safety challenges faced by train travellers with asthma/allergies, depression/anxiety, or reduced mobility, who report higher victimisation rates and stronger feelings of insecurity than other passengers. We turn now to the case study and employed methodology.

12.3 The Case Study

Trains are a vital mode of transport, with passengers averaging 29 kms per trip in 2023 (Transport Analysis 2024). This study's study area comprises 47 rail-bound stations distributed over three railway lines stretching through central and southern Sweden, serving 28 municipalities with a total population of 1.78 million inhabitants (Statistics Sweden 2020). For more details about this case study, see Chapter 4 of this book and the article which is the basis of this chapter, published by Sundling et al. (2024).

12.3.1 Material and Methods

A questionnaire was specially developed for the study (see Chapter 4). After ethical approval, data was collected from May to September 2022 along three major rail routes in Sweden

Table 12.1 Description of the sample

	Asthma/allergy	Depression/anxiety	Reduced mobility
Women	70%	63%	63%
Age span (most common)	40–49 (23%)	19–29 (34%)	50–59 (25%)
No car use (only public transport, bike, or walk)	38%	45%	39%
Feel safe with others	53%	47%	49%
Travel by night	56%	52%	47%

Table 12.2 Victimisation by disability group

Disability	n	%	N	*p*
Asthma/allergy	171	30	576	< 0.001
Depression/anxiety	154	29	531	< 0.001
Reduced mobility	53	32	166	0.009

Significance value at 5%.

(Svealandsbanan and Mälarbanan, Västra stambanan, and Södra stambanan). In all, 47 stations were included in 28 municipalities. The three lines serve approximately 64,000 passengers per day. The questionnaire was delivered as a paper-and-pen survey at train platforms and as a digital survey advertised in the municipalities along these routes. The latter way of advertising was selected to reach even those not travelling by public transport, as we wanted to assess reasons for not travelling by train. The questionnaire is divided into five main parts: control questions, such as how frequently one used the train; questions on perceived safety; questions regarding one's own attitudes and behaviour concerning safety; questions regarding recommendations for improvement; and background questions, such as age, gender, and potential disability. The sample constitutes respondents who identify themselves as having asthma/allergy (N = 595), depression/anxiety (N = 545), or reduced mobility (N = 174). Respondents can be part of more than one group as they have more than one disability (Table 12.1).

In each group, women are in the majority, highest in the asthma/allergy group at 70% (Table 12.1). Respondents with reduced mobility are the oldest with 25% being 50–59 years old. Roughly 50% of the respondents in each group feel safe travelling with others as well as feeling safe travelling by night. However, most commonly in all groups, they travel less than once a month by train.

For both victimisation and perceived safety, chi-square analyses and logistic regression were conducted for each of the three disability groups. However, for victimisation, the logistic regression model was significant only for the group with asthma/allergy and is thus not presented for this group. Victimisation has been set to the last five years and only during a train trip, the "whole trip perspective" (to or from the station, at the station, or on the train). Perceived safety/poor safety concerns "being afraid" or "feeling safe" when travelling by train. A significance level of 5% is used in all analyses. For a more detailed discussion of the methodological steps, see also Sundling et al. (2024).

12.4 Results

12.4.1 Victimisation

Respondents with all three kinds of disability had been significantly more victimised during train travel than other travellers. Around 30% had been exposed to some kind of crime or harassment, the highest for respondents with reduced mobility, 32%, followed by those with asthma/allergy, 30%, and those with depression/anxiety, 29%, see Table 12.2. The most common form of crime and harassment was aggressive panhandling. This was followed by sexual harassment and stalking for respondents with asthma/allergy

Table 12.3 Most common places for victimisation in station areas and most common kinds of crime or harassment in χ^2 analysis

Victimisation	Asthma/allergy	Depression/anxiety	Reduced mobility
Station areas	N = 106	N = 104	N = 34
Platform	55% n.s	69% $p = 0.003$	68% n.s
Underpass	40% n.s	30% n.s	47% n.s
Waiting hall	38% $p = 0.038$	34% n.s	50% $p = 0.01$
Crime/harassment	N = 576	N = 531	N = 166
Aggressive panhandling	21% $p < 0.001$	16% n.s	26% $p = < 0.001$
Sexual harassment	9% $p = < 0.001$	10% $p = < 0.001$	10% $p = 0.006$
Stalking	9% $p = < 0.001$	9% $p = < 0.001$	8% n.s
Threat/hate crime	5% n.s	6% $p = 0.006$	9% $p = 0.01$

Source Sundling et al. (2024), p. 4
Significance value at 5%

and depression/anxiety. However, for respondents with reduced mobility, threats or hate crime was the third most common type of crime/harassment. The other alternatives to choose from were theft, robbery, and violence, and these were experienced to a lesser extent. As seen in Table 12.2, it was significantly more common to have been a victim of these crimes or harassment for all three groups of respondents compared to other respondents.

At the station, the place where the risk of victimisation was highest for all groups is the platform (Table 12.3). Of the victimised respondents in each group, 69% of those with depression/anxiety, 68% of those with reduced mobility, and 55% of those with asthma/allergy had been victimised on the platform. However, this was significantly higher than for other respondents, only for those with depression/anxiety ($p = 0.003$). For all groups, aggressive panhandling was most common.

12.4.2 Perceived Safety

Respondents with disabilities felt more unsafe than others. Those with reduced mobility felt most unsafe (24% at the station at night; $p = 0.002$), but differences between groups were small (for those with depression/anxiety, 23% with asthma/allergy, 21%). Not surprisingly, nighttime travel was perceived to be more unsafe than daytime travel.

In logistic regression, the most important person factor was previous victimisation for those with asthma/allergy and reduced mobility. However, for those with depression/anxiety, identifying as LGBTQI + was most important for perceiving poor safety (Table 12.4). In all groups, women felt more unsafe than men and for those with depression/anxiety, frequent travelling (4–7 days/week) increased the risk of feeling unsafe. The individual factors that were not statistically significant on the 5% level, for any of the groups, were: old age (65 or more), having been born in a foreign country, travel during nighttime (although it was statistically significant on a 10% level ($p = 0.082$) for respondents with depression or anxiety), or only using public transport, cycling, or walking.

Of all place factors, intoxicated people in the station area were perceived, by all groups, to induce a lack of safety the most. Especially travellers with reduced mobility found this frightening; they were more than five times as likely to feel unsafe in such situations as other respondents. This was the strongest predictor for poor safety for all variables (both place factors and person factors) and for all groups. The respondents feel they are left alone with people under the influence of substances:

Table 12.4 Person and place variables impacting safety perceptions at stations for the different disability groups: (a) Asthma/allergy, (b) Depression/anxiety, and (c) Reduced mobility

Y = 1 = Fear of crime	(a) Asthma/allergy				(b) Depression/anxiety				(c) Reduced mobility			
	Exp (B)	CI Lower	CI Upper	Sig.	Exp (B)	CI Lower	CI Upper	Sig.	Exp (B)	CI Lower	CI Upper	Sig.
Individual variables												
Woman	**2.054**	**1.091**	**3.867**	**0.026**	**2.206**	**1.213**	**4.012**	**0.010**	3.929	0.955	16.170	0.058
LGBTQI + / Non-binary/Other	1.681	0.515	5.491	0.390	**4.728**	**1.963**	**11.387**	**0.001**	5.465	0.450	66.299	0.182
Frequent traveller (4–7 days/week)	0.925	0.507	1.689	0.800	**1.827**	**1.084**	**3.080**	**0.024**	0.981	0.264	3.648	0.977
University education	**0.422**	**0.252**	**0.708**	**0.001**	0.705	0.427	1.165	0.173	0.315	0.092	1.073	0.065
Victim of crime	**4.720**	**2.487**	**8.961**	**0.000**	**3.128**	**1.721**	**5.683**	**0.000**	**4.303**	**1.184**	**15.632**	**0.027**
Place variables												
Poor lighting	**2.131**	**1.258**	**3.611**	**0.005**	1.000	0.603	1.658	0.999	0.985	0.270	3.596	0.982
Lack of staff	**1.948**	**1.134**	**3.346**	**0.016**	1.421	0.866	2.333	0.164	**4.930**	**1.375**	**17.669**	**0.014**
Intoxicated people	**2.460**	**1.422**	**4.254**	**0.001**	**2.801**	**1.634**	**4.799**	**0.000**	**5.574**	**1.572**	**19.761**	**0.008**

Logistic regression, significant variables only. Insignificant variables include age (above 65 years old), foreign born, travelling during nighttime, transit captive, restaurant/café at the station, toilet at the station

(a) Nagelkerke R Square = 0.541
(b) Nagelkerke R Square = 0.274
(c) Nagelkerke R Square = 0.309

"In the evening, I sometimes have the feeling that I am the only one waiting for a train. Others are youth gangs who look threatening and older men under the influence of alcohol/drugs".

"It is unpleasant to travel alone as a young woman when they always hang out in waiting rooms or underpasses, etc. I'm in the process of getting my driving licence to avoid being closely influenced every time I travel".

Another respondent commented:

"In the evening, you can be the only adult on the platform. Especially on weekend evenings there are several drunken young people who are going home to (name of place). They know there are no guards there and can be quite rowdy

An important place factor was the lack of staff and the comments address different reasons:

"That drunks and people under the influence and other scumbags are ejected from the place immedi ately"

"When there are traffic problems and so many stations are unmanned today, it can get really scary. No information, no staff to ask [...] No number for the traffic management, etc. This happens more often than you might think, but the management in Gothenburg or Stockholm will never find out because you can't call them".

People with reduced mobility were almost five times as likely to feel unsafe as others if there were no members of staff present. For respondents with asthma/allergy, poor lighting was significantly more important for safety than for other travellers. However, the place factors that were not statistically significant for any of the groups on the 5% level were: if there was a restaurant or café in the area, if there was a toilet at the station, poor maintenance and if the station was isolated. However, the lack of CCTV was significant on the 10% level ($p = 0.062$) for respondents with asthma/allergy and for the presence of underpasses/overpasses ($p = 0.065$) for those with depression or anxiety.

In χ^2 analysis, when different places were compared, underpasses and overpasses were experienced by all groups as the most unsafe places in the station area (Table 12.5).

"The tunnels are scariest as well as desolate obscured stations where bicycles are also parked remotely, etc.".

In every group, more than 50% felt unsafe in such places. This is significantly more than for respondents without disabilities. Bus terminals and car parks were other areas perceived to be unsafe.

Table 12.5 The places in the station area perceived as most unsafe in χ^2 analysis. Significance value at 5%

Most unsafe places	Asthma/allergy	Depression/anxiety	Reduced mobility
Underpass/tunnel	54% $p = < 0.001$ N = 463	53% $p = < 0.001$ N = 447	54% $p = 0.019$ N = 141
Bus terminal	30% $p = < 0.001$ N = 393	28% $p = 0.011$ N = 389	33% $p = 0.008$ N = 120
Car park	29% $p = 0.002$ N = 382	28% $p = 0.026$ N = 383	38% $p = < 0.001$ N = 117

The crimes that most people feared in all groups were theft and violence (Table 12.6). However, for respondents with depression/anxiety, sexual harassment was also feared as much as theft.

At least 90% in each group took precautions of some sort because of poor safety, especially at night, which is significantly higher than for respondents without disabilities. The most common precautions were to stay alert and to avoid certain people. Those with reduced mobility took significantly more precautions because of poor safety, even during the day (63%; $p = < 0.001$).

12.5 Respondents' Preferred Improvements to Increase Safety

More staff was the most desired improvement in all groups (Table 12.7) and significantly more important for the three disability groups than for other respondents. They wanted, for example, "train attendants even at stations for support and for extra vigilance", someone present to keep order:

> "The person who rides a bicycle, skateboard, etc. on the platforms must be told" and they wanted "someone to keep an eye on drunk people who board".

Generally, improvements related to surveillance or alternative ways of getting in touch with staff were found to be more important than other improvements (such as information, frequency, or fewer transfers). Thus, CCTV at the station, an accessible train host, or better police presence at the station were among the most desired improvements. "It is often a place of crime because it is hidden and dark. Want more cameras and guards"; "surveillance even in underpasses, that's the worst place at all stations, tighter spaces where it's hard to escape". Illumination was also deemed a significant improvement. For those with reduced mobility, maintenance was more important than for the other groups: "cleaner and fresher underpasses".

> "Would also need help when the elevator doesn't work, and I have to struggle alone in a wheelchair up the seriously DISGUSTING stairs".

Several of the variables considered to be important were also significantly more important for the participants with disabilities than for other respondents.

12.6 Discussion

The results show that respondents with asthma/allergy, depression/anxiety, or reduced mobility were significantly more victimised and felt more unsafe than other respondents in train travel. This is in line with earlier studies (e.g. Hughes et al. 2012; Yavuz and Welch 2010). Thus, although our groups of respondents have different kinds of disabilities, the higher levels of victimisation and unsafety, compared to other respondents, are still present in all three groups.

Although the platform is the most common place to be victimised, underpasses/overpasses are perceived to be the most unsafe. However, even underpasses/overpasses are among the three most common places to be victimised. In earlier studies, underpasses are often found to be perceived as unsafe (Libardo and Nocera 2012; Coppola and Silvestri 2020; Ceccato et al. 2025a,b), as they are often dark, with limited possibilities for escape. Seeing intoxicated individuals and other "deviant

Table 12.6 The kinds of crime most feared in χ^2 analysis

Most feared crimes	Asthma/allergy	Depression/anxiety	Reduced mobility
	N = 573	N = 531	N = 164
Theft	28% $p = < 0.001$	31% $p = < 0.001$	35% $p = < 0.001$
Violence	27% $p = < 0.001$	29% $p = < 0.001$	36% $p = < 0.001$
Sexual harassment	23% $p = < 0.001$	31% $p = < 0.001$	23% $p = 0.050$

Table 12.7 Desired improvements for safety according to respondents with asthma/allergy, depression/anxiety, or reduced mobility*

	Asthma/allergy	Depression/anxiety	Reduced mobility
	n = 570	n = 530	n = 165
More staff	**61% $p = < 0.001$**	**58% $p = 0.006$**	**61% 0 = 0.027**
Maintenance	35% $p = 0.005$	32% n. s	**44% $p = < 0.001$**
Illumination	**47% $p = 0.001$**	**45% $p = 0.040$**	**49% $p = 0.048$**
Real-time Information	25% n. s	29% $p = < 0.001$	25% n. s
Info. place for arrivals, departures	24% n. s	29% $p = < 0.001$	32% $p = 0.002$
Info on tickets	11% n. s	8% n. s	15% $p = 0.004$
Frequency	26% n. s	27% n. s	29% n. s
Fewer transfers	20% n. s	23% $p = < 0.001$	24% $p = 0.044$
One phone no. for problems	30% $p = < 0.001$	29% $p = < 0.001$	39% $p = < 0.001$
Help button on train	**41% $p = < 0.001$**	**44% $p = < 0.001$**	**46% $p = < 0.001$**
Help button at station	**41% $p = < 0.001$**	**43% $p = < 0.001$**	**41% $p = 0.007$**
Safety app	32% $p = < 0.001$	**40% $p = < 0.001$**	33% $p = 0.047$
Police at station	**48% $p = 0.024$**	**55% $p = < 0.001$**	**49% n. s**
Police on train	27% $p = 0.006$	30% $p = < 0.001$	23% n. s
Accessible train host	**52% $p = < 0.001$**	**48% $p = < 0.001$**	**56% $p = < 0.001$**
CCTV on train	**46% $p = 0.005$**	**44% n. s**	**48% $p = 0.052$**
CCTV at station	**52% $p = 0.023$**	**52% $p = 0.031$**	**56% $p = 0.046$**

*More than one alternative could be marked. Significance value at 5%
Alternatives chosen by 40% or more are in bold

behaviour" is repeatedly found to make a place feel more unsafe (Sundling and Ceccato 2022). Earlier victimisation, found in the present study, is also a fundamental factor affecting safety (de Jubainville and Vanier 2017; Stark and Meschik 2018; Vilalta 2011). People with disabilities were more often victimised, compared to others. Therefore, it is important to include these groups in planning. Thus, it is not surprising that it is significantly more important for our participants that there are staff members present compared to other respondents. Better staff presence was the most important safety improvement, together with different solutions regarding surveillance and access to personal support regardless of the kind of disability (Table 12.7). More staff and surveillance as a highly valued improvement are often found for passengers in general (e.g. Cozens et al. 2004). However, "vulnerable travellers" may be more dependent on accessible staff. According to our results, around 60% of the respondents in the disability groups want more staff at the stations. This is significantly more important than for other travellers. Also, many other desired improvements relating to surveillance staff are significantly more important for our respondents with a disability than for other travellers. However, many stations are small, and staff presence is not realistic, especially not around the clock.

As digitalisation is more and more present, surveillance might be achieved with some kind of "help buttons" in combination with increased police or guard presence, CCTV at the station, and accessible train hosts, which are also among the most desired improvements. What is important is to feel that one is not left alone and that someone is in charge. In many cases, digital alternatives to staff presence can make travellers feel more included. With such digital means, reports can be collected and constitute a statistical base for incidents that take place in a certain area. In this way, decision-making will be easier regarding interventions. In some countries, systems are already in place, such as a help button or app with centralised customer support (Ceccato and Paz 2017). In many other countries, this basic service has not yet been introduced. For a more detailed discussion about the design of future stations and related transit environments, see Chapters 15–17 of this book.

This chapter addresses UN Sustainable Goal 11, "making cities and human settlements inclusive, safe, resilient, and sustainable" (UN 2024), in which "sustainable transport" is included. It is important to underscore that inclusive and secure travel environments need to be created in a "whole journey" perspective, including the way to and from the station or the stop. Thus, municipalities are important actors, together with transport authorities and other transport actors. Variables in our research found to induce a lack of safety at stations such as intoxicated people, poor lighting, underpasses, and isolated places can also be generalised to other public places. Thus, some of the improvements relevant in station areas can also be relevant in other public places. To facilitate activities at different times of the day for different groups of people can be such an example. Similarly, it is important around cities to identify where increased surveillance is needed because of drug dealing or other norm-violating behaviour, to improve sightlines, etc. However, such interventions must be carefully designed so that they do not infringe on individuals' right to privacy or other fundamental human rights.

The chapter also addresses Goal 10, to reduce inequalities. Further research is needed to address the safety needs of different disability groups, which should also be included in planning. Also, if the needs of especially vulnerable groups are considered in planning, safety can be increased for everyone. It is also important to remember that not only the places where the actual crimes occur should be addressed. Places that make people feel unsafe can make people avoid travelling even if most crimes do not occur there.

Certain groups are small for some of the analyses, especially for victimisation. Therefore, these analyses should be interpreted cautiously. Also, causal relationships cannot be determined. Moreover, self-reported data can always be biased in different ways. However, we do not consider that any other method would have been more suitable than asking the travellers themselves about their experiences and points of view. We also tried to achieve a large variation in the sample (see Chapter 4). The body of knowledge on the needs of people with disabilities in public transport must continue to grow; most research still lacks focus on these groups, probably even more so in countries of the Global South. The transport system must be safe and accessible for all passengers. These groups must also be included in urban and transportation planning. According to our results, they are both more victimised and feel more unsafe when they are travelling by train; therefore, improvements for these groups must be prioritised.

12.7 Conclusions and Recommendations

As discussed in previous chapters of this book, adopting a systems thinking approach is fundamental for promoting safe and inclusive transport environments. This study highlights in a small scale the safety challenges faced by train travellers with asthma/allergies, depression/anxiety, or reduced mobility, who experience higher victimisation rates and feelings of a lack of safety compared to other passengers. Tunnels, underpasses and overpasses, often perceived as unsafe due to poor design, emerge as key concerns alongside platforms, which are the most common sites of victimisation. Staff presence and effective surveillance are vital safety improvements. While constant staff presence at smaller stations may not be feasible, technological solutions such as help buttons, CCTV, and accessible digital platforms can offer critical support and data for targeted interventions. All these aspects emerge as critical 'leverage points' for intervention. This chapter addresses the UN's Sustainable Development Goals 10 and 11, advocating for inclusive and safe transport environments with a whole journey approach. Collaboration between munic ipalities, transport authorities, and stakeholders is crucial in order to address safety issues such as poor lighting and intoxicated individuals.

Safety interventions should not follow a one-size-fits-all approach, as different disability groups have distinct needs. Passengers with reduced mobility require reliable boarding assistance and designated safe waiting areas, while those with asthma or allergies need improved ventilation and allergen-free spaces. Installing emergency help buttons with audiovisual signals and CCTV monitoring tailored to detect individuals in distress is essential. One recommendation is to create barrier-free infrastructure, ensuring step-free access in tunnels, underpasses, and overpasses through ramps, elevators, and tactile guidance for visually impaired passengers. Future research must focus on the specific needs of each group to ensure that transport systems are safe and accessible for all.

References

Appleton J (1975) The experience of landscape. John Wiley

Avineri E (2012) On the use and potential of behavioural economics from the perspective of transport and climate change. J Transp Geogr 24:512–521

Bezyak JL, Sabella SA, Gattis RH (2017) Public transportation: an investigation of barriers for people with disabilities. J Disabil Policy Stud 28(1):52–60

Brownridge DA (2006) Partner violence against women with disabilities: prevalence, risk, and explanations. Violence against Women 12(9):805–822

Ceccato V (2015) Ensuring safe mobility in Stockholm, Sweden. In: Proceedings of the Institution of Civil Engineers—Municipal Engineer 168(1):74–82

Ceccato V, Bamzar R (2016) Elderly victimization and fear of crime in public spaces. Int Crim Justice Rev 26(2):115–133

Ceccato, V, Loukaitou-Sideris, A (2020) In: Responding to sexual harassment in transit: towards an agenda for research and practice, Routledge, pp 305–328

Ceccato V, Paz Y (2017) Crime in São Paulo's metro system: sexual crimes against women. Crime Prevent Community Saf 19:211–226

Ceccato V, Sundling C, Gliori G (2024) What makes a railway station safe and for whom? The impact of transit environments on passengers' victimisation and safety perceptions. Eur Transp Res Rev 16(1):1–16

Ceccato, V., Ioannidis, I., & Felson, M. (2025a). Tunnels in the urban fabric: balancing connectivity and safety. *Urban, Planning and Transport Research, 13*(1), 2431514.

Ceccato, V., Ioannidis, I., & Felson, M. (2025b). Graffiti in pedestrian tunnels: A comparison of police records and crowdsourced data in Stockholm, Sweden. *Tunnelling and Underground Space Technology, 159*, 106482.

Coppola P, Silvestri F (2020) Assessing travelers' safety and security perception in railway stations. Case Stud Transport Policy 8(4):1127–1136

Cozens P, Neale R, Whitaker J, Hillier D (2004) Tackling crime and fear of crime whilst waiting at Britain's railway stations. J Public Transp 7(3):23

Cullen M (2006) Improving transport accessibility for all: guide to good practice. OECD Publications

de Jubainville HDA, Vanier C (2017) Women's avoidance behaviours in public transport in the Ile-de-France region. Crime Prevent Community Saf 19(3):183–198

Delbosc A, Currie G (2012) Modelling the causes and impacts of personal safety perceptions on public transport ridership. Transp Policy 24:302–309

Fisher BS, Nasar JL (1992) Fear of crime in relation to three exterior site features: prospect, refuge, and escape. Environ Behav 24(1):35–65

Gabriel U, Greve W (2003) The psychology of fear of crime: conceptual and methodological perspectives. Br J Criminol 43(3):600–614

Henry T (2009) Policy on persons with disabilities. Office of the Prime Minister (Social Services Delivery)

Hughes K, Bellis MA, Jones L, Wood S, Bates G, Eckley L, Officer A (2012) Prevalence and risk of violence against adults with disabilities: a systematic review and meta-analysis of observational studies. The Lancet 379(9826):1621–1629

Iudici A, Bertoli L, Faccio E (2017) The 'invisible' needs of women with disabilities in transportation systems. Crime Prevent Community Saf 19:264–275

Jackson J (2011) Revisiting risk sensitivity in the fear of crime. J Res Crime Delinq 48(4):513–537

Krnjacki L, Emerson E, Llewellyn G, Kavanagh AM (2016) Prevalence and risk of violence against people with and without disabilities: findings from an

Australian population-based study. Aust N Z J Public Health 40(1):16–21
Libardo A, Nocera S (2012) Exploring the perceived security in transit: the Venetian students' perspective. WIT Trans Ecol Environ 155:943–954
Loukaitou-Sideris A (2014) Fear and safety in transit environments from the women's perspective. Secur J 27(2):242–256
Skogan WG, Maxfield MG (1981) Coping with crime: Individual and neighborhood reactions. Sage Publications
Stark J, Meschik M (2018) Women's everyday mobility: frightening situations and their impacts on travel behaviour. Transport Res f: Traffic Psychol Behav 54:311–323
Statistics Sweden (2020) Folkmängden per region efter ålder och kön. Retrieved from https://www.statistikdatabasen.scb.se/pxweb/sv/ssd/START__BE__BE0101__BE0101Y/FolkmDesoAldKonN/
Sundling C (2020) Huddinge, Sweden. In: Transit crime and sexual violence in cities. Routledge, pp 167–175
Sundling C, Berglund B, Nilsson ME, Emardson R, Pendrill LR (2014) Overall accessibility to traveling by rail for the elderly with and without functional limitations: the whole-trip perspective. Int J Environ Res Public Health 11(12):12938–12968
Sundling C, Ceccato V (2022) The impact of rail-based stations on passengers' safety perceptions: a systematic review of international evidence. Transport Res f: Traffic Psychol Behav 86:99–120
Sundling C, Ceccato V, Gliori G (2024) Disability, victimisation, and safety in train travel. Transp Res Interdiscip Perspect 25:101131
Sundling C, Emardson R, Pendrill LR, Nilsson ME, Berglund B (2015) Two models of accessibility to railway traveling for vulnerable, elderly persons. Measurement 72:96–101
Thomas LJ, Rhind DJ, Robinson KJ (2006) Rail passenger perceptions of risk and safety and priorities for improvement. Cogn Technol Work 8(1):67–75
Transport Analysis (Trafikanalys) (2024) Regional linjetrafik 2023. Retrieved February 4, 2024, from https://www.trafa.se/kollektivtrafik/
UN, United Nations (2024) Retrieved August 19, 2024, from https://sdgs.un.org/goals
Uzzell D, Brown J (2007) Conceptual progress in understanding fear of crime in railway stations. Psicologia 21(2):119–137
Vanier C, de Jubainville HDA (2017) Feeling unsafe in public transportation: a profile analysis of female users in the Parisian region. Crime Prevent Community Saf 19(3):251–263
Vilalta CJ (2011) Fear of crime in public transport: research in Mexico City. Crime Prevent Community Saf 13(3):171–186
Wayland S, Newland J, Gill-Atkinson L, Vaughan C, Emerson E, Llewellyn G (2022) I had every right to be there: discriminatory acts towards young people with disabilities on public transport. Disabil Soc 37(2):296–319
WHO, World Health Organization (2023) Disability and health. Retrieved May 23, 2023, from https://www.who.int/news-room/fact-sheets/detail/disability-and-health
Yavuz N, Welch EW (2010) Addressing fear of crime in public space: gender differences in reaction to safety measures in train transit. Urban Stud 47(12):2491–2515
Zegras C, Butts K, Cadena A, Palencia D (2015) Spatiotemporal dynamics in public transport personal security perceptions: Digital evidence from Mexico City's periphery. ICT for Transport: Opportunities and Threats, pp 115–139

13 The Right to the City: The Experiences of LGBTQI+ and Gender Non-Conforming Individuals in Transit Environments

Abstract

Queer individuals, representing LGBTQI+ and gender-nonconforming individuals, face higher levels of discrimination, violence, and crime worldwide compared to the general population. This chapter explores the transit safety of LGBTQI+ individuals in a selection of municipalities in southern Sweden. We discuss victimisation and safety perceptions of these based on 3,407 survey responses, 163 (about 5%) of which identified as LGBTQI+, non-binary, or other. The findings underscore that LGBTQI+ individuals, and particularly gender-nonconforming women, face distinct safety concerns that often stem from societal biases, inadequate infrastructure, and hostile transit environments. Public spaces, including transit systems, are revealed as exclusionary for many within the LGBTQI+ community, perpetuating feelings of fear and insecurity. These experiences highlight broader issues of inequality in access to safe, inclusive public spaces and the need to address systemic barriers. By presenting these insights, the chapter identifies gaps in the current transit safety landscape and provides a platform for discussing actionable strategies to foster inclusivity and equity in public transit systems. The chapter concludes with a summary of key lessons from this study. This chapter offers evidence-based recommendations to improve transit safety for LGBTQI+ individuals, emphasising the importance of creating safer, more inclusive public spaces for all.

Keywords

Queer · Railway · Public transportation · Public space

13.1 Introduction

The right to access public spaces safely is fundamental to a just society. People's ability to use public spaces and public transportation depends on their intersecting identities (such as gender, race, income, or disability) (Crenshaw 1989/2013), which can lead to unequal access to transit services. For example, some groups may face more barriers—such as affordability issues, discrimination, or lack of convenient routes—than others. For many LGBTQI+ and gender-nonconforming individuals, using public spaces often means facing harassment or assault, highlighting the gaps in our commitment to safety and equality (Ceccato et al. 2024; Lubitow et al. 2017a, b; Weintrob et al. 2021). The United Nations has urged countries to reduce discrimination against LGBTQI+ individuals and ensure basic human rights, security, freedom, and dignity, regardless of sex, sexual orientation, gender identity, or perceived "otherness" (UN 2017; OHCHR 2024).

V. Ceccato et al., *Transit Safety for a Sustainable City*, Sustainable Development Goals Series,
https://doi.org/10.1007/978-3-031-97427-4_13

Stations as public environments are important nodes of public life. They should be accessible to all individuals in terms of transportation, yet they are contested spaces regarding inclusion and exclusion. As a risky facility (Clarke and Eck 2007), stations can create a setting conducive to criminal activity, thus heightening the vulnerability of marginalised groups, including LGBTQI+ individuals. LGBTQI+ individuals experience safety differently from those enjoying normative privileges (Doan 2007). Poor safety perceptions of this group can affect them disproportionately, leading to place and time avoidance, or limiting their movements, thus reinforcing exclusionary boundaries in public life (Angeles and Roberton 2020).

We have adopted the framework of "mobility justice", which aims at achieving fairer and more equitable transportation for everyone and is essential for building a more sustainable society (Sheller 2018). Yet, limited research has been conducted to examine the victimisation and safety of LGBTQI+ individuals (Weintrob et al. 2021), and among those that exist, there are calls for new perspectives. Another important concept in this context is that of "transmobilities" suggested by Lubitow et al. (2017a, b). These authors remind us about how transgender and gender-nonconforming individuals experience and navigate mobility in public transportation. They indicate that despite employing strategies to cope with harassment, discrimination, and violence, LGBTQI+ and gender-nonconforming individuals often face negative interactions with transit users, staff, and passers-by, making their experience on public transit distinct and challenging.

Following the same vein of thought (Angeles Roberton 2020), public safety, particularly crime prevention, must adopt a more inclusive, intersectional feminist perspective to better serve gender-diverse communities, as current approaches often fail them. Still, in many parts of the world, incident-reporting systems, for example, fall short because they do not address aggression, hate crimes, and discrimination against LGBTQI+ individuals. Despite underreporting, LGBTQI+ and gender-nonconforming individuals (Mitchell et al. 2024) tend to be overrepresented as victims in transit violence statistics worldwide (Angeles and Roberton 2020; Ceccato et al. 2024; Nourani et al. 2020a). This significantly impacts people's ability to use public spaces, go to school, and work, directly affecting their life opportunities.

In this chapter, we examine the experiences of LGBTQI+ and gender-nonconforming individuals in transit environments. We also highlight how the risks and fears associated with these public spaces contribute to their exclusion and limited mobility in the broader city context. The analysis is based on 3,407 survey responses, of which 163 (about 5%) identified as LGBTQI+, non-binary, or other. In this chapter, "woman" and "man" refer to individuals whose gender identity or expression aligns with societal norms based on their assigned sex at birth. These respondents selected "woman" or "man" in the survey without choosing options like LGBTQI+ , non-binary, or others. Individuals whose gender identity or expression does not align with these norms are categorised as "gender-nonconforming". This group includes those who may identify as transgender, gender-queer, non-binary, gender-fluid, or who challenge traditional gender roles. The term queer and LGBTQI+ encompasses a wide range of gender identities beyond the male–female binary.

Even though Sweden constitutes an interesting case study, as equal rights regardless of sexual orientation, gender identity, or expression are fundamental, LGBTQI+ individuals still face significant discrimination and victimisation (BRÅ 2020). By examining travellers' experiences in Sweden, this study contributes to gendered mobility research in the Nordic context. It also addresses Sustainable Development Goal (SDG) 11 on inclusive and sustainable cities and communities as well as Sustainable Development Goal 10 on reducing inequalities by advocating safer and more equitable public spaces for all individuals.

Earlier chapters of this book have shown that applying a systems thinking lens is key to creating transport environments that ensure both safety and inclusivity. In this framework, the travellers are the end-users or "beneficiaries" of the system, meaning they are those who ultimately gain the most from the system's existence and effective functioning. Transit safety is an essential component of well-functioning public transportation. This chapter first explores the interplay between safety, transit victimisation, and fear of crime for LGBTQI+ individuals, in urban environments. This is followed by an empirical analysis of victimisation patterns and fear of crime, highlighting significant disparities across gender and sexual orientation, with LGBTQI+ individuals reporting higher rates of harassment, stalking, and hate crimes. The study also examines precautionary measures adopted by LGBTQI+ individuals to navigate transit spaces and presents their recommendations for improving safety and accessibility, emphasising the need for structural and cultural changes. The discussion ties the findings to existing theories, including feminist and environmental criminology, and underscores the role of intersectionality in shaping the experiences of LGBTQI+ individuals. The chapter underscores the broader implications for public space equity and urban inclusivity by addressing safety and belonging in transit environments.

13.2 Public Space, Safety, and Belonging of LGBTQI+ Individuals

The concept of the "Right to the City", as articulated by Lefebvre, emphasises public space as a democratic right for all individuals to use and shape. Lefebvre argued that cities should be spaces of participation where every person has access to the urban environment's social, economic, and cultural life (Lefebvre 1968/1996). Similarly, Mitchell (2003) argues that the "right to the city" goes beyond just accessing physical space, it also includes the freedom to live in that space without facing discrimination or exclusion. According to the author, public spaces often provoke fear because they are seen as uncontrolled and fragile (Mitchell 2003).

Stations and public transit environments often occupy a liminal space between private and public domains, where social behaviours and norms are less defined, and crime may happen. The concept of "uncontrolled space" resonates here, as public transit is often perceived as a space where social norms are more fluid (Jacobs 1961; Newman 1972). In such environments, the risk and fear of discrimination, harassment, and violence can be imminent (Angeles and Roberton 2020), limiting people's ability to engage in the public life of the city.

Transgender and gender-nonconforming individuals often face harassment on public transit, limiting their access to safe mobility (Lubitow et al. 2017a, b). Studies show that harassment varies across LGBTQI+ subgroups, with female gender-nonconforming individuals at higher risk (Nourani et al. 2020a). Not being able to move because of risk or fear of crime is considered to be one of the dimensions of LGBTQI+ discrimination (Weintrob et al. 2021), as it limits accessibility for queer individuals using the transportation system, creating barriers such as fear, travel-related stress, place and time avoidance, and heightened vigilance of their surroundings.

Fear of crime in public spaces hinders LGBTQI+ individuals' participation in socioeconomic activities (Ceccato and Loukaitou-Sideris 2020). This fear is primarily influenced by environmental elements, particularly perceptions of the built environment, with a focus on transit spaces (Santoro et al. 2020). However, empirical research focusing on transit environments is rare in the international literature.

Thus, this chapter explores key questions regarding the experiences of LGBTQI+ individuals in transit environments. We investigate whether there are differences in victimisation between LGBTQI+ individuals and other railway travellers and whether LGBTQI+ individuals feel less safe compared to the travellers. Finally, we examine the precautionary measures most commonly adopted by LGBTQI+ individuals to navigate safety concerns during their journeys.

13.3 The Study Case

Sweden is one of the most gay-friendly countries in Europe. According to ILGA Europe (2022) which ranks 49 European countries annually on their LGBTQI+ equality laws and policies, considering 71 human rights, Sweden is placed sixth.

The dataset used for this study was collected via a 22-question survey on the web platform between May and November 2022, approved by the Swedish Ethical Review Authority. It targeted passengers on four train lines, with 3,407 valid responses. The sample was designed for statistical analysis by line, station size, gender, age, and time of day. Of the valid responses, 163 identified as LGBTQI+ , non-binary, or other, making up 5% of the sample. For more detailed information about the survey and the methodology of the study, see Chapter 4 of this book.

13.3.1 Victimisation

LGBTQI+ individuals were more likely to experience victimisation than other travellers inquired in the study. Of the 3,407 respondents, more than a fifth reported incidents compared to a tenth in the rest of the sample. This finding aligns with existing research on victimisation, making LGBTQI+ individuals more likely to experience safety concerns, particularly in public transit spaces. Table 13.1 shows experiences of victimisation by gender and crime type, highlighting the various forms of harm they face. LGBTQI+ individuals reported the highest rates of overall victimisation (22%), significantly higher than heterosexual women (15%) and men (10%). Concerning crime locations, LGBTQI+ individuals (15%) experienced more victimisation than heterosexual women (10%) and men (7%) on the way to the station than in other places. At the station, LGBTQI+ individuals (13%) again reported higher rates compared to heterosexual women (9%) and men (7%), while on the train, LGBTQI+ individuals (14%) faced significantly higher victimisation compared to heterosexual women (8%) and men (5%).

LGBTQI+ individuals reported significantly higher rates of sexual harassment (9%), stalking (12%), and threats or hate crimes (13%) compared to heterosexual women (7%, 7%, and 3%, respectively) and heterosexual men (1%, 2%, and 4%). They also faced higher rates of violence (5%) and robbery (2%) than both heterosexual women and men. There is no significant difference between the groups regarding thefts. As previously mentioned by Lubitow et al. (2017a, b), this pattern of victimisation stems from societal stigma and discrimination, which make LGBTQI+ and gender-nonconforming individuals more vulnerable to crimes such as hate crimes and sexual harassment. Their visibility as a diverse but minority group can increase exposure in public spaces, unlike theft, which appears more situational and identity-neutral.

13.3.2 Safety Perceptions

Of the 3,407 respondents, a third declared fear of crime while in transit. LGBTQI+ individuals were the most fearful, with about half expressing some concern, much less than for heterosexual women and men generally (Table 13.2). People who identify as LGBTQI+ expressed the greatest fear of stalking, unlawful threats or hate crimes, and sexual harassment. In contrast, heterosexual men ranked theft as their primary concern, with sexual harassment and stalking being less of a worry. For heterosexual women, theft was also a concern, followed by fears of sexual harassment, violence, and stalking. The varying concerns—sexual harassment and stalking for heterosexual women and LGBTQI+ individuals versus theft for heterosexual men—reveal how power dynamics shape the types of crimes different groups are most worried about. Traditional crime prevention efforts often focus on property crimes like theft, aligning more closely with male fear patterns. Feminist criminology argues for a shift in focus to address gendered crimes against women and LGBTQI+ individuals (Wonders and Shevat 2024).

General fear of crime is reported by 39% of heterosexual women, 24% of heterosexual men, and 47% of LGBTQI+ individuals. LGBTQI+ individuals consistently report higher levels of fear across all contexts compared to heterosexual men and women, indicating heightened vulnerability. The differences highlight disparities in

Table 13.1 Victimisation by gender, sexual orientation, and crime type 2023

	Hetero Women	Hetero Men	LGBTQI+	Chi-square	*p*
	N = 2,031	N = 1,142	N = 161		
Victim of crime	15%	10%	22%	23.99	0.000
Victim of crime (incl. aggressive panhandling)	23%	22%	32%	8.01	0.018
Victimised on the way to the station	10%	7%	15%	14.39	0.001
Victimised at the station	9%	7%	13%	6.56	0.038
Victimised on the train	8%	5%	14%	17.21	0.000
Victim of theft	4%	5%	7%	4.21	0.122
Victim of robbery	0%	1%	2%	12.14	0.002
Victim of violence	2%	3%	5%	9.35	0.009
Victim of threat or hate crime	3%	4%	13%	39.79	0.000
Victim of sexual harassment	7%	1%	9%	58.80	0.000
Victim of stalking	7%	2%	12%	44.44	0.000
Victim of aggressive panhandling	14%	17%	19%	9.56	0.008

Table 13.2 Fear by gender, sexual orientation, and crime type

	Hetero Women N = 2,029	Hetero Men N = 1,139	LGBTQI+ N = 159	Chi-square	*p*
Fear of crime	39%	24%	47%	84.05	0.000
Fear of crime on the way to the station	29%	14%	35%	95.44	0.000
Fear of crime at the station	28%	17%	36%	61.47	0.000
Fear of crime on the train	17%	10%	25%	38.04	0.000
Fear of theft	24%	17%	25%	22.44	0.000
Fear of robbery	21%	16%	23%	15.48	0.000
Fear of threats or hate crime	18%	13%	26%	26.20	0.000
Fear of violence	23%	17%	24%	17.77	0.000
Fear of sexual harassment	24%	5%	26%	191.17	0.000
Fear of stalking	23%	6%	29%	152.72	0.000

perceived safety relating to both gender and sexual orientation in which heterosexual women report more fear than heterosexual men but less than LGBTQI+ individuals, reflecting previous patterns of victimisation between the groups. These patterns reflect previous well-documented trends in criminological studies that highlight disparities in perceived safety and fear of crime based on gender and sexual orientation (Ceccato et al. 2024; Lubitow et al. 2017a, b; Nourani et al. 2020b; Santoro et al. 2020). These disparities underscore the structural inequalities that shape individuals' experiences of safety.

People feel significantly safer when travelling during the daytime, but LGBTQI+ individuals reported feeling more unsafe during the day compared to other respondents. In the evenings and at night, heterosexual women and LGBTQI+ individuals feel more unsafe in transit than heterosexual men. While there are no significant differences overall, LGBTQI+ individuals report higher levels of poor safety, especially on board the train. Finally, heterosexual women and LGBTQI+ individuals are more likely to feel unsafe around train stations compared to heterosexual men. The pedestrian underpass or overpass to the platform was the most fear-inducing location, followed by the parking area (Ceccato et al. 2024).

13.3.3 Determinants of Safety Perceptions of LGBTQI+ Individuals

LGBTQI+ and gender-nonconforming individuals are 3.4 times as likely to feel unsafe at the station and 3.3 times as likely to feel unsafe on their way to or from the train as others ($p < 0.001$). Similarly, those with prior victimisation in transit are 3.4 times as likely to feel unsafe when commuting ($p < 0.001$). The results in this section build upon prior research (Ceccato et al. 2024), which analysed the relationship between individual factors, environmental factors, previous victimisation, and safety perceptions in transit environments using the 28 municipalities. This study highlights significant safety concerns for LGBTQI+ and gender-nonconforming individuals, especially those with prior victimisation experiences, who face unique challenges and an elevated likelihood of feeling unsafe in transit environments. Our findings show signs of interactions between individuals' predisposition to feel fear and the environment they spend time in. Factors such as a lack of staff, the presence of tunnels, and poor illumination had a negative effect on travellers' safety (see Ceccato et al. 2024). These results call for the need for a comprehensive approach to addressing the safety concerns of LGBTQI+ individuals in public spaces and transit systems, which are discussed in more detail in the following sections.

13.3.4 LGBTQI+ Individuals' Precautions

Figure 13.1 indicates a significant disparity in safety concerns across groups, with LGBTQI+ individuals adopting precautions almost as frequently as heterosexual women (91%) but far more than heterosexual men (73%). Precautions are more common at night (83%) than during the day (48%). They report higher rates of carrying weapons (17%) and avoiding particular types of clothes (22%). More than half declare taking any safety precautions during the morning or daytime (66%) compared with heterosexual women (51%) and heterosexual men (39%).

In contrast, heterosexual women are more likely to stay visible (54%) or rely on phone contact (46%). Over half of LGBTQI+ individuals and heterosexual women try to appear confident, and nearly a quarter of heterosexual women and LGBTQI+ individuals take five or more precautions, compared to less than 10% of heterosexual men. The only precaution with minimal gender difference is avoiding jewellery. The need to take precautions reflects how public spaces are not equally accessible or feels as safe, reinforcing power dynamics that limit the autonomy of heterosexual women and LGBTQI+ individuals. These behaviours illustrate how intersecting identities, such as gender and sexual orientation, amplify vulnerability. Both feminist theory and transmobilities call for structural and cultural transformations to address the unequal experiences of safety and mobility, ensuring equitable access to public transit for all (see Ceccato et al. 2024).

13.3.5 LGBTQI+ Individuals' Recommendations

LGBTQI+ respondents prioritise safety and accessibility measures more than heterosexual men, closely aligning with heterosexual women in some areas (Fig. 13.2). They suggest more staff at stations (51%), better maintenance (44%), illumination (43%), digital timetables (41%), and better signage (38%). They also prioritise higher frequency (36%) and safety features such as reporting apps and police presence, reflecting heightened safety concerns.

Figure 13.2 also shows that heterosexual men consistently recommend measures such as better signage (21%), better ticket information (20%), higher frequency of trains (19%), and train hosts on board (18%)—significantly less than LGBTQI+ and heterosexual women. It is worth noting that heterosexual men's lesser emphasis on features such as safety apps (22%),

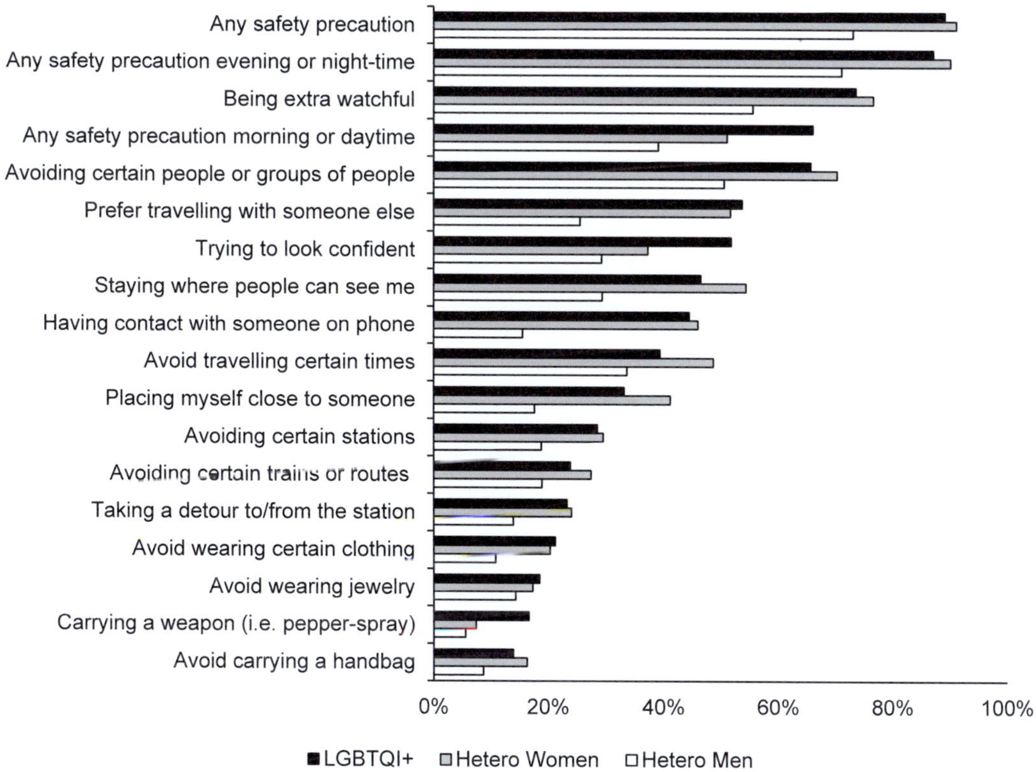

Fig. 13.1 Precautionary measures of LGBTQI+ individuals and gender-nonconforming individuals. $N_{LGBTQI+} = 151$, $N_{Heterosexual\ Women} = 1{,}945$ and $N_{Heterosexual\ Men} = 1{,}029$

police patrols (22%), and station staff (45%) suggests they might feel less vulnerable in transit settings. More than a fourth of LGBTQI+ and gender-nonconforming individuals would wish for higher frequency trains (36%), for a help button to get help on the train (32%) or station (30%), and for more police/guards patrolling the stations (27%) than heterosexual women and men. These findings reveal that LGBTQI+ and gender-nonconforming individuals prioritise measures such as police patrols, help buttons, and increased train frequency due to their unique vulnerabilities in transit settings. This heightened demand for systemic changes in transit environments underscores a lack of structural consideration for LGBTQI+ and gender-nonconforming individuals experiences, an indicative of how power imbalances are embedded in urban and transit planning.

13.4 Discussion of the Results

The data presented in this chapter reveals striking disparities in victimisation across different groups, with LGBTQI+ individuals experiencing significantly higher levels of victimisation compared to their heterosexual counterparts. This finding aligns with previous research on the vulnerability of marginalised groups, particularly those that challenge normative gender roles (Lubitow et al. 2017a, b; Nourani et al. 2020b).

We show how elevated victimisation rates, especially in the form of sexual harassment, stalking, and hate crimes, affect LGBTQI+ individuals and exacerbate their vulnerability in public spaces. LGBTQI+ individuals face many vulnerabilities due to intersecting factors. These factors increase their exposure to targeted harassment and violence, both of which amplify their

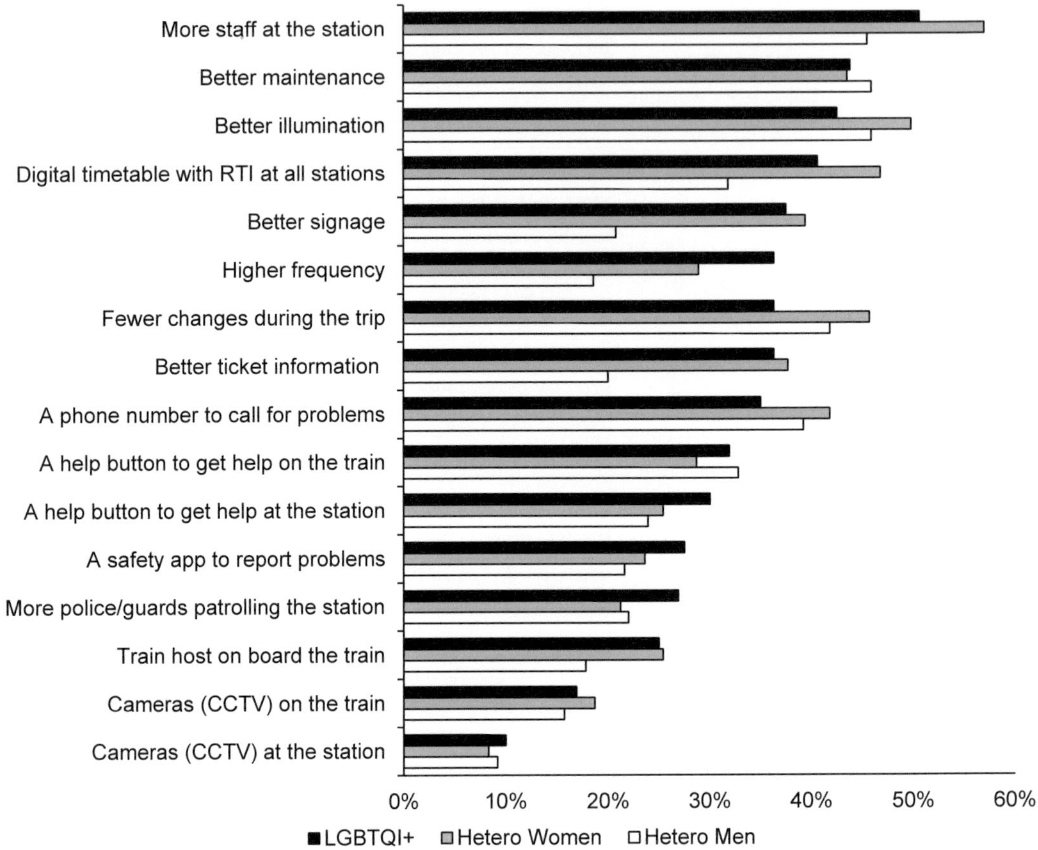

Fig. 13.2 LGBTQI+ individuals' recommendations to improve their safety. $N_{LGBTQI+} = 160$, $N_{Heterosexual\ Women} = 2{,}015$, $N_{Heterosexual\ Men} = 1{,}121$

fear of crime. The survey data suggests that while fear is a concern for all groups, LGBTQI+ individuals are more likely to feel unsafe during the day, indicating that their safety concerns are not confined to nighttime environments but pervade their everyday mobility experiences.

Our study shows that nearly a quarter of LGBTQI+ individuals and heterosexual women take five or more precautions, which reflects the extent to which these groups feel the need to safeguard themselves while navigating public transit. We also show the precautionary measures employed by LGBTQI+ individuals to cope with safety concerns. These measures, including travelling with others, staying in visible areas, and avoiding certain stations or times, reveal the significant impact of fear on their daily lives. The higher prevalence of precautionary measures among LGBTQI+ individuals compared to heterosexual men points to their deep-rooted concerns regarding safety. These precautionary measures also emphasise the safety work involved in mobility for LGBTQI+ individuals, who must constantly monitor their surroundings and make decisions based on their perception of risk (Fig. 13.3). Thus, the experiences of LGBTQI+ and gender nonconforming individuals in transit should be recognised as a "mobility justice" issue (Sheller 2018).

Lefebvre and Mitchell discuss the concept of the "Right to the City", which emphasises the importance of public spaces being accessible to all individuals free from fear and discrimination. In the context of public transit, this means addressing the structural inequalities that disproportionately affect LGBTQI+ individuals, particularly gender-nonconforming individuals. In order to do this, we need to make sure that policing and other actions

Fig. 13.3 The experiences of LGBTQI+ and gender nonconforming individuals in transit should be recognised as a mobility justice issue, essential for safe and sustainable transportation. (*Source* Pixabay (2025))

integrate intersectional feminist frameworks to address the needs of gender-diverse communities effectively. According to Angeles and Roberton (2020), this means having proper records about these incidents, including those that involve verbal harassment and other acts of microaggression. Fear of police power abuse and negligence worldwide has meant that incidents have been under-reported (e.g. Lubitow et al. 2020). Currently, institutional conceptions of hate crimes provide an incomplete picture of who the victims are, neglecting various forms of racial microaggressions in public spaces (Sue et al. 2007). These new frameworks demand more than "just" knowledge about discrimination and violence; they involve personal, professional, and social empathy, which is important when it comes to developing effective relationships and reciprocal exchanges that resonate with mutual responsibilities in public safety among minority groups (Angeles and Roberton 2020). This demands long-term investment in education, training, and professional development for urban planners (Ceccato and Brantingham 2024), police officers and other experts in charge of public safety.

More concretely, creating public spaces that are welcoming and accessible to LGBTQI+ individuals, such as gender-neutral restrooms, safer street lighting, and well-maintained public areas (such as parks), can reduce victimisation and poor safety perceptions. In the city, this may also involve the installation of cameras in high-risk areas, increasing people's presence in areas known to be a concentration of violence against LGBTQI+ individuals. Finally, channels that facilitate co-learning and collaborative action between actors in urban safety planning need to be created. In practice, this means that mutual understanding and collective efforts to address shared challenges (Angeles and Roberton 2020) in terms of discrimination, victimisation, and fear against LGBTQI+ and gender-nonconforming people is a fundamental step.

13.5 Conclusions and Recommendations

This chapter intended to provide insights into the complex intersection of mobility, victimisation, and safety for LGBTQI+ individuals in transit environments. Our results highlight the chronic challenges faced by LGBTQI+ individuals, especially gender-nonconforming individuals, in accessing public spaces safely. The elevated victimisation and fear of crime emphasise the need for inclusive and safer transit environments, ensuring mobility rights for all, regardless of gender or sexual orientation.

Tackling these issues fosters a more just and equitable society where LGBTQI+ individuals can engage in public life without fear of crime or discrimination. For these issues, it is necessary to develop policies that address hate crimes, harassment, and threats targeting LGBTQI+ individuals. This could include improved reporting mechanisms, staff presence, and educational programmes to promote tolerance and inclusion. In Sweden, previous research has shown that using police statistics to understand what happens in the stations can lead to misconceptions, given the fact that only significant incidents of crime are reported to the police (Ceccato 2013). Finally, we need better support if anything happens. While solutions should be tailored to each context, it is important that LGBTQI+ and gender-nonconforming individuals can access help or the reassurance that support is readily available if needed. Finally, as suggested by Sustainability Development Goal 5, addressing victimisation and fear of crime among LGBTQI+ individuals is necessary to promote the elimination of discrimination and violence against vulnerable populations. Our findings also reinforce feminist criminology's call to highlight marginalised voices, address structural inequalities, and challenge traditional crime prevention frameworks to create safer, more inclusive transit environments, as highlighted by Sustainability Development Goal 11.

References

Angeles LC, Roberton J (2020) Empathy and inclusive public safety in the city: examining LGBTQ2+ voices and experiences of intersectional discrimination. Women's Stud Int Forum 78:102313. https://doi.org/10.1016/j.wsif.2019.102313

BRÅ, Swedish National Council for Crime Prevention (2020) Hatbrottsstatistik. https://bra.se/download/18.79079f9d17cc01fce501ad6/1638968596572/2021_17_Polisanmalda_hatbrott_2020.pdf

Ceccato V (2013) Moving safely: crime and perceived safety in Stockholm's subway stations. Lexington books

Ceccato V, Brantingham P (2024) What is the role of architects and urban planners in crime prevention? Secur J. https://doi.org/10.1057/s41284-024-00442-4

Ceccato V, Gliori G, Sundling C (2024) LGBTQI+ safety perceptions in transit environments. J Urban Mobil 6:100087. https://doi.org/10.1016/j.urbmob.2024.100087

Ceccato V, Loukaitou-Sideris A (2020) Transit crime and sexual violence in cities: International evidence and prevention. Routledge

Clarke R, Eck J (2007) Understanding risky facility (problem-oriented guides for police problem-solving tools series, issue

Crenshaw KW (1989/2013) Mapping the margins: intersectionality, identity politics, and violence against women of color. In: The public nature of private violence. Routledge, pp 93–118

Doan PL (2007) Queers in the American City: transgendered perceptions of urban space. Gend Place Cult 14(1):57–74. https://doi.org/10.1080/09663690601122309

Ilga Europe (2022) Annual Review of the Human Rights Situation of Lesbian, Gay, Bisexual, Trans and Intersex People in Europe and Central Asia. https://www.ilga-europe.org/

Jacobs J (1961) The death and life of great American cities. Vintage books

Lefebvre HE (1968/1996) Right to the City. Blackwell Publishing

Lubitow A, Carathers J, Kelly M, Abelson M (2017a) Transmobilities: mobility, harassment, and violence experienced by transgender and gender nonconforming public transit riders in Portland, Oregon. Gend Place Cult 24(10):1398–1418. https://doi.org/10.1080/0966369X.2017.1382451

Lubitow A, Abelson MJ, Carpenter E (2020) Transforming mobility justice: gendered harassment and violence on transit. J Transp Geogr 82:102601

Lubitow A, Rainer J, Bassett S (2017) Exclusion and vulnerability on public transit: experiences of transit dependent riders in Portland, Oregon [Article]. Mobilities, 12(6):924–937. https://www.scopus.com/inward/record.uri?eid=2-s2.0-85009509160&doi=10.1080%2f17450101.2016.1253816&partnerID=40&md5=d86ab19109db4462975ee0a8e3275f1d

Mitchell D (2003) The right to the city: social justice and the fight for public space. Guilford Publications. https://books.google.se/books?id=2NKoBQAAQBAJ

Mitchell M, O'Neill T, Redd C (2024) Queering crime reporting: representing anti-queer violence in LGBTQ news media. Br J Criminol azae078

Newman O (1972) Defensible space. Macmillan New York

Nourani F, Antonello SL, Govone JS, Ceccato V (2020a) 9 Women and LGBTI youth as targets. Crime and Fear in Public Places: Towards Safe, Inclusive and Sustainable Cities

Nourani F, Antonello SL, Govone JS, Ceccato V (2020b) Women and LGBTI youth as targets: assessing transit safety in Rio Claro, Brazil. In: Crime and Fear in Public Places. Routledge, pp 176–193

OHCHR, United Nations Human Rights Office of the High Commissioner (2024) UN expert calls on States to ensure equal electoral participation without discrimination based on sexual orientation or gender identity. https://www.ohchr.org/en/press-releases/2024/11/un-expert-calls-states-ensure-equal-electoral-participa tion-without

Santoro P, Harkot M, Loukaitou-Sideris A, Ceccato V, Sundling C, Romero-Torres J, Agrawal AW (2020) Intersectionality of transit safety. Transit Crime and Sexual Violence in Cities: International Evidence and Prevention

Sheller M (2018) Theorising mobility justice. Tempo Social 30(2):17–34

Sue DW, Capodilupo CM, Torino GC, Bucceri JM, Holder AMB, Nadal KL, Esquilin M (2007) Racial microaggressions in everyday life: implications for clinical practice. Am Psychol 62(4):271–286. https://doi.org/10.1037/0003-066X.62.4.271

UN, United Nations (2017) Leaving noone behind: equality and non-discrimination at the heart of sustainable development. New York

Weintrob A, Hansell L, Zebracki M, Barnard Y, Lucas K (2021) Queer mobilities: critical LGBTQ perspectives of public transport spaces. Mobilities 16(5):775–791

Wonders N, Shevat S (2024) Gendered Violence: A Human Rights Agenda for Criminology. In A Research Agenda for a Human Rights Centred Criminology (pp. 189–206). Cham: Springer Nature Switzerland.

14 Who Are the Non-Train Travellers? An Explorative Analysis in Sweden

Abstract

Public transportation is essential for building a sustainable society, with high train and bus ridership often reflecting good accessibility, reliability, and transit safety. However, some people still use other means of transportation–who are they, and what motivates their choices? This chapter investigates the profile of non-train travellers compared with those of train users to better understand why they are not using trains. This is a small sample study with 91 answers from a survey applied to 28 municipalities in Sweden along three railway lines collected in 2022. Our findings show that non-users tend to be older than the rest of travellers, declare having one or more disabilities, are less educated, have less trust in the police, and feel more unsafe while travelling with others. Non-train travellers differentiate themselves from travellers by indicating the need for more ticket information (how and where to buy) at the station and a wish for fewer transfers during the journey. Having more staff and other personnel in the station is highly recommended by both users and non-users to improve safety. Still, more staff at the station is particularly recommended by train users. The chapter finishes by making suggestions for improving rail-bound transportation to attract non-train travellers in Sweden and expanding some of these reflections to other contexts.

Keywords

Rail-bound transportation · Safety · Crime · Frequency of use · Station characteristics

14.1 Introduction

Making public transportation accessible and safe means that we are also promoting mobility as a basic right for all city users. This chapter investigates the characteristics of non-train travellers compared with train users to understand why they are not using trains. The study assesses answers from a survey applied to 28 municipalities in Sweden along three railway lines collected in 2022. We conclude the article by making suggestions for improving rail-bound transport to attract non-users. We are particularly interested in the environmental conditions of the train stations in demotivating this group of people from using public transportation. At the same time, we unravel differences in their personal characteristics that may also explain why they prefer not using trains or not travel at all. Earlier chapters of this book have shown that applying a systems thinking lens is key to creating transport environments that ensure both safety and inclusivity, which can also be key to attracting new users.

V. Ceccato et al., *Transit Safety for a Sustainable City*, Sustainable Development Goals Series,
https://doi.org/10.1007/978-3-031-97427-4_14

The international literature highlights that people prefer cars and refrain from public transportation due to convenience and comfort. Others indicate issues such as overcrowding, the price of tickets, lack of reliability (often unpredictability of schedules), long waiting times, limited routes, and frequency of trains/buses. Moreover, a lack of privacy and personal space and safety-related issues may also discourage the use of public transportation (Taylor and Fink 2003; Wang and Liu 2015; Sundling et al. 2014; Chowdhury and van Wee 2020). Railways are limited and less flexible than buses, which means that in rural areas or in remote places with infrequent service, public transport may simply not meet the needs of daily commuters (Brons et al. 2009; Brovarone and Cotella 2020). We adopt a systems thinking perspective to understand these interconnections and identify 'leverage points', such as safety improvements, accessibility measures, and inclusive design, that can encourage greater use of public transport. Understanding who these travellers are and why they make these choices may indicate gaps in accessibility, safety and inclusivity in public transportation. For transit operators, this knowledge is essential because it connects to the demands of Sustainable Development Goals (SDGs). In particular, with SDG 10 on reducing inequalities and SDG 11 on building inclusive, safe, and sustainable cities, which underscores the need for transport operators to integrate these insights into planning and daily operations to ensure equitable and secure mobility for all.

The chapter has five parts. First, we present international research on ridership and transit safety, highlighting the potential reasons why people refrain from using public transportation. Then, we present our conceptual model of analysis and our research questions. In the third part, we introduce the study area and methods. Results are presented in section four, and a discussion of the results in section five is followed by conclusions and recommendations in section six.

14.2 Theoretical Background: Ridership in Trains and Transit Safety

14.2.1 What Affects Ridership?

Numerous factors can affect transit ridership or the number of people who use public transportation. The so-called "service quantity variables" are a significant group of factors that influence transit use. These include service coverage and service frequency, which describe the extent and frequency of public transportation services in a specific area. Studies have demonstrated a strong correlation between transit use and service quantity variables, but also the fare and pricing structure (e.g. Taylor and Fink 2003; Wang and Liu 2015). According to studies, higher fares tend to reduce ridership, while lower fares tend to increase it. For example, in a sample of older adults in Sweden, 62% said they would travel more often if tickets were less expensive (Sundling et al. 2014). Other price-related factors, such as the availability of discounts for regular riders or particular demographic categories such as senior citizens or students may also impact ridership.

Service quality is also essential for many people when using public transport. Frequent, reliable, and clean services (Sundling et al. 2014) and friendly and helpful staff can make a big difference in people's overall experience and satisfaction with public transport. In addition, safety and security are major concerns for many public transport users, and measures such as CCTV cameras (Cozens 2002; Yavuz and Welch 2010), onboard security staff (Ceccato et al. 2024), and well-lit stations (Chowdhury and van Wee 2020) can help to ensure that passengers feel safe and secure. It was discovered in a study of passengers' perceptions of their bus journey experience in Edinburgh, Scotland that the main factors discouraging people from using buses are feeling unsafe, preferring to walk or cycle, issues with service provision, preference for a car, cost, disability, and discomfort, and self-image (Stradling et al. 2007). It has been discovered that aspects of service quality, such as customer service, on-street service, and station safety, are more crucial for drawing passengers than fare reductions.

Trust in society, particularly in local and public authorities, significantly influences train ridership. Confidence in local authorities ensures that users believe in the reliability, safety, and maintenance of transportation systems, while trust in police and public officials fosters a sense of security (Lim and Moon 2022). For example, a study in Mexico showed low feelings of safety in one's neighbourhood and low trust in the police to be associated with feeling unsafe when travelling (Vilalta 2011). Visible, fair, and responsive policing deters crime and reassures passengers, whereas perceptions of inefficiency or discrimination can reduce ridership, particularly among vulnerable groups. Additionally, societal trust, including mutual respect among passengers, promotes a positive travel environment. By addressing concerns transparently and equitably, authorities can strengthen trust, making public transport a more attractive, inclusive, and reliable option for all (Owen et al. 2023).

Travellers' personal characteristics or "individual factors" also affect ridership. Age and frequency of bus use appeared to have more influence on the motivational factors than annual household income, car availability, and gender (Stradling et al. 2007). Having a disability can make travelling more difficult. Among older persons, travel frequency has been shown to be a function of a reduction in ability, so having a more severe disability is correlated to lower frequency of use (Sundling et al. 2014).

14.2.2 Does Safety Affect Ridership and How?

Feeling safe while travelling on public transport should be considered a fundamental right, especially as many travellers do not have alternative ways of travelling. However, many travellers feel unsafe on public transport, at least in certain situations (Ceccato et al. 2024, see also findings from previous chapters of this book). There can be many reasons for feeling unsafe in travel situations. For example, there was a contagion risk during the COVID-19 pandemic (Dong et al. 2021). Also, for older travellers or those with disabilities, unpredictability can induce feelings of poor safety, for example, not knowing beforehand if the lift will work or being afraid of falling because of unpredictable driving behaviour (Sundling et al. 2014). However, much research on perceived safety in transport is focused on the fear of crime. Fear can sometimes be a more significant problem than crime itself as it is more common than actual victimisation but still restricts a person's mobility, specially for women (Hale 1996).

Travellers often take precautions to handle a perceived lack of safety when travelling. Those who feel unsafe have been found to use public transport less than others (Jaitman 2020). Many travellers avoid certain times of the day, especially evenings and nights, specific routes or places (Ceccato and Loukaitou-Sideris 2022), or specific travel modes (Jaitman 2020).

A reason for avoiding rail-bound transport is the risk of encountering intoxicated people or sexual harassment, especially for women (Sundling 2020). Therefore, many women also adapt their way of dressing (Ceccato and Loukaitou-Sideris 2022). A large body of research from different continents such as Europe, North and South America, Asia, and Oceania shows different ways of adapting to such situations when travelling. Other ways of adapting one's behaviour because of a perceived lack of safety is avoiding travelling alone (Shirgaokar 2019), being careful to stay alert and to remain aware of one's surroundings (Ceccato and Loukaitou-Siderins 2022), and trying to look confident (Chowdhury and van Wee 2020). Moreover, many travellers use their mobile phones or headphones to be in touch with family and friends (Kim 2021) or to discourage others from initiating contact (Thompson et al. 2012). Others are careful not to have or wear anything valuable (Jaitman 2020). Some also carry a weapon, such as pepper spray (Gopal and Shin 2019). When people avoid public transportation, it imposes costs on society as a whole, from increased traffic congestion and environmental impact to reduced social inclusion. Understanding the factors that discourage use is therefore vital for identifying leverage points or strategies to promote greater public transport use.

Based on the current state of the research in this area, we ask the following research questions:

(1) Who are the non-train travellers?
(2) Do safety perceptions affect an individual's willingness to use public transportation, controlling for individual and other factors?
(3) What are the recommendations to make public transportation more attractive for users and non-users?

14.3 The Study Case

Our study area includes 47 rail-bound stations along three railway lines covering central and southern Sweden and serving 28 municipalities with a combined population of 1.78 million (Statistics Sweden (SCB) 2020). Together, the three railway lines accommodate approximately 64,000 passengers daily (estimate from the Swedish Transport Administration's annual average number of daily passengers per station 2009). Most stations are located in small towns with a population of less than 15,000. However, many are close to big cities, so residents may often find themselves working in nearby municipalities. The three railway routes also pass through several medium-sized urban areas, including Södertälje, Eskilstuna, Örebro, Norrköping, and Linköping. Gothenburg is the only city along the routes classified as a larger city with a population of roughly 600,000. For more detailed information about the study area, see Chapter 4 of this book.

14.3.1 Data

The data basis for this study comes from a survey carried out between May and September 2022 in Southern Sweden. Before this, the survey was approved by the Swedish Ethical Review Authority. We combined answers from two quasi-identical surveys, one printed and one online. The printed version was handed out and then collected onboard the trains and at the stations, while the online version was distributed on municipality websites and Facebook groups and promoted through posters and cards with QR codes at the stations. Researchers participated in radio programmes that promoted the research project and encouraged people to answer the survey. The total sample was composed of 4,690 responses combined ($N_{print} = 2,167$ and $N_{online} = 2,523$). Among these, 91 responses (2%) reported that they never travel by train, classifying them as "non-users". All these responses were gathered through the online survey, as the printed survey was distributed onboard the trains.

The survey consisted of 28 questions designed to capture the importance of the physical and social environment of the stations in the Swedish context. However, in the online version of the survey, respondents who indicated that they had never travelled by train were directed to a different question about the factors that could influence them to start travelling by train. For users, the question was about what factors would make them feel safer during the train journey. A total of 91 non-users responded to the survey, but the number decreased as they progressed through the survey questions, ending with 49 complete responses.

The following questions were used in the analysis:

- How often do you travel by train? *(followed by five alternatives)*
- Indicate which of the following factors would make you travel by train more often. *(followed by 16 alternatives)*

14.3.2 Methods

The survey data was transferred from the web-based platform (Crowdsignal) via a spreadsheet (Microsoft Excel) to a conventional statistical software package, in which the analyses were carried out (in this case, IBM SPSS Statistics, 28.0.1.1). Exploratory data analysis and hypothesis testing, including frequencies, cross-tabulations, and chi-square tests, were performed to identify variables that best describe the characteristics of the non-users and to develop a parsimonious model.

A logistic regression model was used to identify the profile and characteristics of non-users based on the selected variables. Variables with high collinearity (bivariate correlation, $r \geq 0.6$) were excluded to reduce redundancy. For example, education and income level were removed due to their strong correlation. For more detailed information about the methods employed in the study, see Chapter 4 of this book.

14.4 Results

14.4.1 Travellers and Non-travellers: Main Characteristics

Non-train users tend to be older, declare that they have one or more disabilities, are less educated, feel more unsafe where they live, feel insecure while travelling with others, and tend to have less trust in public authorities, such as the police. We also noticed that gender and ethnicity are not significant factors in determining train use for this sample. Below, we discuss these characteristics in more detail (Table 14.2, Appendix A).

Much older adults (60 years or above) tend not to use trains, as almost a quarter declared this, compared to 15% of those younger (Chi-square = 2.942, df = 1, $p = 0.086$). Older age may also be linked to physical and other types of disabilities, so it is not a surprise that non-train users tend to declare that they have one or more disabilities. A third of the respondents with some disability reported not using the train, whereas the share of individuals with a disability who use the train was just less than 10% (Chi-square = 28.362, df = 1, $p < 0.001$). The main difference can be seen for people who suffer from a major disability (i.e. reduced mobility, hearing or vision impairments), as 1% of these use the train while 10% responded that they never travel by train. For those with a minor disability, there is still a significant difference, considering that 8% of the train users declared that they have this condition, and as many as 21% of the non-users (Chi-square = 42.888, df = 2, $p < 0.001$). The most notable difference was observed among people with reduced mobility. Only 5% of train riders reported having reduced mobility, compared to over 15% of non-users (Chi-square = 11.217, df = 1, $p < 0.001$).

Disabilities such as impaired vision or hearing showed no significant differences between users and non-users of trains. Other conditions, for example, relating to age, such as problems with memory, attention, or concentration, were also shown to significantly affect whether people travel by train. As many as 17% of the non-users reported having these conditions, with 9% of users (Chi-square = 4.951, df = 1, $p = 0.026$). Additionally, people with asthma, allergies, or other hypersensitivities indicated to a higher degree that they do not travel by train, as a third of the non-users reported this, but only 17% of the users (Chi-square = 8.196, df = 1, $p = 0.004$).

Education is a key variable in the variation of train usage. As 59% of respondents reported having a university education, compared to 37% of non-train users, the results show that more people with higher education travel by train. In contrast, a higher percentage of those who claimed to have only completed elementary school–27% as opposed to 9%—were not users (Chi-square = 23.526, df = 2, $p < 0.001$). This is correlated with the fact that income follows a somewhat similar trend, as roughly 38% of those with low incomes do not utilise trains compared to 21% of those who do (Chi-square = 6.312, df = 1, $p = 0.012$). Despite this, no significant changes could be identified when comparing low, middle, and high-income groups. Fewer single parents reported not using the train, as there were zero responses from this group compared to the 6% of lone parents who used the train (Chi-square = 3.352, df = 1, $p = 0.067$).

The experience of previous victimisation itself seems not to affect train ridership. Train users are more frequently victims of crime (outside transit areas) than non-train travellers. Several other variables appear to influence train usage. One is about trust in public authorities. As many as 77% of train travellers declare having trust in the police authorities as opposed to 50% of non-train users (Chi-square = 24.148, df = 1, $p < 0.001$). Moreover, feeling safe while travelling with others is undoubtedly essential for non-train users.

As many as 57% of train passengers reported feeling safe travelling with others, compared to only 19% of those who do not take the train (Chi-square = 34.140, df = 1, $p < 0.001$). Note that these results do not control the direction of the causality here. Moreover, train users reported feeling safe in their neighbourhood more frequently than non-users; 86% of passengers reported this compared to 60% of non-users (Chi-square = 29.354, df = 1, $p < 0.001$). These results may also be linked to the age group, which tends to be older, less mobile and often more fearful. To test the effect of these characteristics simultaneously, we report results from the multivariate analysis below.

14.4.2 Modelling the Profile of Users and Non-Users of Train Travel

This logistic regression model (Table 14.1) aims to detect the main characteristics of non-users of train travel and the main reasons for not travelling by train.

Older age is a strong determinant, with those aged 45 or older being 2.5 times as likely not to travel by train as younger people ($p = 0.006$). As expected, this may be related to the fact that older adults are more likely to experience physical or mobility limitations, which was also identified as a critical factor. The odds of individuals with a disability refraining from using trains is three times greater compared to non-disabled individuals ($p = 0.001$). A poor educational level is another important variable. Those who have completed only an elementary school education are 4.5 times as likely to be non-users ($p < 0.001$).

Moreover, those who have expressed discomfort while travelling with others are almost 3.5 times as likely not to travel by train, pointing towards the influence of social comfort in travel behaviour ($p = 0.002$). Non-train travellers are 2.5 times as likely to distrust the police ($p = 0.008$). In contrast, individuals who have experienced victimisation outside transit are less than half as likely to be non-users ($p = 0.045$).

On the other hand, the advocates of better information about tickets (where and how to buy) were 3.4 times as likely to be non-users, reflecting possible frustrations with the current ticket system ($p = 0.002$). Correspondingly, respondents who preferred fewer transfers were almost twice as likely not to travel by train, which shows that route complexity and convenience have a bearing on travel choice, although only significant at the 10% level ($p = 0.077$).

Looking at the recommendations provided by respondents, people who suggested more staff presence were far more like to travel by train, thus making this the strongest predictor for non-users in the model ($p < 0.001$). There were no significant influences of gender and perceptions of neighbourhood safety on train use, indicating that these factors may be less important in distinguishing between users and non-users.

14.4.3 What Should We Do to Attract More Train Travellers?

Respondents highlighted the need to improve social control at the station and during the trip. This is clear when train and non-train travellers underlined the need for more staff, train conductors on board, police and security guards, and CCTV at and on the train (Fig. 14.1). Significant differences were observed in responses related to safety concerns, with train travellers prioritising factors such as increased staff presence at stations, improved lighting, and the installation of CCTV

Table 14.1 Logistic regression results of Y = Non-users of train travel

Y = Non-user = 1, otherwise, 0 = User	Non-user of train (N = 46)			
	OR	95% C.I for OR		Sig.
Man	0.772	0.392	1.523	0.456
45 years or above	**2.461**	**1.295**	**4.677**	**0.006**
Disability	**3.049**	**1.539**	**6.040**	**0.001**
Elementary school (no higher education)	**4.514**	**2.139**	**9.524**	**0.000**
Previous victimisation (outside transit)	**0.434**	**0.192**	**0.981**	**0.045**
Unsafe in neighbourhood	1.416	0.649	3.089	0.382
Distrust in police	**2.525**	**1.273**	**5.011**	**0.008**
Uncomfortable travelling with others	**3.448**	**1.596**	**7.451**	**0.002**
Recommend—More staff	**0.215**	**0.106**	**0.435**	**0.000**
Recommend—Better ticket info	**3.434**	**1.570**	**7.508**	**0.002**
Recommend—Fewer transfers	**1.907**	**0.932**	**3.902**	**0.077***

Cox & Snell R Square = 0.040.
Nagelkerke R Square = 0.233.
OR Odds ratio; *CI* Confidence interval; *Sig.* *p*-value, statistically significant at the 5% level at most.
*Statistically significant at 10% level.

both at the station and on the train. Non-users, on the other hand, placed less importance on these aspects. Non-users placed greater emphasis on fewer transfers and improved train information (such as signage and real-time updates) compared to users. However, the differences between the two groups for these factors were not statistically significant.

Train travellers recommend similar improvements but are significantly more worried by the station's lack of staff than non-users. Non-users would like fewer train transfers and improved train information, including RTI at all stations. As expected, certain aspects of the stations tend to be more relevant to users, such as the quality of lighting, better maintenance, how to get information about tickets, more CCTV (at the station and on the train), and having someone to ask for help (helpline, staff, or police/security guards). Both users and non-users recommended more frequent trains (Fig. 14.1).

Several key themes emerged from the free-text responses provided by non-users. Figure 14.2 shows examples of recommendations from non-users.

14.4.3.1 Affordability and Cost

Respondents raised concerns about the high cost of travel. One participant stated, *"It is too expensive to travel within the municipality"*, and another suggested "Cheaper tickets" to make travel more affordable. From a systems thinking perspective, affordability is a critical leverage point: it shapes travel behaviour and access to services, with effects on congestion and environmental outcomes, linking directly to SDG 10 on reducing inequalities and SDG 11 on building inclusive, safe, and sustainable cities.

14.4.3.2 Infrastructure Accessibility

Access to train services was a common issue. One respondent mentioned, "I don't have access to a train near where I live. If there were train tracks and an opportunity to travel by train, I would certainly do so". Others requested more stops, with one saying, "The train should stop at more places". Strategically, this is also an important leverage point, because by. improving connectivity and stop availability can influence travel patterns, increase public transport use, and enhance social inclusion. Greater access also

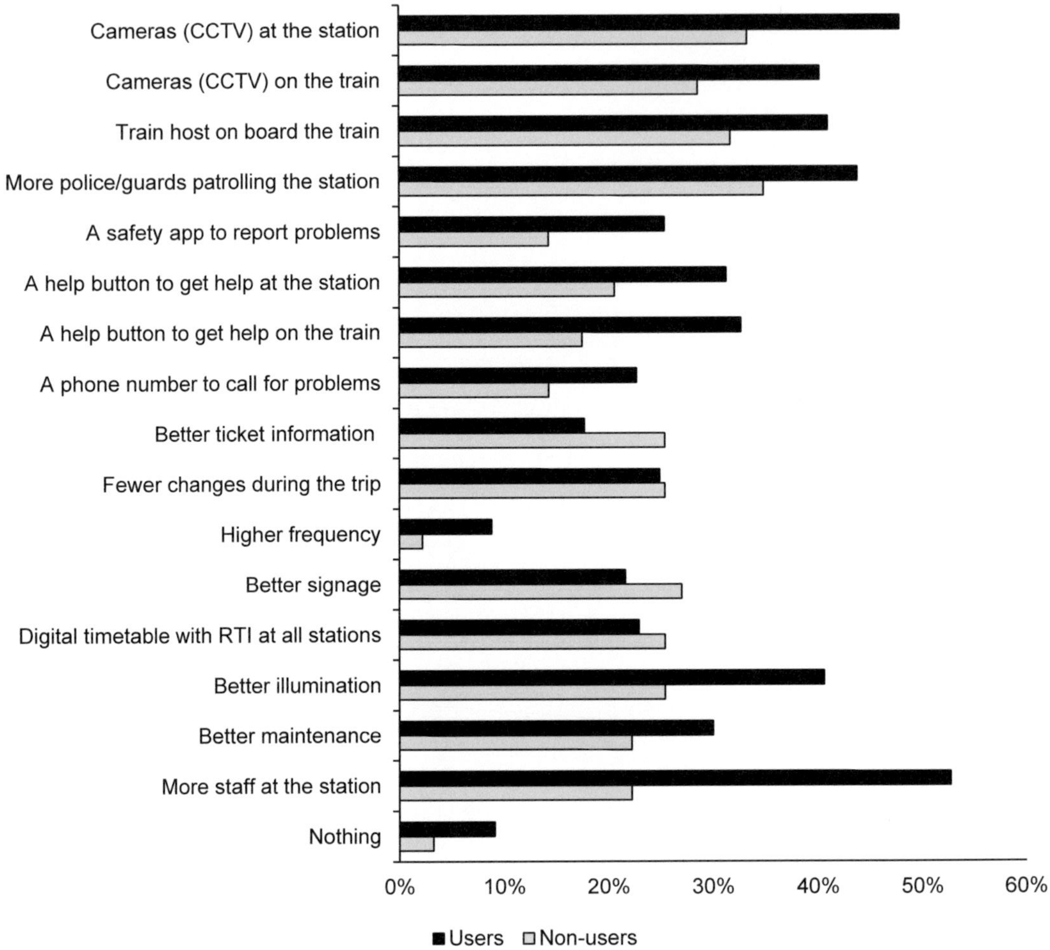

Fig. 14.1 Recommendations: (**a**) Users (Gray): What can be improved to make travel safer? (N = 3,382) and (**b**) Non-users (Black): Indicate which of the following factors would make you travel by train more often (N = 63)

supports sustainable mobility by reducing reliance on private vehicles, contributing to environmental goals, and aligning with SDG 10 on reducing inequalities and SDG 11 on inclusive, safe, and sustainable cities.

14.4.3.3 Fragrance and Allergy Considerations

Health concerns, particularly regarding allergens, were highlighted. One respondent commented, "Nuts and perfumes are the biggest reasons why I cannot take the train, even though I would like to". Another suggested, "Seats or a designated area in a compartment only for asthmatics and respiratory allergy sufferers, with a completely separate carriage clearly labelled on the outside".

14.4.3.4 Safety Issues

Safety, especially for women, was a significant concern. One participant explained, "As a woman, I often experience threatening behaviour from men in public transport, especially in the underground during evenings and weekends". One respondent urged, "Please increase patrolling staff on platforms, stations, and trains". Another safety-related concern involved the location of toilets, with one respondent suggesting, "Toilets should not be located outside the staff's line of sight due to the risk of sexual harassment and violence".

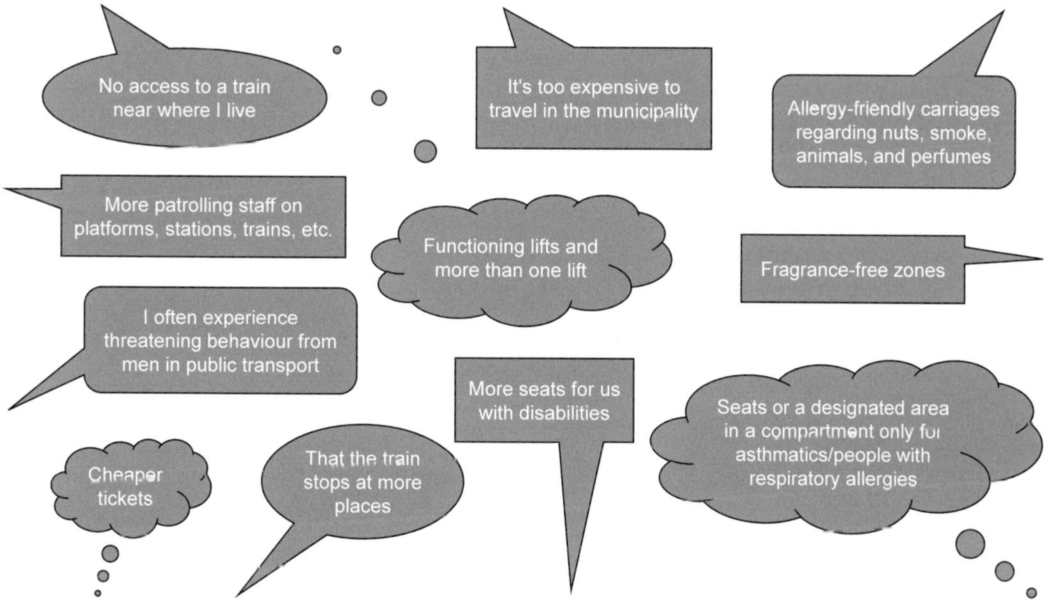

Fig. 14.2 Examples of free-text responses from non-users

14.4.3.5 Accessibility for People with Disabilities

Accessibility issues were also noted. One respondent emphasised, "Functioning lifts and more than one lift [...] We, the chronically ill and disabled, need this!" Concerns about toilets included, "Functional toilets with a basin in the same room as the toilet, with normal lighting, are a requirement for accessibility". Additionally, more seating for disabled passengers was requested: "More seats for disabled people. Also, for those of us with disabilities that are not visible".

14.5 Discussion of Results

Non-train travellers tend to be older, report one or more disabilities, have lower levels of education, feel less safe in their neighbourhoods, experience insecurity while travelling with others, and generally have lower trust in public authorities, including the police. The results show that people who feel trust in other environments and situations use public transport to a greater extent. Earlier studies have shown a link between general trust and perceived safety in public transport (Reichow and Friemel 2020). Specifically, our results show that public transport users have higher trust in the police and feel safer with others and in their neighbourhood. This has also been found in relation to safety. For example, Delbosc and Currie (2012) found that those who felt safe in their street at night and with others also felt safer when travelling. In their research, age and gender were only indirectly associated with safety perceptions. Vilalta (2011) also found a link between safety in one's neighbourhood, trust in the police, and feeling safe on public transport. Jaltman (2020) found low trust in the police to be associated with poor safety, especially for women who do not use public transport. Feeling uncomfortable travelling with others was the strongest predictor of feeling unsafe in a study by Currie et al. (2013). Thus, a lack of trust can be hypothesised to constitute a reason for avoiding public transport. Therefore, rather than just looking at the trip, studies of safety perceptions and travel behaviours need to include the surrounding area and individual factors, as they both may play an important role.

Motivating non-users to start using trains requires addressing key barriers as leverage points and enhancing the overall travel experience. Less mobile and older individuals dominate non-travellers in this particular study. Clear and easily accessible information about train schedules is critical in attracting non-users. Displaying timetables digitally at stations, mobile apps, or interactive websites can help users confidently plan their journeys. Real-time updates on delays or changes further enhance the system's trustworthiness, making trains a more viable choice.

Many potential passengers may view train travel as too costly compared to other transportation modes, particularly private vehicles in more rural areas. Offering competitive and affordable ticket prices is a key strategy for motivating non-users to choose trains. Moreover, train frequency is also relevant. By running trains more frequently, operators can reduce the perceived inconvenience and improve accessibility, ensuring that users can depend on the service for their daily commuting or occasional trips. Delays and unpredictability can create frustration and make train travel seem unreliable. Train services can ensure consistent punctuality by prioritising maintenance, upgrading infrastructure, and improving operational efficiency. Stations are an important part of this. Maintaining clean, well-lit, and safe pathways leading to and from train stations is vital for creating a positive user experience. Neglected or poorly maintained areas can deter potential users due to concerns about safety. CCTV cameras and well-trained staff can help foster a sense of security. Encouraging positive social interactions, such as community-led initiatives or visible staff assistance, can create a welcoming and safe atmosphere, ultimately making train travel more appealing.

Trains should provide specialised amenities to cater to various passenger needs. These include spaces for wheelchairs, pushchairs, and bicycles, quiet zones for relaxation, Wi-Fi and charging points for workers and students, and family-friendly facilities. Offering different service levels, such as standard and premium options, can also accommodate a wider audience.

For rural residents, accessing train stations can be difficult due to limited public transport options. Providing shuttle buses, on-demand transport services, or partnerships with ride-sharing companies can bridge the gap between homes and stations. Ensuring sufficient parking spaces at stations makes it easier for rural dwellers who rely on personal vehicles. Rural travellers, especially women and older adults, may have safety concerns at stations and during late-night travel. Better lighting, security cameras, and visible staff can alleviate these worries. Ensuring clean and well-maintained stations and trains enhances the overall travel experience.

One of the limitations of this analysis was that the non-user group was considerably smaller than the user group, which may have affected the robustness of our statistical comparisons. The dominance of the larger user group could overshadow patterns within the non-user group, making it harder to draw meaningful conclusions. Additionally, with only 53 non-users, we cannot be certain that they represent the broader population, and their opinions may not accurately reflect those of a larger group. This imbalance in sample sizes could introduce bias, particularly when assessing trends across both groups. Another limitation of the study was the lack of questions regarding ticket purchase, the financial implications of travel, and what a free fare would mean to respondents. We believe we need to obtain a better understanding of these factors from a systems thinking perspective, as affordability interacts with access, ridership, and social inclusion.

14.6 Conclusions and Recommendations

The findings of this study underscore the importance of safety perceptions in transit environments and trust in public authorities in delivering public services when the task is to encourage the use of public transport. As previous studies have shown, trust in others, including the police and those around us, significantly influences travel behaviours. Users of public transport tend to feel safer in their neighbourhoods and with other people, which, in turn, leads to a higher likelihood of using public transport. To motivate non-users, mainly those less mobile or older, it is crucial to address key barriers to using trains. Respondents of the survey pointed out a few barriers,

Table 14.2 Descriptives of user and non-users of train travel

	User	Non-user	Chi-square	Sig
Gender orientation	**N = 3,340**	**N = 53**		
Woman	61%	64%	0.215	0.643
Man	34%	32%	0.122	0.727
LGTBQI/Non-binary/Other	5%	4%	0.083	0.773
Ethnicity	**N = 3,299**	**N = 53**		
Swedish born	88%	93%	0.897	0.344
Foreign born	12%	8%	0.897	0.344
Age	**N = 2,732**	**N = 52**		
Young (0–29 yrs)	29%	19%	2.532	0.112
Middle-aged (30–59 yrs)	56%	58%	0.054	0.816
Old (60 + yrs)	15%	23%	2.942	0.086
Household	**N = 3,309**	**N = 53**		
Live alone	23%	25%	0.056	0.813
Lone parent	6%	0%	3.352	**0.067**
Couples	56%	62%	0.871	0.351
Education	**N = 3,335**	**N = 52**		
Elementary school	9%	27%	23.526	**0.000**
Upper-secondary school	32%	37%	23.526	**0.000**
University	59%	37%	23.526	**0.000**
Income	**N = 2,669**	**N = 37**		
Low income (under 250 k SEK/yr)	21%	38%	6.857	**0.032**
Average income (250 k-750 k SEK/yr)	72%	60%	6.857	**0.032**
High income (over 750 k SEK/yr)	8%	3%	6.857	**0.032**
Disability	**N = 3,231**	**N = 49**		
No disability	90%	67%	42.888	**0.000**
Minor disability	8%	22%	42.888	**0.000**
Major disability	1%	10%	42.888	**0.000**
Disabilities	**N = 3,310**	**N = 52**		
Motion sickness	15%	8%	2.309	0.129
Reduced mobility	5%	15%	11.217	**0.000**
Impaired vision	10%	6%	0.989	0.320
Impaired hearing	6%	6%	0.012	0.914
Reading/writing/speaking difficulties	4%	8%	2.330	0.127
Problems with memory/attention/concentration	9%	17%	4.951	**0.026**
Asthma/allergy/other hypersensitivity	17%	33%	8.196	**0.004**
Trouble with anxiety or depression	16%	19%	0.378	0.539
Victimisation	**N = 3,174**	**N = 57**		
Previous victimisation (outside transit environment)	26%	16%	3.060	**0.080**
Trust and safety	**N = 3,408**	**N = 58**		
Safe in neighbourhood	86%	60%	29.354	**0.000**
Trust in police	77%	50%	24.148	**0.000**

such as affordable ticket prices, ensuring reliable service, and improving station infrastructure and safety. Future studies should explore ticket pricing as a key factor affecting non-users' willingness to travel by train, as this study did not directly address this issue. Additionally, future research should examine non-user issues in areas with limited public transport options, such as rural regions where bus and train services are scarce. Understanding the specific needs of these areas could lead to better-targeted solutions for improving accessibility and encouraging train use. More research into the (economic and technological) barriers faced by different groups to access public transportation could also help identify other leverage points to promote more equitable and sustainable public transport use, especially for older adults, as this was the focus in this case study.

Appendix A

(See Table 14.2).

References

Brons M, Givoni M, Rietveld P (2009) Access to railway stations and its potential in increasing rail use. Transp Res Part a: Policy Pract 43(2):136–149

Brovarone EV, Cotella G (2020) Improving rural accessibility: A multilayer approach. Sustainability 12(7):2876. https://doi.org/10.3390/su12072876

Ceccato V, Loukaitou-Sideris A (2022) Fear of sexual harassment and its impact on safety perceptions in transit environments: a global perspective. Violence against Women 28(1):26–48

Ceccato V, Sundling C, Gliori G (2024) What makes a railway station safe and for whom? The impact of transit environments on passengers' victimisation and safety perceptions. Eur Transp Res Rev 16(1):1–16

Chowdhury S, van Wee B (2020) Examining women's perception of safety during waiting times at public transport terminals. Transp Policy 94:102–108

Cozens P (2002) Investigating perceptions of personal security on the Valley Lines rail network in South Wales. Editorial Board 8(1):19–29

Currie G, Delbosc A, Mahmoud S (2013) Factors influencing young people's perceptions of personal safety on public transport. J Public Transp 16(1):1–19

Delbosc A, Currie G (2012) Modelling the causes and impacts of personal safety perceptions on public transport ridership. Transp Policy 24:302–309

Dong H, Ma S, Jia N, Tian J (2021) Understanding public transport satisfaction in post-COVID-19 pandemic. Transp Policy 101:81–88

Gopal K, Shin EJ (2019) The impacts of rail transit on the lives and travel experiences of women in the developing world: evidence from the Delhi Metro. Cities 88:66–75

Hale C (1996) Fear of crime: a review of the literature. Int Rev Victimol 4(2):79–150

Jaitman L (2020) Public transport from a gender perspective: Insecurity and victimization in Latin America. The case of Lima and Asuncion metropolitan areas. J Econ Race Policy 3(1):24–40

Kim H (2021) Service design for public transportation to address the issue of females' fear of crime. Transportation 48(1):167–192

Lim JY, Moon K-K (2022) The implications of political trust for supporting public transport. J Soc Policy 51(1):77–95. https://doi.org/10.1017/S0047279420000707

Owen D, Arribas-Bel D, Rowe F (2023) Tracking the transit divide: a multilevel modelling approach of urban inequalities and train ridership disparities in Chicago. Sustainability 15(11):8821

Reichow D, Friemel TN (2020) Mobile communication, social presence, and perceived security on public transport. Mobile Media Commun 8(2):268–292

Shirgaokar M (2019) Operationalizing gendered transportation preferences: a psychological framework incorporating time constraints and risk aversion. Transp Policy 75:10–18

Stradling S, Carreno M, Rye T, Noble A (2007) Passenger perceptions and the ideal urban bus journey experience. Transp Policy 14(4):283–292. https://doi.org/10.1016/j.tranpol.2007.02.003

Sundling C (2015) Travel behavior change in older travelers: understanding critical reactions to incidents encountered in public transport. Int J Environ Res Public Health 12(11):14741–14763. https://doi.org/10.3390/ijerph121114741

Sundling C (2020) Huddinge, Sweden. In: Transit Crime and Sexual Violence in Cities. Routledge, pp 167–175

Sundling C, Berglund B, Nilsson ME, Emardson R, Pendrill LR (2014) Overall accessibility to traveling by rail for the elderly with and without functional limitations: the whole-trip perspective. Int J Environ Res Public Health 11(12):12938–12968. https://doi.org/10.3390/ijerph111212938

Taylor BD, Fink CN (2003) The factors influencing transit ridership: a review and analysis of the ridership literature

Thompson K, Hirsch L, Loose SM, Sharma-Brymer V, Rainbird S, Titchener K, Dawson D (2012) Riding a mile in their shoes: understanding Australian metropolitan rail passenger perceptions and experiences of crowdedness using mixed-methods research. Road Transport Res: J Austr N Z Res Pract 21(2):46–59

Vilalta CJ (2011) Fear of crime in public transport: research in Mexico City. Crime Prevent Community Saf 13:171–186

Vitale Brovarone E, Cotella G (2020) Improving rural accessibility: a multilayer approach. Sustainability 12(7):2876

Wang D, Liu Y (2015) Factors influencing public transport use: a study of university commuters' travel and mode choice behaviours

Yavuz N, Welch EW (2010) Addressing fear of crime in public space: gender differences in reaction to safety measures in train transit. Urban Stud 47(12):2491–2515

Part IV
The Governance of Stations in the City

15 Making Stations Inclusive and Safer

Abstract

We begin this chapter by presenting and critically looking back at the key findings from the Swedish case study, examining patterns of victimisation and safety perceptions, the influence of the station's urban–rural contexts on people's safety perceptions, temporal variations in transit safety, and the specific safety needs of different traveller groups. We then propose a series of recommendations as 'leverage points' to enhance transit safety while identifying the barriers hindering their implementation. Finally, we discuss these challenges and how they can create opportunities for action by adopting a systems-thinking perspetive. The Swedish example calls for a comprehensive, evidence-based, user-engaged approach to improve transit safety–an approach that can be adapted, improved and rethought to include the specific contexts of other countries.

Keywords

Transit safety · Barriers · Systems traps · Situational crime prevention

15.1 Safety in Stations: Lessons from the Swedish Case Study

Ensuring safety in transit environments is important for sustainable and equitable public transportation. Railway stations, as key nodes in the transport system, play a crucial role in shaping people's daily experiences. Previous research has shown that an individual's concerns about being at risk of victimisation or being in fear when in public places can have long-term consequences (Gardner et al. 2017; Hanson et al. 2010). The physical and social environment of transport nodes, such as stations, plays a crucial role in shaping crime and safety levels (Ceccato et al. 2022; Sundling and Ceccato 2022). Research indicates that open designs with clear sightlines and adequate lighting contribute to safer conditions. Additionally, previous studies suggest that the presence of other passengers, staff, and social interactions can help deter crime, while technological solutions, such as surveillance cameras and emergency alarms, further enhance safety measures.

V. Ceccato et al., *Transit Safety for a Sustainable City*, Sustainable Development Goals Series,
https://doi.org/10.1007/978-3-031-97427-4_15

In this chapter, we present lessons to be learnt from a case study on Swedish railway stations, as presented in Chaps. 5–14 of this book. Based on our analysis, we highlight key findings and propose recommendations to address them. However, implementing these recommendations is not always straightforward. We have identified situations where existing barriers hinder their application and have explored how these challenges can, in turn, create opportunities for action. Through this work, we question short-term, conventional situational crime prevention strategies for transit environments while advocating a systems-thinking approach to promote more integrated and sustainable crime prevention in urban settings.

15.1.1 The Overarching Findings of the Swedish Case

Our findings shed light on important safety concerns in train travel, highlighting both experiences of victimisation and safety perceptions from the home context to stations and beyond. The results reveal that one-third of respondents (34%) feel unsafe while travelling by train, whereas under one-sixth (14%) have been victims of crime in transit during the past five years, based on 4,893 survey responses. We found that those who have experienced victimisation before are more likely to feel unsafe, with specific groups, locations, and times being particularly associated with higher risks. These findings have implications for transport operators, law enforcement, municipalities, and other stakeholders responsible for ensuring public transportation safety, which will be discussed in more detail below.

The most common offences reported were stalking and sexual harassment, with violence and theft also occurring. Despite higher victimisation on trains and platforms, most passengers felt safe, revealing a gap between perceived and actual safety. Frequent travellers experienced higher victimisation and lower safety perceptions, although familiarity can sometimes improve safety. Police data showed challenges in mapping crimes at stations, as only major incidents are reported, leading to the underreporting of minor offences and becoming a poor base for safety interventions. Vandalism was the most common crime reported to the police, followed by theft and issues with drug markets and drug- and weapon-related crimes.

While individual factors such as gender and prior victimisation are major factors in shaping victimisation and perceptions of safety, our study also confirms that a station's physical and social environment plays a role in both. We found that safety perceptions are more strongly influenced by the physical and social characteristics of station environments than by actual individual victimisation. A lack of staff presence and poor station maintenance were linked to lower safety perceptions and increased victimisation of certain offences. Women, especially those identifying as LGBTQI+, as well as individuals with disabilities, more frequently reported feeling unsafe. The results highlight differences between passenger groups and emphasise the need for tailored measures to create safer transit environments, based on gender-neutral policies and interventions.

Sweden's railway stations generally maintain high standards, offering real-time train information and good accessibility for people with disabilities. However, the absence of service facilities (kiosks, restaurants, or cafés) was identified as an issue. The most important concern raised by travellers is the lack of staff availability for assistance or emergencies. Unsurprisingly, safe stations tend to be larger. However, there are variations even within larger ones, and we found that many stations feature both safe and unsafe features, as illustrated in Fig. 15.1, where a platform is well-lit and visible, but a dark, poorly lit tunnel lies underneath. Passengers generally feel safest on the trains and platforms but express higher insecurity in tunnels, on their way to/from the station, and in parking areas. We cannot expect all stations to be large. However, some features follow their perception of quality, such as staff presence, adequate lighting, service facilities, and clear signage and waiting areas. In contrast, unsafe stations have poor maintenance problems, such as open drug markets and have insufficient lighting.

Fig. 15.1 Many stations have safe and unsafe features, where a platform is well-lit and visible inside and outside the station, and a dark, poorly lit tunnel lies underneath. (*Source* Authors)

Long, dark winters can become a safety challenge. Our findings indicate the need for tailored safety measures to address the fact that many respondents feel less safe in the evenings when it is dark–a concern especially pronounced among women. We also observed variations in victimisation patterns across different days of the week and seasons, underscoring the importance of considering temporal factors in safety strategies. According to our results, poor lighting and maintenance have a detrimental effect on feelings of safety in station areas. The importance of these factors is often described in research. Lighting is often one of the most influential factors affecting safety around the station (e.g. Bivina et al. 2019). Lighting influences safety in various ways; by improving visibility, reducing hiding places, and making a place more welcoming (Lorenc et al. 2013), although lighting needs brightness and evenness, the latter to avoid the "fishbowl effect" of being seen but having dark surroundings (van Rijswijk and Haans 2018). Poor maintenance is often found to have a detrimental effect on feelings of safety (Loukaitou-Sideris 2014). Places that are poorly maintained make people feel unsafe, especially infrequent users (Vanier and de Jubainville 2017) who are not as familiar with the place as more frequent users. Other risks involve tunnels and places where drugs are sold or used.

Our findings highlight the need to adopt a whole journey approach to transit safety, considering how victimisation and safety perceptions are shaped by the broader context—from the user's place of residence to departure and destination stations. Socioeconomic and criminogenic conditions influence these experiences, sometimes consistently along the trip but often varying between locations. Similarly, safety perceptions differ across the rural–urban continuum. While young women may not feel more unsafe than their urban counterparts, they take precautionary measures more frequently or feel unsafe in certain environments on their way to stations. These results underscore the importance of tailored safety measures and the importance of considering larger boundaries of analysis beyond the place of residence and the stations.

In summary, we have highlighted the fact that several factors are crucial in shaping transit safety, including staffing and supervision, and environmental and individual factors. These factors often interact with time, as we have explored daily, weekly, and seasonal variations. A comprehensive understanding of what happens at each station is essential for stakeholders, as this knowledge is important for selecting appropriate interventions and monitoring their effectiveness over time. To address these issues, interventions must be tailored to the specific context of each station and user group. Achieving this requires an overall understanding of the broader system and detailed knowledge of local dynamics, including when and where victimisation occurs and the underlying causes. Coordination is important when it comes to bringing these elements together. This is why we have consistently emphasised the importance of unified reporting systems, which can support and create a continuous, evolving body of knowledge which can be used for targeted, context-specific interventions.

Previous research highlights the challenge of fully understanding and addressing safety issues in complex systems such as a railway network with multiple municipalities, stations, and thousands of travellers. Instead, we must identify key leverage points for targeted interventions, requiring collaboration and active participation from stakeholders and users. Next, we offer recommendations for Sweden, which are also potentially relevant for similar systems elsewhere.

15.1.2 Recommendations for the Swedish Case

Based on our findings, we propose the following recommendations to improve safety conditions at train stations and their surrounding areas. These measures are designed explicitly for existing stations that require targeted intervention. In the next chapter, we explore how our findings can guide the planning and development of future stations, ensuring safer transit environments.

15.1.2.1 Enable Contact with Staff Even in the Absence of Their Physical Presence

Our findings show that the primary request on the part of travellers for improvement is the ability to communicate with staff, ideally station staff. Travellers do not want to be left entirely alone when they need help or when they feel victimised or unsafe. It has been shown repeatedly in research from around the world that a lack of staff strongly contributes to victimisation and fear (e.g. Sundling and Ceccato 2022). Many of these stations, however, are small, making it difficult to have staff available around the clock. In such cases, new technologies could serve as an alternative, partially compensating for the absence of staff and enhancing supervision. For instance, users have expressed a desire for more CCTV and help buttons. From earlier research, it has been shown that it is important that someone is monitoring them (Thomas et al. 2020) because the bottom line is to have confidence that there will be help if there is any trouble (Uzzell and Brown 2007). Inspired by successful national and international examples, we propose implementing digital solutions that enable contact with station staff even when they are not physically present. For instance, Swedish SL (Storstockholms Lokaltrafik, the public transportation authority in Stockholm) uses a safety hotline (SL 2025). The safety hotline (and app) is a service to ensure the safety and well-being of passengers and staff using the transport system. It allows people to report accidents, harassment, unsafe situations, or other safety concerns they may encounter while travelling, allowing quick responses to safety issues. Other systems elsewhere, such as in countries such as Brazil, which allow users to see and communicate with staff via screens in bus stops, serve as effective models of communication (see for instance, Sérvio 2023). Additionally, integrating help buttons as those found, for example, in the Netherlands and the UK could facilitate prompt communication with station staff.

15.1.2.2 Consider the Safety Needs of Individual Groups of Travellers

What works for older adults may not necessarily work for young people. In our study, we demonstrate how gender intersects with other individual dimensions, such as sexual orientation, age, and disability. Women, young people, and LGBTQI+ individuals are at a higher risk of being exposed to victimisation than men, particularly crimes of a sexual nature, such as stalking and sexual harassment, but also threats or hate crimes. Our study also focused on women in urban and rural areas and shows that the risk of being subjected to crime or harassment during train journeys varies depending on age. For women under 18 years old, the risk is 3.5 times higher than for older women. When comparing young women in rural areas and cities, the young women in rural areas feel safer than those living in cities. However, they face a greater risk of being exposed to crime or harassment at the station and more often take precautionary measures. These findings call for gender-age-sensitive mobility policies, taking into account the intersectionality of transit safety and reflecting women's and girls' safety needs in particular rural contexts. This includes ensuring that infrastructure is accessible, installing safety features such as adequate lighting and emergency assistance, and offering tailored services to support these passengers. By prioritising their needs, transit systems can foster a safer, more inclusive environment for all travellers.

15.1.2.3 Improve the Physical and Social Environment of the Stations

Amenities such as restaurants, cafés, and kiosks contribute to transit safety by generating activity and facilitating informal surveillance. Our findings indicate that the presence of these establishments in and around stations not only boosts actual safety but also enhances the overall feeling of security among travellers. Therefore, we recommend promoting various service offerings and activities to create a livelier, better-monitored environment. However, we understand that many stations are too small, merely constituting a platform and shelter. In small stations, lighting conditions can be especially relevant, particularly in Scandinavia, where long winters and extended periods of darkness can exacerbate poor safety. Investing in an appropriately designed lighting scheme can significantly enhance passengers' perceptions of safety at stations, and this does not necessarily require exceeding the lighting standards currently used in many stations. Properly implemented, the adapted lighting can improve visibility and create a more secure atmosphere without the need for additional resources (Ceccato et al. 2024b) by improving visibility and promoting natural surveillance. It is also important to consider how specific crimes affect particular stations. Our results further indicate that some areas around the stations are perceived as unsafe, highlighting the need to examine individual contexts in detail, as discussed below.

15.1.2.4 Prioritise the Areas Surrounding Stations that Are Perceived as Unsafe

Although crimes in underpasses and on bridges are relatively rare, many travellers in our study who report feeling fearful in transit identify these areas as particularly problematic. Our findings align with broader research, including a recent survey conducted in 29 Swedish municipalities, showing that poor lighting is the most significant safety concern in tunnels, followed by poor visibility, isolated locations, and graffiti and vandalism. Notably, there is often a lack of consensus when it comes to distinguishing between artwork and physical damage (Ceccato et al. 2025b). To address these issues, we recommend improved maintenance practices by municipalities, including regular cleaning, upkeep of lighting, and removal of what is perceived as physical damage (Ceccato and Ioannidis 2024). Key strategies include identifying the most problematic tunnels, considering their multifunctionality, enhancing lighting and visibility, particularly near stations, and involving users in crime prevention efforts in high-risk areas. Transport operators could also contribute by ensuring that stations are well-lit and maintained, providing clear signage, and investing in security infrastructure, such as CCTV cameras and emergency contact systems.

Moreover, some stations are isolated and inherently more vulnerable, creating an elevated crime risk and/or a perceived lack of safety compared to others, depending on the type of crime. The surrounding areas, shaped by socioeconomic factors and criminogenic conditions (especially open drug markets), can make travellers, particularly women and other vulnerable groups, feel the need to take extra precautions when using public transportation. Increasing visibility through regular patrols, especially during the more problematic hours, can help deter drug selling around stations. By fostering a safer and more secure environment, transport operators can help build passenger confidence and reduce the need for extra precautions.

15.1.2.5 Collaboration and Coordination

By promoting collaboration and considering the entire station area as a unified whole, security and safety measures can be tailored to meet passengers' needs. This includes addressing areas such as bus terminals and car parks, which may be perceived as unsafe, and implementing appropriate measures to improve security in these locations. A pivotal factor in any transport system is the coordination of responsibilities. Cooperation alone is insufficient; a clear, coordinated goal, and a shared vision must support it. Without a shared vision, there is a risk of "accidental adversaries" (Meadows 2008), in which different actors do what is best for their part, with no one having the whole picture and therefore risking undermining each other. Also, this can result in an uneven distribution of resources, for example, if different actors, because of increased crime, simultaneously increase the number of guards in different places, in the station, at the platform, and the area surrounding it because of the fact that there are different owners of the premises. We turn to these issues in more detail in Sect. 15.1.3.

15.1.2.6 Temporal Factors

It is essential to consider how crime incidents vary daily, weekly, and seasonally. Understanding these fluctuations allows authorities to anticipate when criminal activity may peak, which can inform resource allocation, targeted interventions, and long-term planning to address and reduce crime more effectively. Our results show that most crimes are reported to the police late at night, and travellers feel less safe at night. Weekends are the most crime-intense days. The frequency of different types of crime varies across the seasons. Over a more considerable period, crime also develops and takes new paths, so the transport systems, regardless of country, consistently need to adapt to new realities. Although the countries discussed in this book, Sweden, and to lesser extent, Denmark, and the Netherlands, are organised differently and have their distinct characteristics, they share some of the challenges associated with constantly evolving threats typical of the North European context. Poor safety perceptions also peak during the darker hours of the day, showing some indications of variation also in certain seasons.

15.1.2.7 Consider the Whole Journey Perspective

Our findings show that travellers' safety perceptions vary at various stages of their railway journey, highlighting significant concerns, particularly at night. A majority, 63% of respondents, report feeling unsafe at the station and on their way to or from it, while 40% feel unsafe on the train. Even during the day, many travellers express unease, with 30% feeling unsafe at the station, 29% on the way to or from the station, and 17% onboard the train. These findings underscore the importance of improving safety measures across all transit environments along the trip, particularly during nighttime. The way to and from the station may be a barrier in itself and therefore, factors must be considered that impact people positively, such as wide sidewalks, easiness of crossing roads, the presence of trees, and limited traffic noise (Herrmann-Lunecke et al. 2021). The "whole journey approach" to safety is key to achieving the United Nations' 2030 Agenda and Sustainable Development Goals. Understanding travellers safety patterns across rural and urban areas is crucial for ensuring safe travel.

15.1.3 Recommendations According to Problem Type

These recommendations provide general guidelines and emphasise the need to tailor measures for each unique station and its context. A station represents a complex environment where safety is shaped by individual characteristics of that transport node and the surrounding setting. Factors such as fear and vulnerability are influenced by complex interactions from the micro to the macro level subsystems, from a platform to the whole station. Understanding each subsystem is crucial, but it is equally important to consider the entire system, meaning that the interaction between different components of the system must be considered to enhance security and safety effectively.

15.1.3.1 Homelessness and Panhandling in and Around Stations

Aggressive panhandling has also given rise to a new debate about who has the right to public spaces. Panhandling around train stations is complex, intersecting with social welfare and public safety. Previous research shows that while asking for money in itself is not necessarily a crime in most municipalities in Sweden, the presence of individuals engaging in panhandling or occupying station areas can reduce feelings of safety among travellers. Some municipalities have imposed local bans on begging, while others have implemented permit systems. As of October 2024, the Swedish government initiated an inquiry into the feasibility of implementing a national ban on begging, which sparked significant debate, with critics arguing that such a ban could infringe on human rights and the right to public places (Bryant 2024). The police and security guards could work with transport operators and social services to implement alternative interventions. This could include setting up designated areas for support services near stations rather than allowing station spaces to become informal gathering points for vulnerable individuals. To address challenges related to aggressive begging and individuals under the influence of alcohol or drugs at stations, targeted measures that consider specific groups could be implemented. Identifying these groups and tailoring long-term responses for each through collaboration with social services and other organisations is recommended to address issues related to homelessness and alcohol dependency.

Current research shows that cities around the world have adopted either "punitive strategies", in which criminalisation, policing and enforcement of laws and codes of conduct prevail or "outreach-related strategies", which aim towards providing help and support to unsheltered individuals (Ding et al. 2022). The authors conclude that there is limited research on the outreach strategies of transit agencies but that most transit agencies involve both punitive and outreach strategies. The authors highlight an example of how public library staff assist people who experience homelessness by making libraries serve as makeshift shelters and offering access to services and special programmes. However, staff face challenges due to inadequate training and a lack of formal partnerships with shelters and social services. Similar experiences were also found in libraries in the Stockholm capital (Ceccato et al. 2023, 2024a, c). Because homelessness and panhandling are widespread in transit systems around the world, we echo Ding et al. (2022) by stressing the fact that transit operators and municipalities have a social duty to ensure their services are accessible to those riders and to assist them in accessing support. We call for better documentation of their experiences and needs (most of those living on the streets are harassed and badly treated) and for information from those working in transit environments to improve inclusive and safe responses.

15.1.3.2 Measures to Combat Sexual Harassment and Stalking

To address sexual harassment and stalking at stations, recommended measures include activating locations with the most problems while considering all hours of the day, ensuring good lighting for better visibility, launching safety campaigns to raise awareness among passengers, and encouraging them to report suspicious behaviour or to know when is it safe to intervene in incidents as they happen. Additionally, investigations should be conducted to determine which technologies are best suited to addressing different issues—for example, surveillance cameras may be appropriate for parking areas. Furthermore, tailored measures should be developed to address the needs of particularly vulnerable groups (e.g., young women) and stations.

15.1.3.3 Measures to Combat Theft and Robbery

To combat theft and robbery at public transport stations, several measures can be implemented to create a safer environment and protect passengers' belongings. Recognising seasonal variations in theft and robbery is crucial for developing adaptive strategies over time. Awareness campaigns should be directed at passengers to highlight the risks of theft and provide practical tips on safeguarding their belongings. A key measure is to identify specific locations, such as platforms, where improved surveillance is necessary. Especially at smaller and more isolated stations, monitored parking areas can effectively reduce the risk of theft. Strategically placing seating to allow passengers to observe their surroundings can deter potential thieves. Additionally, bicycle parking should be in high-activity areas with good visibility to enhance natural surveillance. These measures aim to reduce theft and robbery and may also interest researchers studying crime prevention and environmental design. For parking lots, consider incorporating activity support through kiosks or other engaging features that enhance foot traffic and optimize the flow of people throughout the area.

15.1.3.4 Measures to Combat Violence, Threats, and Hate Crimes

Our study shows that LGBTQI+ individuals, particularly gender nonconforming women, face unique safety concerns compared to the general population. While theft and robbery are common fears for general travellers, LGBTQI+ individuals are more concerned with threats, hate crimes, and sexual harassment. These findings highlight the distinct vulnerabilities of LGBTQI+ individuals in public spaces, emphasising the exclusionary nature of current transit infrastructure and the denial of their right to fully access public spaces (Ceccato et al. 2024b). Our study also shows that people with disabilities are more likely to be victimised in transit than those with no disability and declare feeling more unsafe than other travellers. Commonplace factors for poor safety are a lack of staff and intoxicated people present at the station (Sundling et al. 2024). Chapter 11 of this book also emphasises the importance of addressing the needs of travellers with allergies, a topic that is underrepresented in the international literature. It is challenging to come up with solutions that tackle all types of safety needs of all these groups of travellers. In general terms, several measures are recommended to enhance security at Swedish public transport stations to prevent and manage potential conflicts and violence. Installing emergency buttons or telephones in strategic locations, such as platforms, trains, and waiting areas, can prompt a swift response and assistance. However, using emergency buttons to call for help may not be suitable if the threat is in close proximity or with the individual. Training station staff to act proactively and de-escalate potential conflicts is crucial for creating a safer environment, yet this demands personell training. Surveillance cameras should be used to monitor critical areas, such as platforms and transit spaces, while the presence of security guards and police officers should be increased, especially during high-risk periods, such as between 20:00 and 00:00. However, earlier research from different countries, stress that real-time monitoring is essential for users to feel safe, as they do not want to feel left alone (e.g. Sundling and Ceccato 2022). To promote natural surveillance and improve visibility, it is essential to create or enhance sightlines within the station. Additionally, transparent protective barriers in waiting areas can contribute to an increased sense of safety and security for passengers. These measures are adapted to the specific conditions and needs of Swedish public transport stations.

15.1.3.5 Measures to Combat Drug-Related Offences in and Around Stations

Results from our study show that observed drug use or drug selling at the station was one of the most important predictors of victimisation and fear of crime. Addressing drug-related offences at train stations requires a combination of enforcement, environmental strategies, and social interventions. Stations often serve as hotspots for drug activity due to their accessibility and relative anonymity. While policing and security patrols play a crucial role in deterring drug-related offences, a more practical approach involves continuous monitoring of crime trends, ensuring that security efforts are focused on stations and times with the highest risk. This includes collaboration between transit security personnel, local law enforcement, and station staff to disrupt drug activities. Enhancing surveillance through CCTV monitoring, real-time reporting systems, and increased visibility of security staff can also serve as deterrents. Strategic environmental design, such as reducing hidden areas, improving station lighting, and limiting loitering spots, can make stations less attractive for drug-related activities.

Beyond enforcement, addressing the root causes of drug-related activity requires outreach and harm-reduction strategies. Instead of solely relying on punitive measures, transit agencies should collaborate with social services to implement outreach programmes at key stations. This could involve designated outreach teams trained to engage with individuals, offering referrals to support services rather than immediate displacement. Experiences from other urban transit environments indicate that a combination of policing and social support leads to more sustainable safety outcomes (Ding et al. 2022; Ceccato et al. 2024b).

15.1.3.6 Artwork or Physical Damage?

Managing graffiti in public spaces is challenging because not all graffiti is perceived as art; some forms of grafitti are mere tags, yet the reasons behind public complaints about graffiti as damage to property remain unclear. The perception of graffiti is subjective; its acceptance often depends on the viewer's perspective, regardless of whether it was created with permission (Ceccato et al. 2025b). In an incident in Stockholm, police were called to investigate vandalism in a public underpass, but upon arrival, they discovered that the graffiti was actually a municipality-commissioned summer project created by young people as part of a youth job programme. This case highlights the contrasting perceptions of graffiti in public places and the challenges of balancing the maintenance of underpasses and ensuring that interventions are inclusive and align with urban sustainability goals. One possible measure could be the creation of designated legal walls for artistic graffiti, as seen in Stockholm.

Maintenance and prompt repair of property damage should be implemented to tackle physical damage/vandalism and prevent minor issues from escalating into extensive destruction. Measures such as surveillance cameras and good lighting in areas particularly vulnerable to vandalism, such as tunnels, stairwells, and platform connections, can help reduce incidents. Local community initiatives, such as "adopt a station" or "adopt a pedestrian tunnel", where local residents are engaged in improving the design and safety of these environments, could be encouraged. Using art as a preventive measure and introducing materials resistant to graffiti and vandalism are additional strategies to consider; for more details, see, for example, (Ceccato et al. 2025a). Implementing tagging systems and databases for the locations most affected can also be an effective strategy for monitoring and managing these challenges in an organised yet inclusive manner.

15.1.4 Pre-requisites for Measures in Safety Planning

A prerequisite for evidence-based crime prevention is having national guidelines supporting knowledge gathering through a unified reporting system and monitoring of stations and the city. In the Swedish case, the National Board of Housing, Building and Planning (Boverket), the National Transport Administration (Trafikverket), and the National Council for Crime Prevention (Brå) are key actors in developing this framework. Collaboration between national and local stakeholders needs an interface at the regional level to build knowledge and deploy interventions that could be assessed in the long term. Cooperation with local authorities involving the public and private sectors is fundamental to addressing specific local challenges; in the next section, we discuss recommendations in more detail.

As mentioned in Chap. 3, following the recently adopted Act on Municipalities' Responsibility for Crime Prevention (2023:196), Swedish municipalities need support to develop comprehensive plans to prioritise evidence-based crime prevention locally. For transit environments, municipalities can seek support from safety experts, urban planners, architects, and law enforcement for a station that already exists and needs safety intervention. Following the steps suggested by Ceccato and Newton (2024), the steps we suggest for applying systems thinking to crime and safety problems at stations are:

Envision a Desirable Transit System: This step involves imagining an ideal, safe, and functional transit system. The stakeholders involved could include station owners and managers, transport operators, and other actors within civil society, including representatives of travellers' organisations and vulnerable groups.

Get "the Beat of the System": This refers to understanding the system's dynamics and rhythm. It involves examining how the station (as a node) and city interact, the flow of people, and how different factors (e.g. the time of day, weather, and social conditions) affect safety and crime patterns.

Focus on Leverage Points: It is challenging to tackle all issues at once, therefore selecting the most important leverage points is essential. They are points within the system where small changes can have a significant impact. By identifying and addressing these points, such as improving lighting or enhancing surveillance, interventions can be more effective in reducing crime and improving safety. This step involves designing interventions that can impact the system as a whole, rather than focusing on isolated issues.

Create Well-Informed Models of the Problem: Developing accurate, data-driven models is crucial for identifying and understanding the complexities of crime problems. These models should account for all environmental and social factors and help frame the problem in a way that supports targeted interventions. Data and information are a precondition.

Promote Resilience, Self-organisation, and Hierarchy: Systems should be designed to adapt and recover from disruptions. A structured hierarchy helps coordinate efforts and ensures that key stakeholders can work together effectively. This step involves looking for example, beyond the immediate environment of the station and considering how broader factors–such as the surrounding neighbourhoods, transportation systems, or even cultural dynamics–can influence safety. Expanding the boundaries allows for a more comprehensive understanding of the system and its challenges.

Include the Beneficiaries: In systems thinking, involving those affected by the changes is also important. This means consulting with passengers, local communities, and other stakeholders to understand their concerns, needs, and ideas for improvement. Their input is crucial for designing relevant and effective interventions.

Assess Consequences and Impact: Finally, assessing the consequences and impact of actions helps identify what has worked, what has not, and how improvements can be made. This is an ongoing process that will inform future decisions and adjustments.

These steps create a place-based framework that allows for an interconnected understanding of crime problems in transit environments and supports the development of interventions that address causes while considering the broader system context. The following section explores the challenges of implementing and evaluating safety interventions in station and railway system interventions.

15.1.5 Current Barriers to Safety Governance of Transit Environments in Sweden

If a traveller is mugged, they can report the incident, allowing it to be collected alongside thousands of others. This enables transport operators to systematically address these issues, assess whether interventions are effective, and determine if the problem has been resolved. If not, new investigations and improved knowledge are expected to lead to new interventions, creating an ongoing cycle of improvement (Fig. 15.2a). In an ideal world, this process would function almost automatically. As discussed in Chap. 3, this model is called SARA and form the base for evidence-based planning. However, in Sweden one of the main current challenges in the railway system is the lack of comprehensive information about safety incidents occurring while people are in transit–whether on the train, at stations, or on their way to them. If a traveller is mugged on a station platform, there is often no immediate support available beyond calling the police. This means that the whole evidence-based planning system falls apart because this crucial link does not exist in the system (Fig. 15.2b).

When single stakeholders collect data, it is often fragmented across different transit operators, making it difficult to obtain a complete picture of what happens across the system. Police data, as previously mentioned, cannot reveal what happens during the trip or in transit environments. While some incidents are reported to the police, many—mainly minor ones, such as panhandling or harassment—are never or rarely officially recorded. This disproportionately impacts specific groups, making them more vulnerable to these incidents than the general population. As a result, there is a limited understanding of what happens in the railway system, including how stations interact with the rest of the city, how people move, and how various factors, such as the time of day and social conditions around the stations, influence safety and crime patterns. In other words, we are currently unable to capture "the beat of the system", as we suggested earlier, even when stakeholders come together and share a vision to improve the safety conditions for train users. This issue is not only important for planning, as the lack of staff and limited access to assistance have emerged as some of the most significant causes of poor safety in transit environments, according to this study. A potential solution to the lack of a unified incident-reporting system could be the implementation of a National Survey of Transit Safety. This would be modelled on Sweden's existing National Victimisation Survey, specifically on public transportation, offering a continuous monitoring and evaluation framework.

As suggested in Chap. 3, another problem is a lack of coordination of relevant stakeholders, making safety interventions arising from planning experiences fragmentary. Today, we lack a national coordinative body with a clear set of security and safety guidelines for the railway system that can be applied consistently across sectors. While municipalities may revise comprehensive plans, existing guidelines from the National Transport Administration and the National Board of Housing, Building and Planning, although

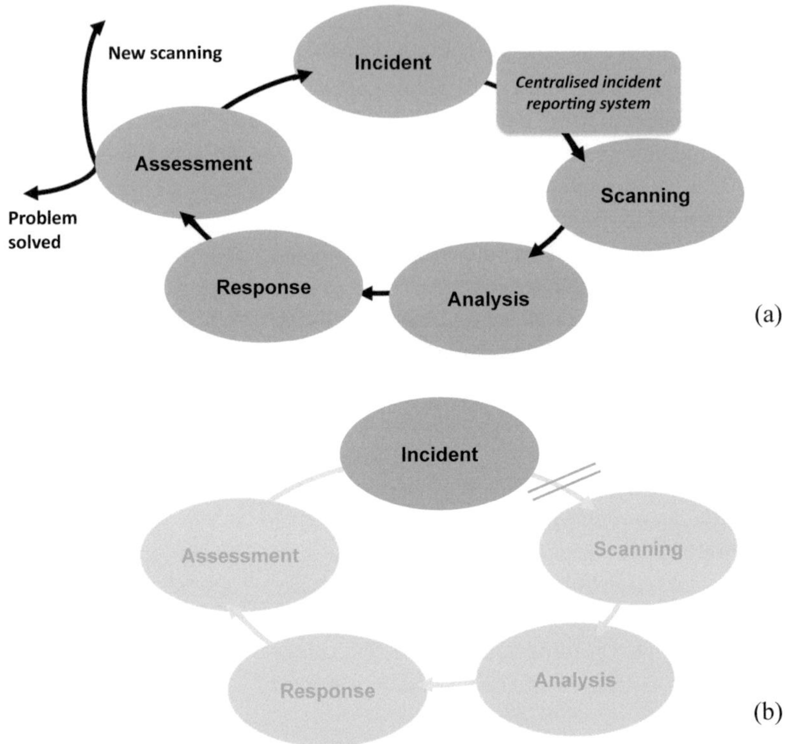

Fig. 15.2 Barriers against evidence-based planning: the railway systems in Sweden

important, are often recommendatory when interventions are done locally. This leads to discrepancies in approaches to address similar challenges, or the other way around. The deregulation of the railway sector has led to the involvement of multiple new partners with responsibilities for different parts of the system, including how a station is maintained. As discussed in Chap. 3, these changes have brought opportunities, such as improved efficiency and investment in modern infrastructure (Boyne 1998), but also fragmented responsibilities, reduced public oversight, and inequalities in service access (Hodge et al. 2010; Klijn and Teisman 2000). At the local level, these issues are reflected by the fragmentation of responsibilities, which hinders the development of coordinated long-term interventions by obscuring the need for an overview of what happens at the station level and across the entire system.

This structural fragmentation is not the only barrier to sustainable governance; both public stakeholders and authorities, including municipalities, face a significant loss of institutional memory (Schein 2010). When skilled planners, experts, or key personnel leave, they often take valuable knowledge, insights, and experience with them which are difficult to replace. This gap in expertise can lead to a lack of continuity in planning and decision-making processes, hindering the effectiveness of long-term projects. Furthermore, the loss of institutional memory can result in repeating past mistakes or overlooking important historical contexts that could provide information for better, more sustainable approaches to urban development and crime prevention.

Institutional memory loss is also triggered by many interventions relying on "quick fixes" for chronic problems, often leading to poor long-term solutions. Short-term funding often limits the long-term perspective essential for effective safety interventions. For example, research on homelessness in the US highlights how short-term approaches can worsen the issue over time by ignoring systemic factors such as unemployment, lack of affordable housing, and mental health challenges (Stroh 2015). Some quick fix policies, like "zero-tolerance" policing, may lead to mass arrests but also fuel cycles of violence. A significant challenge is the short time to evaluate crime prevention interventions—early assessments may overstate success, while delayed evaluations risk missing long-term trends. A broader, long-term approach is needed to capture sustainability, feedback loops, and system adaptability. The EMMIE framework (Chap. 3) highlights how unexpected outcomes may benefit some groups initially but harm others over time.

By nature, a challenge is the persistent siloed approach. Architects, urban planners, designers, and engineers as "the shapers of urban space" (Brantingham 1989) often view crime and safety as beyond their scope, considering it the responsibility of the police, particularly in stations and other transit environments (Paulsen 2013). However, their decisions inevitably shape criminogenic conditions. They can enhance safety by integrating crime-reducing features into new developments or by modifying existing areas through situational crime prevention, which reduces crime by making offences more complex, risky, or less rewarding. Without shared knowledge and resources, meaningful progress remains challenging, as constant short-term focus hinders lasting change. Currently, these fields are mainly separate. Architects and urban planners often lack exposure to crime prevention concepts, while safety experts focus on social prevention, and urban planners take a situational approach. On the other hand, law enforcement and security experts focus on preventing criminal behaviour. This divide is especially evident in Scandinavia. Bridging this gap requires better education and collaboration between these groups (Ceccato and Newton 2025).

Another barrier is the "seeking the wrong goal dilemma", which occurs when a solution to a problem fails to tackle its underlying cause or worsens the situation. For example, installing signage at a train station to raise awareness about pickpocketing may seem like a good strategy to reduce theft. However, it inadvertently helps offenders by providing clues about where passengers store their valuables, making it easier for them to target specific individuals. Thus, the intervention worsens the problem rather than solving it (Ceccato and Newton 2025).

Moreover, despite having a shared mission to combat safety problems in stations, urban planners and police officers may become "accidental adversaries" (Meadows 2008) due to conflicting strategies and misaligned goals. Urban planners often envision stations as part of the broader public realm—accessible, welcoming, and integrated into the surrounding city. The design approach tends to emphasise openness, connectivity, and transparency. On the other hand, law enforcement and security experts tend to prioritise control and surveillance to ensure the safety of passengers and prevent crime. These measures aim to reduce risks but may create a sense of exclusion, making travellers feel as though they are in a restricted space of freedom of movement. Planners and law enforcement professionals may unintentionally find themselves in opposition, as the focus on maintaining openness can clash with the need for security.

In addition, in Sweden, situational crime prevention measures have always been viewed with overall scepticism compared to social crime prevention, which are traditionally better aligned with the prerogatives of welfare but a market-oriented state model. A common criticism is that interventions in specific areas, such as a station, may merely displace crime—geographically, temporally, or according to target, tactic, or even crime type (Clarke 1998). However, a review of research revealed that displacement occurs in only a small fraction of cases and is often overshadowed by the diffusion of benefits, where the positive effects of interventions spread beyond the targeted area (Guerette and Bowers 2017). Even when displacement does occur, it tends to be less significant than the overall reduction in crime resulting from the intervention (Ceccato and Newton 2025). Another reason for this scepticism is how interventions (are perceived to) impact different groups in society. While some individuals directly benefit from safety measures, others may experience no effect or even negative repercussions. For instance, installing CCTV cameras in public spaces is intended to enhance overall safety and assist in crime prevention (see also the case of LGBTQI+ groups discussed in Ceccato and Bogacka (2025)). However, the intended beneficiary (the general public) may not perceive a direct personal advantage from such surveillance. Instead, the primary beneficiaries are often law enforcement agencies, who use these tools for crime detection and prevention. Finally, most of the situational interventions have largely focused on large urban areas (the model has been highly "urbancentric"), where crime opportunities are more concentrated. Yet, rural areas also face threats that demand special knowledge. Adapting these strategies to rural contexts can help better understand the needs of travellers using public transportation in small, more remote municipalities in Sweden.

Similar to those seen in transit environments worldwide, current barriers to safety governance in Sweden persist and hinder efforts to improve safety conditions. The challenge lies in balancing these competing goals while ensuring security without compromising the public's sense of openness in stations or restricting access to those who use these spaces. Understanding these dynamics helps bridge the gap between public aspirations and governance, fostering more sustainable urban planning that effectively addresses crime and safety.

15.2 Conclusions

We have highlighted key findings and proposed recommendations to tackle the most important challenges found in the Swedish railway system. The quality of the station environment plays a crucial role in the overall safety of individuals and should not be underestimated; we argue that recognising the importance of transit environments is key to encouraging more people to use trains and other sustainable transport options. We have limited information about how travellers are affected by the station environment, as they often have no one to turn to in case of an incident. The lack of staff presence is one of the most significant issues across the system and negatively impacts travellers' safety. Regardless of the safety problems, implementing these recommendations is not always straightforward. We have identified several barriers that hinder their application and explored how these challenges can create opportunities for action and improvement. Our recommendations offer a roadmap for overcoming these barriers, but successful implementation demands a commitment to a long-term systems-thinking approach.

In the next chapter, we shift our focus to stations that do not yet exist, even though they may already have been planned, exploring future rail projects in Sweden and internationally. We critically discuss how these new developments and planned expansions can shape the future of public transportation as part of the land use planning of cities and regions.

References

Bivina GR, Gupta A, Parida M (2019) Influence of microscale environmental factors on perceived walk accessibility to metro stations. Transp Res Part D: Transp Environ 67:142–155

Boyne GA (1998) Competitive tendering in local government: a review of theory and evidence. Public Admin 76(4):695–712

Brantingham PL (1989) Planning, development designs and crime. In: Paper presented at Designing Out Crime: Crime Prevention Through Environmental Design (CPTED), Hilton Hotel, Sydney

Bryant M (2024) Swedish government accused of trying to 'outlaw poverty' over begging ban plans. The Guardian. https://www.theguardian.com/world/2024/oct/02/swedish-government-accused-of-trying-to-outlaw-poverty-over-begging-ban-plans?utm_source=chatgpt.com

Ceccato V, Ioannidis I (2024) Brott i tunnlar: Kriminogena egenskaper hos tunnlar i stadsmiljöer. https://doi.org/10.13140/RG.2.2.20515.05927

Ceccato V, Newton A (2024) Systems thinking for sustainable crime prevention in risky places. Routledge

Ceccato V, Bogacka E (2025) Queering the rural: Tales from Poland and Sweden. In: Ceccato V, d'Arbois de Jubainville AH (eds) Crime, peoples and places: perspectives on rural safety and justice. Routledge

Ceccato V, Newton A (2025) Systems thinking for sustainable crime prevention: planning for risky places. Taylor and Francis

Ceccato V, Gaudelet N, Graf G (2022) Crime and safety in transit environments: a systematic review of the English

and the French literature, 1970–2020. Public Transport 14(1):105–153
Ceccato V, Ercin E, Hazanov J, Elfström S, Sampaio A (2023) Safety in a public library: the perspective of visitors and staff. Libr Manag 44(3/4):229–245
Ceccato V, Ariel B, Ercin E, Sampaio A, Hazanov J, Elfström S (2024a) Changing environments to promote safety in libraries. Eur J Criminol. https://doi.org/10.1177/14773708231213157
Ceccato V, Besenecker U, Höglund M, Eizadi H (2024b) Ljusförhållanden för trygga tunnelbanestationer: Experiment i Fridhemsplan station och i virtuell verklighet (VR). In KTH Royal Institute of Technology
Ceccato V, Ercin E, Sampaio A, Hazanov J, Elfström S (2024c) Crime at micro-places in public libraries. Secur J 37(3):686–711
Ceccato V, Ioannidis I, Felson M (2025a) Graffiti in pedestrian tunnels: A comparison of police records and crowdsourced data in Stockholm, Sweden. Tunnelling and Underground Space Technology Incorporating Trenchless Technology Research
Ceccato V, Ioannidis I, Felson M (2025b) Tunnels in the urban fabric: balancing connectivity and safety. Urban, Plann Transport Res 13(1):2431514
Clarke RV (1998) Situational crime prevention: Successful case studies (2nd ed.). Harrow and Heston
Ding H, Loukaitou-Sideris A, Wasserman JL (2022) Homelessness on public transit: a review of problems and responses. Transp Rev 42(2):134–156. https://doi.org/10.1080/01441647.2021.1923583
Gardner N, Cui J, Coiacetto E (2017) Harassment on public transport and its impacts on women's travel behaviour. Austr Planner 54(1):8–15
Guerette RT, Bowers KJ (2017) Assessing the extent of crime displacement and diffusion of benefits: a review of situational crime prevention evaluations. In Crime opportunity theories, pp 529–566
Hanson RF, Sawyer GK, Begle AM, Hubel GS (2010) The impact of crime victimization on quality of life. J Trauma Stress 23(2):189–197
Herrmann-Lunecke MG, Mora R, Vejares P (2021) Perception of the built environment and walking in pericentral neighbourhoods in Santiago, Chile. Travel Behav Soc 23:192–206
Hodge GA, Greve C, Boardman AE (2010) International handbook on public-private partnership. Edward Elgar Publishing
Klijn E-H, Teisman GR (2000) Governing public-private partnerships: Analysing and managing the processes and institutional characteristics of public-private partnerships. In: Public-private partnerships, Routledge, pp 102–120
Lorenc T, Petticrew M, Whitehead M, Neary D, Clayton S, Wright K, Renton A (2013) Fear of crime and the environment: systematic review of UK qualitative evidence. BMC Public Health 13:1–8
Loukaitou-Sideris A (2014) Fear and safety in transit environments from the women's perspective. Secur J 27:242–256
Meadows DH (2008) Thinking in systems: a primer. Chelsea Green Publishing
Paulsen DJ (2013) Crime and planning: building socially sustainable communities. Routledge
Schein EH (2010) Organizational culture and leadership (Vol. 2). John Wiley and Sons
Sérvio G (2023) Novo ponto de ônibus que 'conversa' e faz companhia chega a SP. https://olhardigital.com.br/2023/08/08/carros-e-tecnologia/novo-ponto-de-onibus-que-conversa-e-faz-companhia-chega-a-sp-veja/
SL, Greater Stockholm Local Transport (2025) Security Centre. Region Stockholm. https://sl.se/en/help-and-contact/sl-security-centre
Stroh DP (2015) Systems thinking for social change: A practical guide to solving complex problems, avoiding unintended consequences, and achieving lasting results. Chelsea Green Publishing.
Sundling C Ceccato V (2022) The impact of rail-based stations on passengers' safety perceptions. A systematic review of international evidence. Transport Res f: Traffic Psychol Behav, 86:99–120. https://doi.org/10.1016/j.trf.2022.02.011
Thomas G Bennie J A De Cocker K Castro O Biddle S J (2020) A descriptive epidemiology of screenbased devices by children and adolescents: a scoping review of 130 surveillance studies since 2000. Child Indic Res 13(3):935–950.
Transport for London (2025). Help from staff. Transport for London. https://tfl.gov.uk/transport-accessibility/help-from-staff
Trip JJ (2007) What makes a city? Planning for 'quality of place': the case of high-speed train station area development (Vol. 12). IOS Press
Uzzell D, Brown J (2007) Conceptual progress in understanding fear of crime in railway stations. Psicologia 21(2):119–137.
Vanier C d'Arbois de Jubainville H (2017) Feeling unsafe in public transportation: A profile analysis of female users in the Parisian region. Crime Prev Community Saf 19(3):251–263.
van Rijswijk L Haans A (2018) Illuminating for safety: Investigating the role of lighting appraisals on the perception of safety in the urban environment. Environ Behav 50(8):889–912.

16 Planning for New Stations in the City

Abstract

A safe public transportation system is essential for a sustainable future. Drawing on examples from Sweden and beyond, this chapter explores how future railway station designs are being planned to enhance accessibility and safety. We apply foresight strategies to anticipate future trends, external change drivers, opportunities, and risks. Considering diverse transit users, this discussion is intended to set the foundation for sustainable urban mobility in different contexts. These efforts align with the Sustainable Development Goals, particularly SDG 11 (Sustainable Cities and Communities), by promoting safer, more inclusive public transport environments.

Keywords

Future station · Design · Foresighting · Norrbotniabanan · Travellers · Sustainability

16.1 Introduction

A safe public transportation system is a key element of a sustainable future. In this chapter, we explore how current railway station designs and related features aim to create accessible and safe environments. By critically examining real-world examples from Sweden and beyond, we illustrate how these projects seek to optimise passenger flow, enhance natural surveillance, and integrate other sustainable elements such as energy-efficient systems. These features are focused on the short-term challenges; however, in the long term, predicting the future becomes more complex and uncertain. They tend to become more complex, especially with shifting demographics, rapid technological advancements, and revealing inequities in travellers' safety. Innovations such as automated trains, smart infrastructure, and the integration of stations with other forms of mobility influence the future railway system and stations as nodes, places, and hyperspaces (see Chapter 3). These innovations also introduce potential risks (e.g. cybersecurity concerns and privacy issues) and create new challenges in managing complex, interconnected systems. This chapter applies foresight strategies to proactively shape sustainable transport systems by anticipating future trends, external change drivers, opportunities, and risks.

We argue that technological advancements may not yet have been accompanied by improvements in the social dimensions of trips by train. One example of this concerns the inequities in safety for vulnerable groups. As illustrated in this book, many travellers feel unsafe while travelling by train or when in transit between their homes and the stations, especially during nighttime. This is not unique to Sweden; in many countries, safety concerns make individuals consider other less

V. Ceccato et al., *Transit Safety for a Sustainable City*, Sustainable Development Goals Series,
https://doi.org/10.1007/978-3-031-97427-4_16

sustainable transportation forms, such as automobiles (for a review, see Ceccato et al. (2024). Furthermore, in the event of an incident, passengers may feel isolated, as many stations lack on-site personnel and remote support systems, leaving them without immediate assistance and creating a sense of vulnerability. Public transportation and future stations should not be affected by these challenges. How can we envision a future where the quality of train travel becomes a key factor in attracting people?

Thus, we shed light on the technical and social dimensions of station planning and set the stage for ways of thinking about governance at various levels, from local to national. We will devote time to reflecting upon future travellers' needs, embracing the different types of transit users from an intersectional perspective. We wonder whether the strategies employed in Sweden can be adapted to fit the unique context of other cities, considering different urban dynamics, station designs, and community characteristics. We also report on a few international experiences with the project "Stations of the Future" and include other examples from countries in the Global South.

Cities must adapt to evolving technologies and changing societal preferences while maintaining reliable and safe infrastructure. This chapter reviews insights into station planning that we hope can be helpful elsewhere. It demonstrates that well-designed, efficiently managed stations can be uplifting for urban and rural communities by catering to diverse user needs. This engagement and strategy aligns with the Strategic Development Goals, particularly SDG 11 (Sustainable Cities and Communities), by fostering safer, more inclusive urban environments.

The chapter provides two-level examples at national and international level by combining a national case study (Norrbotniabanan) with international perspectives (UITP 2023). Norrbotniabanan illustrates specific station features (e.g. safety barriers, tactile signage, inclusive design), while the UITP shows examples not only of specific station features but also of strategies and policy frameworks.

16.2 Foresighting Stations in the Future

Stations and railway terminals can be criminogenic places; as discussed in Chapter 2, they can be crime generators and attractors. They may radiate crime in the morning or absorb crime at other times (Bowers 2014; Brantingham and Brantingham 1995). Stations behave as self-contained entities but open to the public and frequently open 24 h a day, making passengers and the stations potential targets for criminal activity, including rare terrorist incidents. Trains offer amenities similar to those of hotels, such as temporary accommodation, restaurants, and lounges (Neal 2020). Yet, the design and management of the stations can be maximised to reduce risks of crime or other safety incidents. As we will discuss in the next section, station design can be optimised for better passenger flow, with clear signage, well-maintained facilities, and integrated services, such as retail and information hubs, to make the experience more seamless and enjoyable. Additionally, sustainable features, such as energy-efficient systems and green spaces, can be incorporated, making stations functional and environmentally friendly.

As we navigate the challenges of shaping tomorrow's cities, we can look at the innovations of past generations to provide a blueprint for crafting the future. Another alternative is to embrace uncertainty.

In Chapter 3, we showed how Trip's (2007) transport planning model can be used to identify four main decision-making arenas: public administration, transport, design, and real estate. The inclusion of policing and safety planning as a fifth arena underscores the importance of addressing security concerns, both in station design and operation, as an integral part of the system. This fifth arena focuses on crime prevention partnerships and collaboration among various stakeholders to ensure that stations remain safe spaces for passengers and to build new ones that will attempt to be safe from the start. Can this be possible if we do not know what the future holds?

Planning new stations requires coordination in all these arenas to ensure that each aspect of station design, from accessibility to security, is thoroughly considered. Challenges in terms of collaboration and coordination will most likely remain for some time. Involving local, regional, and national actors will ensure that stations are well-integrated into the broader urban transport network and meet the needs of diverse populations. Collaboration across different sectors is essential for creating stations that attract passengers and become thriving, multifunctional spaces that support the city's growth and mobility. This means that we must address the future planning of cities and transportation nodes evolving to keep up with a continuously changing world. Peter Senge, more than three decades ago, wrote that the managers (of urban systems) must have the capacity to shift "from seeing parts to seeing wholes, from seeing people as helpless reactors to seeing them as active participants in shaping their reality, from reacting to the present to creating the future" (Senge 1990).

16.3 Research Design

This chapter adopts a systems-thinking approach, using foresight strategies to navigate future uncertainties and design sustainable transport systems. According to the APA (2025), foresight, or strategic foresight, involves anticipating future trends, understanding external drivers of change, and preparing for potential successes or challenges. Figure 16.1 illustrates the analytical model used in this chapter to assess the documentation from various case studies.

Foresight begins with the future and reverse-engineers what needs to happen today to achieve the most desirable outcome in the future. Planners can enhance thinking about possible alternatives for the future by applying foresight in iterative cycles. This approach enables planners to proactively shape the future of transport systems, anticipating both opportunities and risks (APA 2025). In the following section, we apply insights from our study on railway systems in southern Sweden to explore a new project in northern Sweden. We assess the proposed plans with the potential risks and opportunities, focusing on design

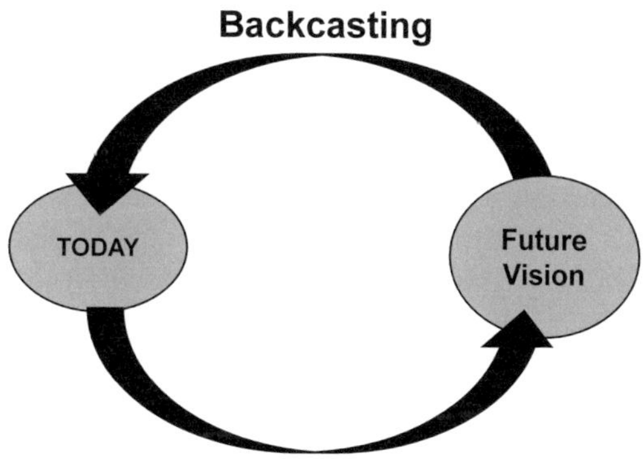

Fig. 16.1 Foresighting: an analytical model to assess the features of various case studies. (*Source* Authors)

and station management features. Additionally, this evaluation considers factors that may not yet be visible but could influence the project's success. The same procedure is also used to analyse other international projects, allowing for a broader perspective on potential challenges and solutions.

16.4 Case Studies

16.4.1 The Swedish Norrbotnia Line

Norrbotniabanan is a new 270 km coastal railway connecting the cities of Umeå and Luleå in northern Sweden. It is designed to make travel and transport faster, safer, and more environmentally friendly. Once completed, it will bridge Sweden's most densely populated region without a commuter train service. The project is extensive and highly complex, requiring the construction of 250 bridges and 550 km of new roads, and significant land modifications. The Norrbotniabanan railway is expected to improve travel times, support business and industrial growth, enhance safety, and contribute to a greener future by reducing emissions and shifting transport from road to rail.

The Norrbotnia Line is important to many actors who are involved in this new railway project. Close cooperation with the municipalities along the route is fundamental, and a comprehensive planning approach is needed to ensure that the railway is as accessible to all users as possible. Along several parts of the route, the planning process involves train stations and larger travel hubs (bus terminals), which may include restaurants, shops, and hotels. The Swedish Transport Administration is responsible for tracks, platforms, and platform connections. At the same time, partnerships involving municipalities and the private sector will handle other aspects, such as station buildings, connecting streets, parking areas, pedestrian and bicycle paths, and public transport connections.

Now, let us turn to the Norrbotniabanan case to assess this project's safety opportunities and risks, and how these are intertwined with long-term planning choices, sustainability goals, and the integration of transport infrastructure into broader city and regional development.

16.4.1.1 Station Design

Existing stations in this part of the country vary in size and have also been built in different styles depending on the ideals prevailing at the time (many stations are from the 1800s). When new stations are planned, some of the architectural characteristics that were relevant at the time an old station was built, such as nooks and crannies, are avoided today and now impose challenges to safety. Thus, foreseeing the future is challenging. Today, unrestricted sightlines are required, which implies, for example, large glass areas. These glass surfaces enhance visibility, allowing for natural surveillance and reducing hidden corners that could otherwise feel unsafe. However, while glass façades contribute to safety by improving sightlines, they also introduce potential risks. As society changes and security concerns evolve, particularly regarding terrorism, large glass areas may present vulnerabilities in the event of an attack. Extensive glass surfaces are visible from the outside, compromising travellers' privacy and, in extreme cases, creating a 'fishbowl effect' (see Chapter 8 for a more detailed discussion on the topic).

The stations along the Norrbotnia Line are designed with extensive use of glass to enhance openness, visibility, and natural surveillance (Fig. 16.2). While large glass surfaces contribute to perceived safety, they also present new challenges.

Good maintenance is essential (as indicated in Chapter 9), as signs of disorder and desolation seriously decrease safety (see Chapter 10). Especially at rural stations, maintenance contributes to better-perceived safety and less victimisation. Likewise, lighting is one of the most important factors for safety. Good sightlines are especially important when people are not feeling safe. Good even lighting can, therefore, enhance safety. Appropriate lighting is needed especially in rural and mixed areas, but it is also important in urban areas (as suggested in Chapter 10). Rural stations, as our results show, are experienced as less safe than urban stations, and people tend to be also more victimised (see Chapter 7). Therefore, in the case of a new station, the station itself and surrounding areas must be designed with better visibility, ensuring proper lighting. This presents an opportunity to design stations with open, well-lit spaces and clear sightlines from all access points. Also, the connectivity with the surrounding area should be seamless, making it easy to change between modes of travel and well-integrated into the surrounding area.

Station pedestrian tunnels or "underpasses" can prevent accidents as people do not need to cross the tracks. However, our findings indicated that they are perceived as the most unsafe places at the station (see Chapter 7). It is thus an example of a conflict of goals between traffic security and safety from crime. Tunnels can be dark with poor sight. They can be curved, thus not possible to see through and they can give the impression of being trapped if something happens. This problem can be generalised to other environments as well. In some cases, pedestrian bridges can serve as a better alternative than tunnels, offering visibility and safety. Where tunnels are unavoidable, design improvements—such as wider passages, increased height, and lighter colours—can make them feel safer.

For the Norrbotnia Line, where tunnels are primarily avoided, semi-open underpasses provide an effective alternative, as shown in Fig. 16.3. The clear visibility from multiple directions, along with the presence of activity and

Fig. 16.2 Example of design with glass from the station on the Norrbotnia Line. (*Source* Dissing + Weitling)

movement, reduces the feeling of isolation and entrapment that is often associated with tunnels. Such designs can maintain the safety benefits of grade-separated crossings while mitigating the risks of crime and fear associated with enclosed tunnels. However, semi-open underpasses on the Norrbotnia Line would be welcome to improve visibility and reduce isolation, but they still pose risks. Limited surveillance during off-peak hours, harsh weather conditions, poor maintenance, and potential loitering can impact safety in these areas if tunnels and lighting are not well maintained.

16.4.1.2 Technical Potentialities, Staff, and Formal and Informal Surveillance

The Norrbotniabanan coastal railway connecting Umeå and Luleå is expected to have around six to seven stations along its route. These stations will serve key locations, enhancing travel and transport in the region. We can apply current knowledge to evaluate the design and feasibility of Norrbotniabanan. As we have learned in previous chapters, staff and effective surveillance are among the most essential factors for safety and victimisation regardless of which group is studied. Train users in particular want more staff at the station; they want to know that they can get in touch if something happens that makes them feel unsafe, which is also important for preventing crime. This need is especially pressing during late-night hours and weekends when crime rates tend to be higher (Chapter 9). Rural stations, in particular, experience more significant challenges, with higher victimisation rates and lower levels of perceived safety (Chapter 7). The upcoming Norrbotnia Line, which will traverse remote areas with small stations, presents a potential threat to travellers safety due to the lack of on-site personnel. Addressing this issue requires combining digital surveillance solutions, increased activity in and around stations, and a national approach to incident reporting and monitoring.

Rapid technological development will offer new possibilities for reaching staff even if they are not physically present. However, currently, there are already successful examples of digital solutions in Sweden and internationally. In Sweden, public transport in the Stockholm Region has a security centre accessible to users via telephone, app or chat (SL 2025). In countries such as Brazil, seeing and talking to a person on a screen is possible. In the Netherlands, users can contact staff via help posts, WhatsApp, or text messages

Fig. 16.3 Visualisation of a planned station along the Norrbotnia Line and its tunnel design. (*Source* Dissing + Weitling)

(NS 2025). Help posts can also be found on underground platforms in London (Transport for London 2025).

However, these systems are sometimes well developed in parts of the country, such as in the capital city, but are lacking or uneven implemented in the rest of the country. This is the case in Sweden, where certain operators and stakeholders in parts of the country provide these resources but not other operators. Passengers should be provided with a single point of contact, rather than multiple numbers depending on their location within the transport system, whether on the train, in the station, or elsewhere. It should not be the responsibility of passengers to discern ownership or jurisdictional boundaries within the transport infrastructure; instead, they should be able to rely on a seamless, integrated system that ensures consistent and accessible reporting (See Chapter 8).

Integrating such digital solutions from the start will be essential for the Norrbotnia Line, where many stations will be unmanned. Remote assistance centres, interactive help screens, and CCTV surveillance should be incorporated into station infrastructure to provide security without requiring a constant staff presence. A coordinated national incident-reporting system as part of a Decision Support System (DSS), as suggested in Chapter 3, would ensure that passengers can always get help, no matter where they are.

Increasing activity around stations is also important for ensuring both actual and perceived safety. Stations that are active and well-frequented tend to feel safer, as the presence of people naturally increases surveillance. Establishing restaurants, cafés, and other service-oriented businesses within or near stations can help create a lively environment and deter criminal activity. Our results indicate that such service facilities in or near stations positively impact perceived safety and can prevent crime (see Chapters 5, 8 and 9), particularly during evenings and nights when safety concerns are highest. This is especially important for the Norrbotnia Line, where some stations may be built on the outskirts of towns (Fig. 16.4). If these stations remain empty and inactive outside commuting hours, they risk becoming isolated and unsafe.

Therefore, a recommendation is to facilitate and create different activities and services and that the environment should be lively at different times of the day. Note that the evenings constitute the riskiest conditions for both victimisation

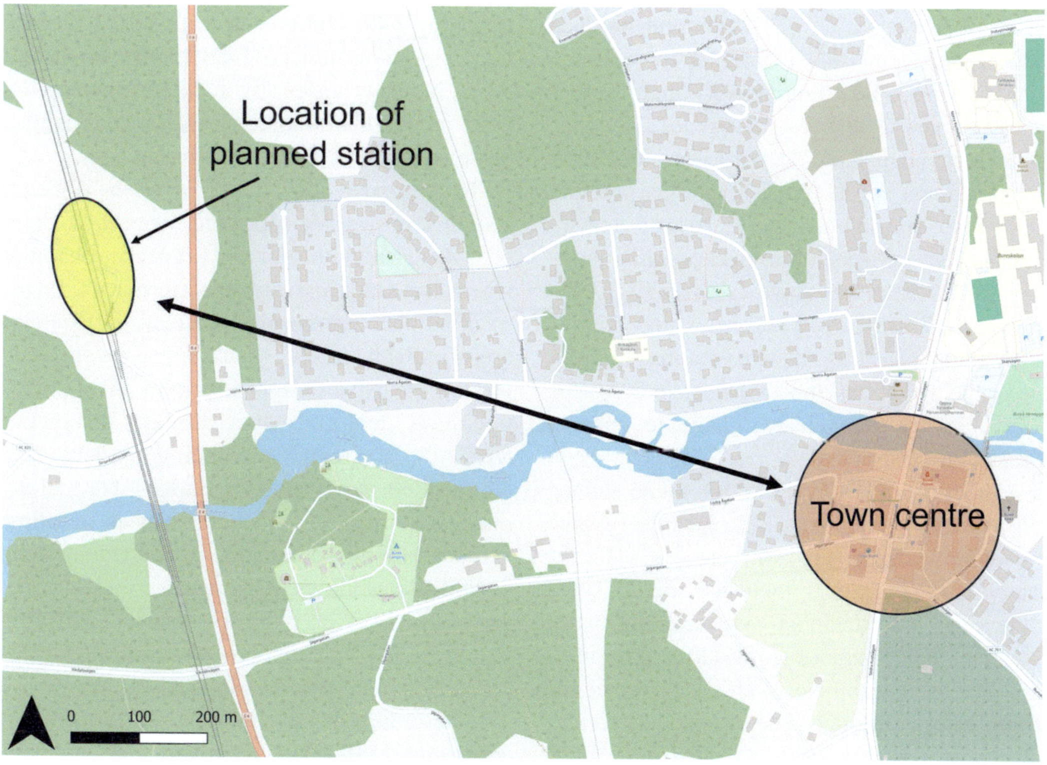

Fig. 16.4 The new station planned in relation to the town centre. (*Source* Authors)

and the feeling of safety; it is, therefore, especially important to facilitate activity at these times. At the Norrbotnia Line stations, a key strategy is to create active and vibrant environments that attract people and promote natural surveillance. Integrating shops, cafés, and public spaces designed for gathering people together aims to increase positive activity and discourage illicit behaviour. However, this approach also requires careful planning to ensure increased foot traffic does not unintentionally invite new security challenges. Open sightlines, well-lit areas, and clear territoriality can help maintain a balance, encouraging activity while minimising opportunities for unwanted behaviours. Addressing these issues requires collaboration between stakeholders, including police, municipalities, and property managers, to ensure that station areas remain safe and welcoming. Furthermore, the personnel of these establishments, particularly those who share some responsibility for managing stations, must receive continuous training to ensure adequate preparedness in handling safety incidents.

16.4.1.3 Traveller's Safety Needs

As illustrated in this book, groups that need to be focused on include women, LGBTQI+ individuals, frequent travellers, rural residents, young travellers, and people with disabilities. Interventions and planning should not follow a one-size-fits-all approach, as different groups of travellers (and non-travellers) have distinct needs. For example, although a large body of research shows that women (Chapter 10) feel more unsafe than men (e.g. Sundling and Ceccato 2022), most transport policies remain gender-neutral. A non-gender-neutral approach might involve designing with specific gender-based needs in mind (e.g. improving lighting in particular areas that women perceive as unsafe).

Some of the solutions may include what is known as universal design. Universal design focuses on creating environments and products that are accessible and functional for as many people as possible, regardless of their gender, mobility, age, or other characteristics. The challenge here is making environments accessible and functional for as many people as possible, while also considering the needs of specific groups. These two goals are not necessarily mutually exclusive. An example is the design of a station along the Norrbotnia Line, which incorporates features such as shelters, seating, lighting, accessible ticket machines, wheelchair spaces, safety barriers, and clear signage to enhance accessibility and safety.

Although guidelines exist for some vulnerable groups, many others lack specific support, and improvements are still needed. Continuous staff education, inclusive design, and addressing intersectional vulnerabilities are essential. This includes considering non-users and addressing conflicts, such as allergens and crime safety, through policy and awareness campaigns. Ensuring safety for all requires identifying marginalised voices among travellers, reducing barriers, and improving infrastructure for vulnerable and general travellers.

In many countries, there are guidelines for some vulnerable groups, such as for those with reduced mobility or visual impairments, but for many other groups, these are lacking. Moreover, despite the improvements that have been made, for example, for those with reduced mobility, the environment is not adapted to enable them to travel safely (Chapter 12). Thus, the work must continue. For example, staff have an important role to play if something happens. Also, even if guidelines and legislation exist, there may be discriminatory. Therefore, continuous educational interventions for staff are important. It is also important to continue incorporating inclusive and accessible design for individuals with reduced mobility, especially in rural and other small municipalities. In this way marginalised voices will be in the spotlight, and structural inequalities will be addressed, while challenging traditional crime prevention. Also, it should be remembered that most people can be "vulnerable travellers" at certain times, thus making this group even larger. For example, travelling with children or with heavy luggage may impose barriers not met on other occasions. Also, in many cases, by reducing barriers for these groups, travelling can be facilitated for other passengers, for instance, by offering a seat.

To improve movement between different station levels, grade-separated connections have been designed with stair and lift packages, ramps where necessary, and interactive terminals. Tactile track number signs and directional signage also assist individuals with visual impairments, reinforcing inclusive design principles. At information hubs, real-time train displays, interactive terminals, and clear wayfinding markers help all travellers to navigate the station—particularly those who require guidance, such as first-time visitors, elderly passengers, or individuals with cognitive disabilities. A designated meeting spot for assisted travel is clearly marked, ensuring that those needing support can easily find assistance. For more detailed suggestions tailored to specific groups of travellers, such as women and girls, and LGBTQ+ individuals, see previous chapters of this book.

16.4.1.4 Whole Journey approach

When considering safety in public transport, a "whole journey" approach is essential. This means that not only the stakeholders within the transport system, but also the municipalities are responsible. Also, it is not only places where the crimes are committed but also places where people feel unsafe that need to be considered before interventions are initiated, as poor safety impacts travel behaviour and wellbeing. Moreover, to be able to travel, every piece of the travel chain, from door-to-door, must be sufficiently safe; otherwise, the trip might not even be started if there is an alternative to public transport.

For the Norrbotnia Line, we note that the station design aims to support seamless travel, where connections between transport modes are safe and intuitive. Another aspect to consider is that wayfinding should be clear so passengers can navigate stations and transfers without confusion. Visual contact between waiting areas and key station points also contributes to a sense of safety. The bus waiting room, for example, is strategically positioned to provide clear sightlines to both the bus stops and the station entrance, reducing hidden spaces that can feel unsafe (Fig. 16.5). Additionally, ensuring easily accessible and visible travel information helps passengers navigate their journey.

16.4.1.5 Governance, Coordination, and Financial Constraints

A well-functioning collaboration between different actors provides a basis both for the planning stage and for later management at different levels. Guidelines need to be established at a national level, based on knowledge of people's safety needs and preferences. These must offer clear guidance for safety management and interventions across the railway system, down to the local level. In Sweden, these are partially missing (Chapter 8). It is unclear how national regulations differ in their approach to safety management at the local level and how these variations impact the overall security and operational efficiency of railway systems. A key question for Sweden is how improved coordination among stakeholders from local to national levels can be better implemented to enhance the safety conditions of travellers across railway systems. A risk stemming from poor organisation among different stakeholders at a local level is the creation of "accidental adversaries" in which each actor only has an overview of their own work, with no one having the big picture.

Regardless of goodwill, financial constraints restrict both planning and management. In planning, a risk is that the costs will be higher than expected because of unexpected factors. The planning and building of Norrbotnia Line are conducted over several years and involve several actors. Given the complexity and multiple stakeholders involved, significant uncertainty remains regarding costs and timelines.

16.5 Stations of the Future: The UITP Platform

Turning our gaze to international organisations addressing future stations, we can report the work of the UITP - Union Internationale des Transports Publics (2023). Established in 1885, this organisation constitutes a network for public transport stakeholders. Their project "Stations for the future" (UITP 2023), conducted with the assistance of the company Kone, outlines global trends in how society, technology, and the economy are evolving and what challenges and solutions are for making stations lively hubs. Their motivation for this project is that from a global perspective, it is pointed out that, as the world population grows, the demand for public transport will also grow and must be planned for. The ageing population is also growing globally and may constitute, for example, a more even distribution during the day, with a larger part of the passengers not commuting and thus travelling at times other than peak hours. Below, we discuss in detail the components of the stations of the future as envisioned by UITP.

16.5.1 Design and Lighting

Designing a station involves a comprehensive approach incorporating functional aspects such as lighting, accessibility, and user comfort and safety while ensuring that the space is efficient, safe, and visually appealing. The UITP suggests that light needs to be bright enough for users to feel safe, although no clear guidance is provided. Areas of darkness or shadows should be avoided, as should flickering. They also recommend that colour temperature be chosen to be perceived as welcoming. Light should also be comfortable when moving from one place to another and only change gradually, creating a more unified space. Stations should be easily identifiable and stand out clearly, aiding passenger wayfinding (UITP 2023).

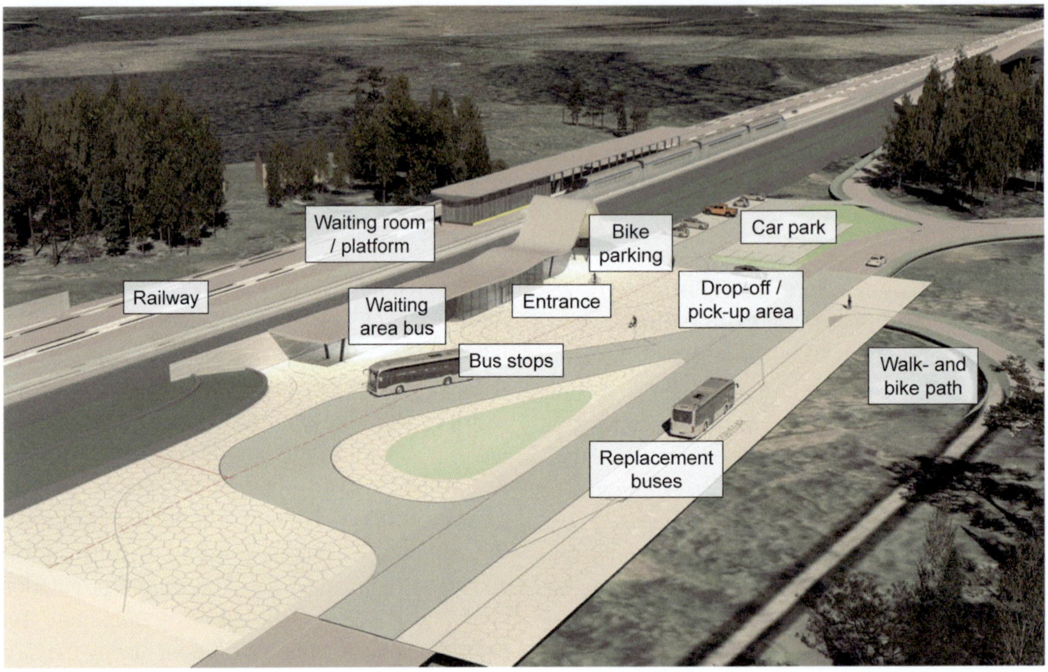

Fig. 16.5 Overview of a planned station along the Norrbotnia Line with well-connected spaces for different transport modes. (*Source* &Rundquist)

16.5.2 Technical Potentialities, Staff, and Formal and Informal Surveillance

Digitalisation is rapidly progressing, providing opportunities to develop services based on collected data. It can allow a more direct communication between users and operators, for example, making it possible for users to report damage. However, cyber security is a challenge and the users' trust is essential for them to be willing to use digital services.

Lack of staff is also a challenge; however, technology, automation, and self-service are seen at possible ways forward, where the UITP sees the possibility of releasing staff from some of their work tasks, enabling more visible human presence at stations. However, users should have direct contact with staff in cases of emergency or for information. As an alternative to help points, they recommend robots or holograms as these can be more interactive. It should, for example, be possible to get in touch with staff when ticket vending machines do not function or there are other problems with tickets. Also, even if technical solutions can replace staff in some cases, and are preferred by some travellers, everybody will not like, or will not be able to use, these tools, and for them more personalised assistance will be important. Because of that, user centres at major hubs are recommended. Moreover, instructors may be needed to teach travellers how to use the digital tools. They can also help with wayfinding or organising queues. Mobile stands with staff at certain occasions is another idea.

The UITP argues that stations should not only be for those travelling but also for others, and proposes different solutions for bringing about more activity which will attract different kinds of people to the place. Stations, centrally located as they often are, can be a place to sit and work for people who can work away from the office. Another idea is to place parcel pick-up points at stations, as online shopping is a noticeable part of retail sales today. Shops can also be placed at stations, and surveys can be used to identify the kind of shops most people want. The stations can also be used for art exhibitions, attracting more people to the location.

16.5.2.1 Travellers' Safety Needs

Making stations fully accessible for as many people as possible is an important task for the future. Installing more lifts and having university students employed part-time at stations to help passengers board and alight are two of many presented examples that can be used in future. Also, making personalised information available before entering the station (real-time information about the availability of facilities such as toilets, lifts, services, and other amenities or alter native arrangements if needed) helps people to be more in control. This is already being used and will be continuously developed with technical advancements. Real-time information is already common and should be extended in cases where it is missing. Knowing the waiting time, for example, until a lift arrives if it is not visible, makes users feel more in control and shortens their perceived waiting time. AI will continuously develop ways to provide personalised recommendations. Different digital solutions can be used and continuously developed to plan new stations.

An interacting example from Singapore is the SMRT—the Enhancing dementia-friendly Transport Programme. This project enhances service provision for those with dementia as they travel in Singapore, acting as "Go-To-Points", locations to be relied upon when an issue with someone who has dementia arises. These "Go-To Points" serve as resource centres and safe return points to help people with dementia return home.

16.5.2.2 The Whole Journey Approach

The first and last mile may be a weak link in the safety chain and if so, it is essential in improving connectivity with the surrounding area. The UITP addresses mobility models, such as car sharing and other ways of buying transport as a service, which can complement public transport and make the station more accessible and safer. Also, combining public transport with, for example, cycling needs drop-off areas. Change between different modes of transport needs to be seamless.

Stations are landmarks within a city. To effectively fulfil this function, stations should be more seamlessly integrated into the urban fabric, and transport services should be better connected with onward travel options. In order to gather insights and solutions, workshops and brainstorming sessions were conducted in various locations, including Singapore, Jacksonville, and Delhi, as well as online, bringing together transportation industry experts and key stakeholders (UITP 2023). Another example is the Metro de Madrid, which launched its accessibility and inclusion plan in 2016 to enable full autonomy for those with reduced mobility and/or sensory/cognitive difficulties by installing lifts and improving supplementary measures (UITP 2023).

16.5.2.3 Governance, Coordination, and Financing

The UITP concludes that cooperation between stakeholders and governance of station development are critical and complex issues. They argue that the development of larger stations should be seen as making them destinations instead of places to pass through. Collaboration should be fostered, not only between transport stakeholders but between service stakeholders.

By using data analysis, stakeholders can better understand users' needs. Local inhabitants can be engaged in increasing safety through community outreach programmes for immigrants or homeless people, such as neighbourhood watch programmes, or safety campaigns, to foster a stronger sense of community. Early cooperation is needed in planning new stations, including engaging the local stakeholders and civil society in general. An example from Japan was Takanawa Gateway City, a partnership with state and city authorities, the railway industry, public transport operators, and others. The project is part of a broader initiative to create a Transit-Oriented Development model, and the recently opened new station has already become one of Tokyo's busiest rail nodes with significant potential. In the UK, St Pancras station in London uses a real-time digital twin to enhance customer experience and station performance. The system helps attract and retain passengers. The digital twin optimises wayfinding, ads, and crowd management and it also monitored social distancing during the pandemic (UITP 2023).

16.6 Other Examples Around the World

In Brazil, the Guarded Bus Stops project ("Abrigo Amigo") enhances safety at bus stops, particularly for women travelling at night (Sérvio 2023). The project transforms traditional bus shelters into interactive safety spaces with digital screens, microphones, night-vision cameras, and internet connections. Between 8 PM and 5 AM, passengers can connect to a service centre, providing them with company until the bus arrives. During the day, the screens display regular advertisements. The project has expanded to 100 locations in São Paulo, Rio de Janeiro, and Campinas, with support from brands such as Santander, Vivo, Bradesco Seguros, Sempre Livre, Diageo, and Anhanguera, as well as backing from municipal, state, and federal governments.

Green Park Station will connect to Nairobi's ongoing transport developments in Africa, including the Nairobi Commuter Rail System and the bus rapid transit (BRT) system. This integration will help create a seamless, efficient transport network, making it easier for residents and visitors to move around the city. The station is being built with a strong focus on sustainability. It will feature green spaces, energy-efficient systems, and eco-friendly infrastructure that contribute to reducing the environmental impact of the project (NMS, n.d.).

In Asia, the Beijing New Airport Railway Station is currently under construction and expected to be completed in 2025. It is designed to handle high passenger volumes, offering advanced services such as intelligent ticketing, real-time travel information, and automated check-in, ensuring an efficient and smooth travel experience (Beijing Daxing International Airport, n.d.).

While the examples presented in this chapter are not exhaustive, they illustrate how innovation can be leveraged to enhance the safety and sustainability of public transportation. These cases demonstrate various approaches, from technological advancements to design interventions, contributing to safer and more accessible railway stations. However, a more in-depth analysis of these examples is necessary to fully assess the opportunities they present and the potential risks and challenges they may entail. Despite these limitations, we hope this exploratory discussion serves as a first step towards understanding how future railway stations can be designed to better meet diverse users' needs.

16.7 Conclusions

In this chapter, we attempted to explore how the design of future railway stations can be planned to enhance accessibility and safety, drawing on examples from Sweden and other international contexts. We applied foresight strategies to anticipate future trends and discuss opportunities and risks related to public transportation. Even if we do not know how the future will evolve, we can plan for what needs to be improved in the short term. In a longer-term perspective, it is more difficult to foresee. Building new transportation infrastructure requires significant investment, especially in more remote areas. However, it will probably be important to consider differences in safety and victimisation over time, place, and group of travellers when planning new stations, as discussed in this chapter. We are seeing some interesting examples from around the world regarding public transportation and the planning of new stations. Cities are reimagining how to improve mobility, efficiency, and sustainability with innovative designs and technologies.

References

APA, American Planning Association (2025) Shape the future: 2025 Trend report for planners. https://planning-org-uploaded-media.s3.amazonaws.com/publication/download_pdf/2025-Trend-Report-for-Planners-r1.pdf

Beijing Daxing International Airport (n.d.) Beijing New Airport Railway Station. http://www.bdia.com.cn

Bowers K (2014) Risky facilities: crime radiators or crime absorbers? A comparison of internal and external levels of theft. J Quant Criminol 30:389–414

Brantingham P, Brantingham P (1995) Criminality of place: crime generators and crime attractors. Eur J Crim Policy Res 3:5–26

Ceccato V, Sundling C, Gliori G (2024) What makes a railway station safe and for whom? The impact of transit environments on passengers' victimisation and safety perceptions. Eur Transp Res Rev 16(1):21. https://doi.org/10.1186/s12544-024-00641-5

Neal A (2020) How to stay safe on your daily train commute. Safety risk. https://www.cardinus.com/insights/security/how-to-stay-safe-on-your-daily-train-commute/

NMS, Nairobi City County (n.d.) Nairobi Railway City Project. https://www.nairobi.go.ke

NS, Nederlandske Spoorwegen (2025) Safe train travel with NS. Nederlandse Spoorwegen. https://www.ns.nl/en/travel-information/facilities/safety-on-the-train-and-at-the-station.html

Senge P (1990) The fifth discipline. Currency Doubleday

Sérvio, G (2023) Novo ponto de ônibus que 'conversa' e faz companhia chega a SP. https://olhardigital.com.br/2023/08/08/carros-e-tecnologia/novo-ponto-de-onibus-que-conversa-e-faz-companhia-chega-a-sp-veja/

SL Storstockholms Lokaltrafik (2025) Security Centre. Region Stockholm. https://sl.se/en/help-and-contact/sl-security-centre

Sundling C, Ceccato V (2022) The impact of rail-based stations on passengers' safety perceptions. A systematic review of international evidence. Transport Res f: Traffic Psychol Behav 86:99–120

Transport for London (2025) Help from staff. All our staff are trained to offer you assistance and support. Transport for London. https://tfl.gov.uk/transport-accessibility/help-from-staff

Trip JJ (2007) What makes a city? Planning for 'quality of place': the case of high-speed train station area development (Vol. 12). IOS press

UITP Union Internationale des Transports Publics (2023) Stations for the future. How to make stations better, more inclusive transport hubs. Report. Brussels, Belgium

Responding to Challenges in Transit Safety: Towards an Agenda for Research and Practice

17

Abstract

In this chapter, we summarise the findings from the analyses presented in the preceding chapters to highlight the need for further knowledge in this area and forward-thinking policies at both national and local level. The results are grouped into the research agenda, changes to the design of the transit environment of cities and changes to planning and policy. Our aim is for this chapter to serve as a guide for researchers calling for interdisciplinary collaborations and emphasising the role of sustainable transport in supporting broader sustainability and development goals.

Keywords

Railway stations · Transit safety · Research · Governance · Systems thinking

17.1 Introduction

Cities, as hubs of social, economic, and cultural life, rely on a well-functioning and safe public transport system to connect people, reduce social inequalities, and promote sustainable development (UN-Habitat 2025; United Nations 2015). With this book, our aim has been to illustrate the importance of safe public transport environments in fostering sustainable cities. We have examined how environments impact transit safety and have provided strategies to address the diverse safety needs of city users. Drawing on insights from the Swedish case, we offer broader recommendations that we hope can be considered locally and, if context-relevant, elsewhere.

Our aim for this chapter is for it to serve as a guide for researchers calling for interdisciplinary collaborations and to emphasise the role of sustainable transport in supporting broader sustainability and development goals. Safe public transportation is important in achieving several of the Sustainable Development Goals. At the same time, innovation and infrastructure improvements support SDG 9 while reducing car dependency to address SDG 13 in mitigation and adaptation. Moreover, addressing challenges collectively strengthens stakeholder collaboration. It demands well-prepared coordination, contributing to the advancement of SDG 17, which is relevant in a governance system that is currently, to various extents, fragmented. Finally, SDG 11 emphasises the need for universal access to safe, inclusive public spaces, particularly for women, and individuals with disabilities (United Nations 2017), which has been the focus of this book.

By presenting lessons to be learnt from the past in Chap. 15 and looking ahead in Chap. 16, we lay the foundation for a discussion on how public transportation can become a more attractive alternative and a safer and more inclusive mode of travel. This chapter focuses first on establishing

V. Ceccato et al., *Transit Safety for a Sustainable City*, Sustainable Development Goals Series,
https://doi.org/10.1007/978-3-031-97427-4_17

a research agenda that might also have implications for practice, then on designing changes to the transit environment of cities and rethinking overall planning policy.

17.2 Sustainability and the Governance of Transit Safety

In stations and other transit nodes that attract crime more than anywhere else, the role of planning must "take into account a variety of aspects of society with good touch" (Forsberg 2019, p. 12). The challenge lies in interpreting democratically made decisions, addressing the diverse needs and expectations of different groups, and integrating them into long-term, sustainable solutions. Incorporating a sustainability agenda into this framework is crucial, as it encompasses the structures and processes of authority that govern social, political, and economic relationships. This includes formal and informal institutions and private forms of authority (UNODC 2020).

Sustainability as a concept is increasingly recognised across various fields; although crime and safety have not traditionally been part of the sustainability discourse, there are signs of change which address environmental crimes, ecological justice and restorative justice, which focuses on repairing the harm caused by crime. A more direct link between crime and sustainability is environmental design and social cohesion, which is crucial for a community's long-term stability and growth (Cozens 2007; Gamman and Thorpe, 2009). A sense of safety is closely tied to community sustainability, where social engagement and participation in activities contribute to overall well-being.

Sustainability is also linked to the societal costs of crime, including the costs of processing and preventing crime. Paulsen (2013) split these categories into economic, social, and environmental costs. Economic costs involve direct losses, such as reductions in property values or government spending. Social and emotional costs are harder to quantify, but can be seen in how fear of crime affects mobility and behaviour. Vulnerable groups bear a higher price due to unequal levels of victimisation. Environmental costs include, for example, carbon dioxide emissions arising from crime prevention efforts. Other costs stem from the environmental impact of construction and the renovation or demolition of buildings and residential areas. According to Ceccato and Newton (2025), one way of addressing these costs and challenges is through systems thinking, which requires a clear, shared vision that is continually discussed and refined. The 2030 Sustainable Development Goals (SDGs) aim to create safer places and develop sustainable crime prevention. In the next section, we put forward a set of suggestions for research in this area.

17.3 Suggestions for Further Research

A research agenda for transit safety must embrace systems thinking (Ceccato and Newton 2025; Meadows 2008), recognising that transit environments are embedded within broader urban, regional, and global systems. As discussed in previous chapters, we must in our research move beyond the fragmented approach that currently characterises transportation planning and governance. The reason for this is that if we are to seriously address the unequal conditions faced by vulnerable groups, such as women, older adults, and people with disabilities, who often experience heightened safety concerns in public transportation, we must extend our focus beyond transit nodes and neighbourhoods first, because safety conditions in stations can differ significantly from those experienced or perceived along the journey to and from them. Second, ensuring safety from home to destination involves multiple stakeholders who do not always coordinate their actions effectively. Bridging these gaps requires a more integrated approach considering the entire mobility chain and fostering collaboration across sectors.

There is no doubt that the "whole journey approach" to safety is essential for aligning with the United Nations' 2030 Agenda and its sustainable development goals. Current data limitations restrict our ability to access real-time information or disaggregate data according to individual characteristics. This gap hinders our capacity to fully understand the variations in safety risks along the trip, particularly in the "first and the last mile" (Roy et al. 2024). By examining safety patterns across time and space, we can support travellers, helping them to experience safer travel from door-to-door. Our findings show that the conditions around the stations seem to affect the safety perceptions of the travellers we studied more than the context of the place of residence and the context of the station of destination. Different types of crimes occur in varying environmental conditions, emphasising the need for research into safety risks throughout the entire journey, particularly for groups that are more frequently victimised or feel less safe. In this study, we were unable to investigate the full journey, as we only had data on the journey from home to the destination station. Future research should explore safety on the return journey, which is often during the evening hours, a time generally perceived as less safe compared to daytime travel. Our findings corroborate recent calls for more contextualised evidence-based studies on first- and last mile connectivity (Roy et al. 2024). More research is needed to investigate the dynamic space–time dimensions of people's routine activities, which is essential for effectively deploying interventions.

More research on the physical and social qualities of the stations and surrounding areas is needed. Our findings suggest that while individual characteristics significantly impact train travellers' victimisation, their safety perceptions are also influenced by the environment of the stations and surrounding areas. Larger stations often tend to be perceived as safer; they are also often located in bigger cities. They tend to have good lighting conditions, amenities such as cafés, restaurants, and kiosks, which support natural surveillance, positively impact safety perceptions, but well-maintained stations and the presence of CCTV cameras also further serve to enhance safety. However, stations vary in their design, and it is not uncommon that stations display both good and poor design features. Further research is needed to assess individual elements of the environment in more detail, such as the location and type of illumination and its impact on travellers' comfort and safety. If the presence of staff is important, how they should be deployed across the system to ensure that travellers get the support they need remains an unanswered question. More research is needed on homelessness and panhandling in transit environments in order to understand the impact of different interventions, including enforcement and outreach strategies. Better documentation on the experiences of affected individuals and what can be learnt from current projects is also necessary.

We need to learn about the challenges and opportunities involved in coordinating actions among these stakeholders to improve station safety in more isolated, remote areas. Smaller stations, often located in remote rural areas without staff, are frequently perceived as less safe by travellers. Stations situated in areas with open drug markets, panhandling activities, and poorly maintained underpasses have a significantly negative impact on safety perceptions. We should further investigate what factors contribute to the perception of these environments as unsafe, and how they can be addressed through design or other interventions. These spaces have become "grey zones", where responsibility is unclear, and no party takes responsibility. There is also a lack of coordination among place managers and other actors, such as transport operators, municipalities, and law enforcement, that leads to the creation of environments that are neither accessible nor safe for travellers.

A centralised incident-recording system is vital for making informed decisions and improving the safety of train travellers. The fact that travellers often cannot report incidents anywhere presents a significant barrier to evidence-based interventions aimed at enhancing the safety of train travellers. This raises the question of how such a system could be implemented to improve decision-making and create consistent, informed strategies across stations. Further research is needed to investigate the challenges of sharing data across different stakeholders, and how these challenges could be overcome. More research is needed to understand how to utilise this information in a long-term perspective to improve the system, including identifying the organisational changes required, as well as the technological challenges and opportunities that a centralised system would present. In particular, as has been discussed in Chap. 3, we should further investigate how Decision Support Systems (DSS) can identify and predict safety concerns or patterns of risk in transportation environments and how real-time processing and automated alerts could improve incident response times, potentially enhancing user confidence in public transport. We could also consider whether we should integrate what-if modelling for scenario simulation to help forecast the effects of various changes and interventions on user safety.

Even though our study shows that women are overrepresented as victims of transit crime, women are far from a homogenous group; and this has an impact for both research and policy implications (Loukaitou-Sideris 2016). We found in our study that LGBTQI+ individuals, particularly gender nonconforming women, have distinct safety concerns. While theft and robbery worry the general public, LGBTQI+ individuals are more affected by threats, hate crimes, and sexual harassment. Research on sexual harassment in public spaces requires detailed data collection. Many crime-coding systems fail to capture the full spectrum of sexual offences, from serious crimes such as rape and sexual assault to coercive acts such as inappropriate touching and verbal harassment. We echo previous research on sexual crime in transit to call for further research that is needed to identify patterns across contexts, including variations according to offence type, transport mode, and time, such as differences between weekdays and weekends or peak and off-peak hours (Ceccato and Loukaitou-Sideris 2020). It is also essential to develop robust methodologies for data collection and analysis that consider the different areas within stations and between bus stops, as well as the journey to and from these transit points. Without this contextual approach, policymakers risk overlooking key opportunities for effective intervention for these specific groups.

The impact of victimisation and fear in public spaces significantly shapes people's behaviour, often leading to avoidance of certain areas or stations or altered travel patterns. Understanding these effects is important for developing strategies to enhance safety and ensure that public spaces are accessible and inclusive for all individuals. For instance, over half of both LGBTQI+ people and women preferred not travelling alone, but only a quarter of men. About half of women and LGBTQI+ individuals stayed in visible areas of the station, compared to a third of men. Over half of LGBTQI+ individuals tried to appear confident, while only a third of women and men (not belonging to this group) did so. Future research could investigate what factors influence the decision of LGBTQI+ individuals and women to adopt safety measures, such as avoiding travel alone or staying in visible areas within stations. A limitation of our study was the inability to capture all gender status categories due to low response rates, resulting in the merging of identities into a single category. Future research should maintain separate gender categories to better understand the specific safety concerns of these groups.

The intersectionality of transit safety demands more research. For instance, this study shows that people with disabilities are more likely to be victimised in transit than those with no disabilities; they also declare feeling more unsafe than the rest of the travellers. These results raise important questions about how safety regulations can be improved, and which stations require staff presence the most, as well as at what times of the day. For example, smaller stations with the size of bus stops may not realistically accommodate on-site staff, but alternative ways of contacting staff could be explored. This is particularly important for people with disabilities. For larger stations, it is worth considering whether increasing staff presence in the evenings would suffice. Another key issue is the perception that train hosts are not readily available onboard, prompting the need to investigate how this could be improved. Similarly, for those who suffer from allergies, our research shows that "allergens" were the most feared factor when taking a train trip, especially onboard the train. This demands more detailed data collection along the trip as well as testing of designated areas in carriages for individuals suffering from a high degree of asthma or allergies. Further research is needed in each setting and temporal context to understand where the problem lies and how to deal with it. In cases of health-related risk caused by allergens, for example, there may be variations in transportation mode as well as temporal variations such as differences between weekdays and weekends and between seasons.

Improving safety in transit environments requires an interdisciplinary approach that bridges urban planning, transportation studies, criminology, and social sciences. As illustrated in this section, calls for research that address inequities in mobility cannot be confined to changes within the transit system alone; they must extend to transformations in housing policies, urban development, regional planning, and global transportation networks. Mobility, justice, and safety are deeply intertwined, demanding sector-overarching changes that integrate transport accessibility with broader policies on social inclusion and spatial justice. This demands a shared vision and national guidelines to empower stakeholders and support coordination and collaboration from national to local levels, across all sectors. However, none of this development is possible without the knowledge and evidence necessary to guide intentional change.

17.3.1 Limitations of the Book

As in any piece of research, this book has its limitations. While it covers a wide range of potential threats in transit environments, including violence, theft, and health-related risks for individuals with allergies and disabilities, it does not address terrorism in railway systems as a distinct threat. Additionally, it does not cover cybercrime or cyberattacks targeting transit systems and passengers. Future research should focus on these areas. The interactions between different transport modes deserve more attention, as safe transitions between rail, bus, and other forms of mobility are necessary for reducing safety risks. Despite the book's effort to incorporate international perspectives and examples from around the world, much of the focus remains on the Nordic context, particularly Sweden. This is both a limitation and a strength, as it allowed for an in-depth examination of transit safety within this context. Future studies could expand comparative research by analysing transit safety challenges and solutions in diverse geographical and socioeconomic contexts, particularly in regions with different transportation agencies, policing and governance models.

17.4 Suggestions for Changes in the Design of Transit Environments

Research has long supported the fact that certain environmental characteristics have a positive effect on reducing crime rates and enhancing feelings of safety, while others have adverse impacts and seem to relate to higher crime incidence and/or fear (Ceccato et al. 2022; Sundling and Ceccato 2022). Crowded rush hours might just provide the right opportunity for pickpocketing or sexual crimes. Meanwhile, an empty station or train in the late evening hours may just offer the anonymity that an offender needs to commit a robbery. Based on the Swedish case and the international evidence from the two literature reviews discussed in Chap. 2, we propose a number of changes and improvements in these transit environments to ensure more favourable safety conditions.

Our study shows that railway stations tend to blend well-planned design features with poorer design features. Larger stations in big cities feel safer due to good lighting, amenities, maintenance, and CCTV, but staff presence remains the most important element for their safety. However, ensuring staff presence at all stations is neither feasible nor economically viable. Additionally, investing in any single element, such as lighting, surveillance, or amenities, does not automatically guarantee that the safety conditions are ensured. Addressing these incongruences requires a strategic approach that balances different interventions based on the station's context. For instance, in smaller or less busy stations where permanent staff presence is not an option, alternative measures such as enhanced remote surveillance by CCTV or similar forms of surveillance, improved visibility, community engagement initiatives, and targeted maintenance efforts can be alternatives. While these potential solutions may be implemented in isolation, the station's specific context and temporal variations must be considered (see Chap. 15 for an in-depth discussion). Thus, a tailored, multi-layered approach rather than a "one-size-fits-all solution" is necessary to enhance safety across diverse station environments.

As previously indicated, aggressive panhandling raises questions about the right to public space and impacts travellers' feelings of safety at train stations. Our results indicate a low tolerance for panhandling, suggesting the need for targeted interventions. Addressing homelessness and substance dependency requires tailored responses, integrating both enforcement and outreach strategies. Lessons learnt from public libraries, which serve as informal shelters but struggle with inadequate training and partnerships, highlight the importance of structured support systems. Since homelessness and panhandling are common in transit systems worldwide, we echo calls for transit agencies to balance security with social responsibility in any kind of intervention.

To construct new stations, a balance between accessibility and safety is important so they can be adapted to their surroundings while promoting sufficient social interaction that encourages conditions for natural surveillance. For indoor environments, the choice of materials and the design used in construction could align with energy consumption targets while creating spaces perceived as safe and welcoming. Providing features promoting wayfinding, such as adequate lighting and clear signage, can help enhance accessibility. For outdoor environments, it is advisable to avoid constructing areas that may be perceived as inaccessible or unsafe, particularly by minimising the use of underpasses or areas with limited visibility. Having residential areas with views of the platforms can reduce isolation and foster greater interaction between residents and travellers. Achieving these goals requires an interdisciplinary design approach involving urban planners, architects, transport authorities, and local communities.

Safety interventions must address diverse disability needs. Passengers with reduced mobility need boarding assistance and step-free access, while those with asthma or allergies require better ventilation. Emergency help buttons, audiovisual signals, and CCTV for distress detection are essential. Barrier-free infrastructure, including ramps, elevators, and tactile guidance, improves accessibility. Since constant staff presence at smaller stations is unfeasible, technology such as apps, help buttons, CCTV, and digital platforms can provide crucial support and data for targeted interventions. To make travel safe for those suffering from allergies, regular evaluations of train stations and transit routes conducted with affected users could help reduce these threats and raise awareness. This could include an assessment of foods and drinks served on board.

Transit operators worldwide implement tailored actions to address specific challenges, such as improving accessibility for people with disabilities. Examples include installing elevators and ramps, adapting buses, and offering assistance to older passengers. Initiatives also reinforce courteous behaviours to enhance the travel experience for individuals with specific needs. Future studies should explore how designated seats and campaigns encourage acts of kindness in transit environments, examining who performs these acts, and how they affect safety. Alternatively, future research should investigate how acts of incivility, such as not giving up a seat or being rude to another fellow passenger, are perceived and how incivility affects the overall safety travel of those witnessing these acts.

17.5 Suggestions for Policy Changes

The introduction of a centralised reporting system for crimes and incidents at railway stations is necessary for making events visible and enabling thorough follow-ups. The lack of a central system in Sweden for collecting data on incidents in the railway system means that safety interventions are currently not evidence-based, which undermines all good shared visions and intentions to improve transit safety in train travel. While some operators have station-specific databases, these are rarely shared, limiting the ability to generalise solutions across Sweden since the system also lacks coordination across operators and other stakeholders.

We would envision a centralised reporting system that is part of a global Decision Supporting System (DSS), as discussed in Chap. 3. This system should cover not only criminal incidents but also non-criminal events that contribute to insecurity, addressing actual crime and other incidents as separate issues with their dynamics. It is worth testing digital solutions for reporting to increase accessibility and efficiency. By introducing opportunities to report issues via apps, help buttons, a shared phone number, or other digital platforms, the lack of staff and limited assistance options can be partially mitigated, especially in small stations, and more remotely located stations. By having a centralised system, it could be possible to monitor developments and evaluate whether interventions have solved the safety problems in the system. Evidence-based planning is central to ensuring that the measures achieve the desired results following a systems thinking approach. A reassessment of actions should be considered if they prove ineffective. As a tool for evaluation and follow-up, we also propose the introduction of a national transit safety survey. Respondents' answers would also be linked to the stations they travel from, potentially providing valuable insights into where actions should be implemented within the public transport system. This survey would be conducted regularly or whenever needed.

Creating educational campaigns and policy frameworks is crucial. Public transport authorities should launch awareness campaigns to reduce transit crime, encouraging collaboration between institutions, organisations, and the public. Additionally, training for architects, engineers, and urban planners should include the relationship between travellers' victimisation and fear of crime in transit environments, fostering sensitivity to gender-based safety concerns in urban design. For people with disabilities and, in particular, for travellers with allergies, educational campaigns and policy frameworks are essential. Public transport authorities should launch awareness initiatives, encourage collaboration, and implement strategies to minimise allergens.

It is also important to promote collaboration between local and regional actors in order to tailor measures to specific issues and contexts (but note that collaboration between national and international levels is also important, see Chapter 8). Better coordination between these levels is essential for creating cohesive and customised strategies for safety and security at railway stations. For this to happen, legislators and the criminal justice system must take transit crimes in public spaces seriously. A comprehensive, systems thinking, multi-pronged response is needed, involving transport operators, the police, legislators, educators, municipal authorities, NGOs, and civil society to ensure people feel safe in public spaces.

We have learnt that transit environment qualities, such as train frequency, accessibility, cleanliness, ticket prices, real-time communication and safety, make people choose trains over other modes of transportation. However, attracting users is about improving the service and addressing broader societal challenges, including financial constraints, technological advancements, and competition with other transport modes. At the same time, public transportation systems must be adaptable. As we discussed in Chap. 16, future stations may include further investments to maximise the role of stations as multifunctional spaces, allowing travellers to spend time online or for leisure; this is, for instance, a way of making trains more attractive than the competition which is less sustainable, such as cars and aeroplanes.

17.6 Conclusions

Our findings underscore the need for a comprehensive approach to improving safety for individuals in public spaces and transit systems. This requires targeted policies and broader societal efforts to foster a safer and more inclusive environment for all, regardless of, for instance, gender identity or sexual orientation. Prioritising women and LGBTQI+ /non-binary individuals, especially nighttime travellers and those with a disability, is essential. While solutions must be context-specific, having the option to contact someone in case of an incident remains a valued safety measure. The book also shows that data, methods, and research must be tailored to enable well-informed action.

Ensuring equitable mobility requires acknowledging and addressing the wider societal inequalities that shape our daily experiences—poor transit safety can be one of them. This demands targeted interventions and structural changes in legislation and governance to integrate safety, accessibility, and inclusion into urban and transport policies. Some recommendations demand long-term investments and systemic changes in transport infrastructure, governance, and public attitudes, while others require more immediate localised interventions in transit environments. Additionally, emerging technologies and innovations in transportation play and will play a more critical role in shaping the future of urban mobility, influencing how cities, transit hubs, and mobility networks evolve. Preventing crime in dynamic spaces such as stations is vital for ensuring public safety. By anticipating and addressing current safety challenges, we can ensure that the future of mobility is safer and more sustainable.

Acknowledgements We would sincerely like to thank all travellers and non-travellers who responded to our May to September 2022 survey. We also appreciate the experts from transit agencies, municipalities, and other organisations interviewed in Sweden, the UK, Denmark, and the Netherlands. Special thanks to the Swedish Transport Administration (Trafikverket) for supporting and funding this research (the Effect of the Station Environment on Crime and Passenger Safety, 2021–2023, TRV 2020/22903).

References

Ceccato V, Loukaitou-Sideris A (2020) Transit crime and sexual violence in cities: international evidence and prevention. Routledge

Ceccato V, Newton A (2025) Systems thinking for sustainable crime prevention: planning for risky places. Taylor & Francis

Ceccato V, Gaudelet N, Graf G (2022) Crime and safety in transit environments: a systematic review of the English and the French literature, 1970–2020. Public Transport 14(1):105–153

Cozens P (2007) Planning, crime and urban sustainability. In: WIT Transactions on Ecology and the Environment, p 102

Forsberg G (2019) Samhällsplaneringens teori och praktik. Liber. http://urn.kb.se/resolve?urn=urn:nbn:se:su:diva-177379

Gamman L, Thorpe A (2009) Less is more: what design against crime can contribute to sustainability. Built Environ 35(3):403–418

Loukaitou-Sideris A (2016) A gendered view of mobility and transport: next steps and future directions. Town Plann Rev 87(5):547–565. https://doi.org/10.3828/tpr.2016.38

Meadows DH (2008) Thinking in systems: a primer: chelsea green publishing

Paulsen DJ (2013) Crime and planning. Planning

Roy S, Bailey A, van Noorloos F (2024) Understanding the barriers affecting women's mobility in the first- and last-mile stretches in low-and middle-income countries: a systematic review. J Transp Geogr 121:104036

Sundling C, Ceccato V (2022) The impact of rail-based stations on passengers' safety perceptions. A systematic review of international evidence. Transport Res f: Traffic Psychol Behav 86:99–120

UN-Habitat (2025) Healthier cities and communities through public spaces: A guidance paper. https://unhabitat.org/sites/default/files/2025/01/final_public_space_and_urban_health.pdf?trk=public_post_comment-text

UN, United Nations (2015) Transforming our world: The 2030 Agenda for Sustainable Development. https://sds.un.org/2030agenda

UN, United Nations (2017) New Urban Agenda. Habitat III. https://haitat3.orf/the-new-urban-agenda/

UNODC, United Nations Office on Drugs and Crime (2020) UNODCITY Urban safety governance initiative. https://www.unodc.org/documents/Urban-security/Safety_Governance_Assessment_Guidance_final.pdf

Appendix—Survey (in English)

To Promote Safe Stations

A research survey from KTH (Royal Institute of Technology) and Södertörn University in collaboration with the Swedish Transport Administration.

1a. Which railway station are you departing from?	**1b. Which railway station are you travelling to?**
2a. What is your postal code?	**2b. What is your city?**
3. How often do you travel by train? a) 4-7 days a week b) 1-3 days a week c) 1-3 days per month d) Less than 1 day a month	**4. At what times do you usually travel by train?** Multiple options can be selected. a) In the morning (ca 06-09) b) During the day (ca 09-17) c) In the evening (ca 17-22) d) At night (ca 22-06)

5 In the past 5 years, have you been victim to or witnessed any of the following on the train, at the station, or on the way to/from the station?

	Yes, I have been victim to	Yes, I have witnessed	No, never
Theft			
Robbery			
Violence			
Threat or hate crime			
Sexual harassment			
Stalking			
Aggressive panhandling			

V. Ceccato et al., *Transit Safety for a Sustainable City*, Sustainable Development Goals Series,
https://doi.org/10.1007/978-3-031-97427-4

	Way to/from the station	Platform	Restroom	Waiting room	Bus terminal	Parking lot	Tunnel/ bridge	On the train
Theft								
Robbery								
Violence								
Threat or hate crime								
Sexual harassment								
Stalking								
Aggressive panhandling								

5d If you indicated that you did nothing, explain why? Multiple options can be selected. Skip the question if you have not experienced or witnessed anything.

- I was afraid
- I was embarrassed
- I thought it wouldn't lead to anything
- I didn't think the incident was important
- I wanted to avoid creating more problems
- None of the above

6a. Do you feel safe during the day when you...?

	Always	Sometimes	Never	Not applicable
Are on the train				
Are at the station you depart from				
Are on the way to/from the station				

6b. Do you feel safe when it is dark (evening)?

	Always	Sometimes	Never	Not applicable
Are on the train				
Are at the station you depart from				
Are on the way to/from the station				

7. Tänk på den station du reste ifrån, brukar du känna dig trygg på följande platser?

	Yes	No	Not applicable
On the platform			
In the restroom			
In the waiting room			
At the bus terminal			
In the parking lot			
In the tunnel/footbridge			

5c What did you do about the crime(s) you have experienced or witnessed? Multiple options can be selected. Skip the question if you have not experienced or witnessed anything.

	Reported to police	Reported to staff/other	Tried to help	Did nothing
Theft				
Robbery				
Violence				
Threat or hate crime				
Sexual harassment				
Stalking				
Aggressive panhandling				

8. Indicate which of the following factors affect your sense of safety at the station you usually depart from. Multiple options can be selected.

Heavy traffic/noise	
Poor lighting	
No or few other passengers	
Poorly monitored (e.g., guards, surveillance cameras)	
Underground passages or tunnels	
Intoxicated/drug-affected people present	
Feeling of confinement/lack of oversight in the environment	
Isolated/desolate	
Difficult to buy tickets	
No or too few staff	
No weather protection	
Crowded	
Drug use or dealing	
Graffiti/vandalism/littering	
Poor information (departures, signage, etc.)	
No emergency phone	
None of the above	

9. When you travel by train, do you often feel afraid of being subjected to the following? Multiple options can be selected.

	Yes, on the way to/from the station	Yes, at the station	Yes, on the train	No
Theft				
Robbery				
Violence				
Threat or hate crime				
Sexual harassment				
Stalking (being followed)				
Aggressive begging				
Disturbing behaviour (drinking alcohol, obscene language, etc.)				
Disturbing smell				
COVID-19 infection				
Train accident/risk of being hit				
Terrorist attack				

10. Indicate which of the following statements about safety/unsafety apply to you when travelling by train during the day or evening. Multiple options can be selected.

	Daytime	Evening
I avoid carrying a handbag		
I place myself where I can be seen		
I try to look confident		
I avoid certain stations		
I place myself near another person		
I avoid wearing jewellry		
I am extra vigilant		
I avoid travelling at certain times		
I dress in a certain way		
I avoid certain trains or routes		
I take a detour to/from the station		
I avoid certain people/groups		
I carry a weapon (e.g., pepper spray)		
I prefer to travel with someone else		
I stay in touch with someone on the phone		

11. Indicate what could make your train journey safer. Multiple options can be selected.

- More staff at the station
- Better maintenance (cleaning, graffiti removal, broken glass cleanup, bush clearing)
- More police/guards patrolling the station
- Fewer transfers during the journey
- More frequent trains
- Better lighting
- Cameras (CCTV) at the station
- Cameras (CCTV) on the train
- A single phone number to call for problems
- A security app on the phone to report problems
- Digital timetable with real-time information at all stations
- Better information on where and how to buy tickets
- Better information on where trains and buses depart/arrive
- A "help button" to get assistance at the station
- A "help button" to get assistance on the train
- Available train conductor onboard the train
- None of the above

12. Are there any other aspects, besides those mentioned in the questions above, that could make you feel safer when travelling by train? If so, what are they?

13 Apart from what you may have mentioned earlier, have you been subjected to any type of crime in the past 5 years? Multiple options can be selected.

- Yes, personal crime (e.g., robbery, assault, threats, sexual offenses)
- Yes, property crime (e.g., burglary, car theft, bicycle theft)
- No
- Don't know/Prefer not to say

14 . Indicate which of the following statements apply to you. Multiple options can be selected.

I only use public transport, bicycle, or walk	
I feel safe travelling with people I don't know	
I feel safe in my residential area	
I trust my neighbours	
I trust the police	
I can count on the police's help if something happens to me	
None of the above	

15. How do you identify?
Multiple options can be selected.
a) Woman
b) Man
c) Non-binary
d) LGBTQI+
e) Other

16. How old are you?

____

17. Are you born in another country than Sweden?
a) Yes, which?________________________
b) No

18. How do you live?
a) Alone
b) Alone with children
c) With spouse/partner and children
d) With spouse/partner
e) With parents/siblings/other adults

19. What is your highest level of completed education?

a) Primary school
b) Secondary school
c) College/university

20. Vad är din sammanräknade årsinkomst?

a) 1-249 KSEK
b) 250-499 KSEK
c) 500-749 KSEK
d) 750+ KSEK
e) Don’t know/Prefer not to say

21. Do you have any disability?
- No
- Yes, I have a minor disability
- Yes, I have a major disability
- Don't know/Prefer not to say

22. Do you have any of the following? Multiple options can be selected.
- Problems with attention, memory, or concentration
- Reading, writing, or speech difficulties
- Anxiety, depression, or other emotional issues
- Bowel or urinary problems
- Reduced mobility (including balance or coordination problems)
- Reduced vision or vision impairment
- Asthma/allergies or other hypersensitivities
- Motion sickness
- Reduced hearing or hearing impairment
- None of the above

Appendix—Survey (in Swedish)

Trygg Stationsmiljö

En forskningsenkät från KTH (Kungliga Tekniska högskolan) och Södertörns högskola i samarbete med Trafikverket.

1a.Vilken järnvägsstation reste du ifrån?	**1b. Vilken järnvägsstation reser du till?**
2a.Vad är ditt postnummer?	**2b. Vilken är din postort?**
4. Hur ofta åker du tåg? a) 4-7 dagar i veckan b) 1-3 dagar i veckan c) 1-3 dagar i månaden d) Mer sällan än 1 dag i månaden	**4. Vilka tider reser du vanligtvis med tåg?** Markera ett eller flera alternativ. a) På morgonen (ca 06-09) b) Dagtid (ca 09-17) c) Kvällstid (ca 17-22) d) Nattetid (ca 22-06)

5 Har du under de senaste 5 åren *varit med om eller sett* något av följande på tåget, på stationen eller på väg till/från stationen Markera ett eller flera alternativ.

	Ja, jag har själv varit med om	Ja, har sett någon vara med om	Nej, aldrig
a) Stöld			
b) Rån			
c) Våld			
d) Hot eller hatbrott			
e) Sexuella trakasserier			
f) Stalking (bli förföljd)			
g) Aggressivt tiggeri			

V. Ceccato et al., *Transit Safety for a Sustainable City*, Sustainable Development Goals Series,
https://doi.org/10.1007/978-3-031-97427-4

5b Om ja, vänligen ange *var* detta skedde? Hoppa över frågan om du inte varit med om eller sett något.

	På väg till/från staionen	På plattformen	På toaletten	I väntsalen	Vid bussterminalen	På parkeringsplatsen	I gångtunnel /på gångbro	På tåget
a) Stöld								
b) Rån								
c) Våld								
d) Hot eller hatbrott								
e) Sexuella trakasserier								
f) Stalking (bli förföljd)								
g) Aggressivt tiggeri								

5c Om ja, vad gjorde du? Markera ett eller flera alternativ. Hoppa över frågan om du inte varit med om eller sett något.

	Jag anmälde till polisen	Jag anmälde till personalen/någon annan	Jag försökte hjälpa	Ingenting
a) Stöld				
b) Rån				
c) Våld				
d) Hot eller hatbrott				
e) Sexuella trakasserier				
f) Stalking (bli förföljd)				
g) Aggressivt tiggeri				

5d Om du markerat att du *inte* gjort något, förklara varför? Markera ett eller flera alternativ. Hoppa över frågan om du inte varit med om eller sett något i fråga 5.

a) Jag var rädd
b) Jag var generad
c) Jag trodde att det inte skulle leda till något
d) Jag trodde inte att händelsen var viktig
e) Jag ville undvika att skapa mer problem
f) Inget av ovanstående

6a Känner du dig trygg *på dagtid* när du…?

	Alltid	Ibland	Aldrig
a) Är på tåget			
b) Är på stationen du åker från			
c) Är på väg till/från stationen			

6b Känner du dig trygg *när det är mörkt* när du…? (kvällstid)

	Alltid	Ibland	Aldrig
d) Är på tåget			
e) Är på stationen du åker från			
f) Är på väg till/från stationen			

7 Tänk på den station du reste ifrån, brukar du känna dig trygg på följande platser?

	Ja	Nej	Inte aktuellt
På plattformen			
På toaletten			
I väntsalen			
Vid bussterminalen			
På parkeringsplatsen			
I gångtunnel/på gångbro			

8 På den station du reste ifrån, markera vilka av följande faktorer som påverkar din trygghet?
Markera ett eller flera alternativ

a) Inget väderskydd	
b) Dålig belysning	
c) Isolerat/ödsligt	
d) Ingen eller för lite personal	
e) Inga eller få andra passagerare	
f) Ingen nödtelefon	
g) Trängsel	
h) Mycket trafik/buller	
i) Känsla av instängdhet/brist på överblick i miljön	
j) Klotter/vandalisering/nedskräpning	
k) Dålig information (avgångar, skyltning, etc.)	
l) Svårt att köpa biljetter	
m) Underjordiska passager eller tunnlar	
n) Dåligt bevakat (t.ex väktare, övervakningskameror)	
o) Berusade/påverkade personer närvarande	
p) Narkotikaanvändning eller handel	
q) Inget av ovanstående	

9 Tänk på resan du gör just nu, känner du dig rädd för att utsättas för följande? Markera ett eller flera alternativ.

	Ja, på väg till/från stationen	Ja, på stationen	Ja, på tåget	Nej
a) Stöld				
b) Rån				
c) Våld				
d) Hot eller hatbrott				
e) Sexuella trakasserier				
f) Stalking (bli förföljd)				
g) Aggressivt tiggeri				
h) Störande beteende (dricka alkohol, obscent språk, etc.)				
i) Störande lukt				
j) COVID-19-smitta				
k) Tågolycka/risk att bli påkörd				
l) Terroristattack				

10 Markera vilka av följande påståenden om trygghet/otrygghet som stämmer in på dig när du reser med tåg dagtid respektive kvällstid? Markera ett eller flera alternativ.

	Dagtid	Kvällstid
a) Jag reser helst med någon annan		
b) Jag undviker vissa tåg eller sträckor		
c) Jag undviker vissa stationer		
d) Jag undviker vissa personer/grupper		
e) Jag placerar mig där jag syns		
f) Jag placerar mig nära en annan person		
g) Jag har kontakt med någon på telefon		
h) Jag är extra vaksam		
i) Jag klär mig på ett visst sätt		
j) Jag undviker att bära smycken		
k) Jag undviker att bära handväska		
l) Jag har något vapen på mig (t.ex pepparspray)		
m) Jag försöker se självsäker ut		
n) Jag tar en omväg till/från stationen		
o) Inget av ovanstående		

11 Markera vad av följande som skulle kunna göra din resa med tåg tryggare? Markera ett eller flera alternativ

a) Mer personal på stationen
b) Bättre underhåll (städning, klottersanering, krossat glas, buskröjning)
c) Bättre belysning
d) Digital tidtabell med realtidsinformation vid alla stationer
e) Bättre information om var tåg och bussar avgår/ankommer
f) Bättre information om var och hur man köper biljetter
g) Högre turtäthet
h) Färre byten under resan
i) Ett enda telefonnumer att ringa till för problem
j) En "hjälpknapp" för att kunna få hjälp på tåget
k) En "hjälpknapp" för att kunna få hjälp på stationen
l) En säkerhetsapp på mobilen för att rapportera problem
m) Fler poliser/vakter som patrullerar på stationen
n) Fler poliser/vakter som patrullerar på tåget
o) Tillgänglig tågvärd ombord på tåget
p) Kameror (CCTV) på tåget
q) Kameror (CCTV) på stationen
r) Inget av ovanstående

12 Finns det andra aspekter än de som tagits upp i frågorna ovan som skulle kunna få dig att känna dig tryggare när du reser med tåg? Vad är det i så fall?

Slutligen kommer några frågor om dig själv

13 Om du bortser från det du eventuellt tagit upp i fråga 5, har du blivit utsatt för någon typ brott under de senaste 5 åren?

a) Ja, personbrott (t.ex. rån, misshandel, hot, sexualbrott)
b) Ja, egendomsbrott (t.ex. inbrott, bildstöld, cykelstöld)
c) Nej
d) Vet ej/vill ej svara

14 Markera vilka av följande påståenden som överensstämmer med Dig? Markera minst ett alternativ

a) Jag använder enbart kolletivtrafik, cykel eller går	
b) Jag känner mig trygg att resa med människor jag inte känner	
c) Jag känner mig trygg i mitt bostadsområde	
d) Jag känner förtoende för mina grannar	
e) Jag känner förtroende för polisen	
f) Jag kan räkna med polisens hjälp om något händer mig	
g) Inget av ovanstående	

15 Hur identifierar du dig? Flera svar möjliga

f) Kvinna
g) Man
h) Ickebinär
i) HBQTI+
j) Annat

16 Hur gammal är du?

____

17 Är du född i ett annat land än Sverige?

c) Ja, vilket________________________
a) Nej

18 Hur bor du?

a) Ensam
b) Ensam med barn
a) Med make/maka/sambo och barn
b) Med make/maka/sambo
c) Med föräldrar/syskon/andra vuxna

19. Vad är din högsta avslutade utbildning?

a) Grundskola
b) Gymnasieskola
c) Högskola/universitet

20. Vad är din sammanräknade årsinkomst?

a) 1-249 Tkr
b) 250-499 Tkr
c) 500-749 Tkr
d) 750+ Tkr
e) Vet ej/Vill ej svara

21 Har du någon funktionsnedsättning?

a) Nej
b) Ja, jag har en mindre funktionsnedsättning
c) Ja, jag har en större funktionsnedsättning
d) Vet ej/vill ej svara

22 Har du något av följande? Markera ett eller flera alternativ.

a) Åksjuka
b) Nedsatt rörelseförmåga (även problem med t ex balans eller koordination)
c) Nedsatt syn eller synstörning
d) Nedsatt hörsel eller hörselstörning
e) Läs-, skriv- eller talsvårigheter
f) Besvär med uppmärksamhet, minne eller koncentration
g) Astma/allergi eller annan överkänslighet
h) Besvär med oro, ångest, eller nedstämdhet
i) Tarm- eller urinbesvär
j) Nej, jag har inget av ovanstående

Glossary

This book uses specific terminology that may need to be clarified. The following terms are explained in the context of this book and their application in everyday practice. They may not fit other contexts, but can be particularly useful for anyone seeking to engage with the key topics discussed in this book.

Alcohol-selling premises—In Sweden, there are two main categories of premises for the sale of alcoholic beverages: government-owned establishments (Systembolaget) and licensed premises (e.g., bars, restaurants, and shops with special licences). The main differences between these two types of premises stem from their ownership, the kind of alcohol they are allowed to sell, opening hours, and age restrictions for alcohol purchases.

City or municipality—For the purpose of this book, these two terms are used interchangeably to reflect the urban areas of municipalities and sometimes both urban and rural areas. In Swedish, municipality ("kommun") is the official administrative unit, while city ("stad") is often the urban area. A municipality ("kommun") in Sweden is an administrative unit responsible for local governance and services, with the authority to make decisions on local matters. There are 290 municipalities in Sweden, each with its own elected council.

City users—Individuals who use public spaces and services, including public transportation. This group is diverse in demographics, socioeconomic status, daily needs, and travel patterns. While some are regular public transit users, others may not use it frequently, but they all interact with public spaces in various ways.

Crime—A crime is defined as an anti-social act that violates a law and for which a punishment can be imposed by the state or in the name of the state (UNHSP 2007). A crime is an act that is punishable under the Swedish Penal Code. According to the Swedish Penal Code (1994:458), a crime is "an act described in this Code or in another law or regulation, for which the prescribed punishment is specified below. Unless specifically prescribed, an act shall be considered a crime only if it is committed intentionally."

V. Ceccato et al., *Transit Safety for a Sustainable City*, Sustainable Development Goals Series,
https://doi.org/10.1007/978-3-031-97427-4

Fear of crime—Fear of crime is an emotional response related to potential victimisation, general worry about crime, or symbolic representations of crime (Ferraro 1995). Fear of crime is, therefore, influenced by individual experiences, social contexts, and environmental factors, and it can occur even when the likelihood of victimisation is low. Gabriel and Greve (2003) suggest two complementary dimensions of fear of crime. The first dimension considers fear as a trait. This reflects individual differences (e.g., age, gender, ethnicity) in the tendency to experience fear. This dimension is also called disposition-based fear. The other dimension considers fear as a state of fear or situational fear, which refers to a temporary emotional state. For example, a person may be afraid to walk through a dark street but feel safe there during the day. Gabriel and Greve (2003) also note that these two types of fear influence each other.

Gendered safety—This refers to how perceptions of safety and experiences of victimisation can differ based on gender identity and status. It may also involve how gender norms and expectations play a significant role in shaping the way individuals experience and navigate public spaces.

Non-travellers – Here, non-users of public transportation are individuals who do not use buses, trains, subways, or other forms of public transit for their daily commuting or travelling. These people might rely on private vehicles (cars, bicycles, etc.), walk, or use other forms of transportation such as taxis or ridesharing services (Uber, Lyft), or they may even avoid travelling altogether. In this book, non-travellers declare that they do not use rail-bound public transport, as analysed in Chap. 14.

Public transportation—This is used as a synonym for public transport, public transit, and mass transit. It includes buses, trains, metros, trams, ferries, and other forms of mass transit, and government agencies or private entities usually manage it under public regulation. In this book, this term is often linked to the use of buses but more frequently linked to the use of trains and railway systems.

Place/station manager—This person or a group of people is responsible for the overall maintenance, preservation, and improvement of one or more facilities, such as a station or school. The owner may or may not contribute to the management of the place or station (may need to be adjusted).

Public spaces—Areas that are open and accessible to all individuals, typically owned and managed by government or public authorities, but they may also include privately owned spaces, such as stations or shopping malls. These spaces are intended for use by the general public, and their primary purpose is to facilitate social interaction, recreation, and access to services or transport. Public spaces are typically free from restrictions, although they may be subject to specific regulations or rules of conduct. Németh (2012) suggests that ideally, public space means space legally open and accessible to all that does not require permission from anyone else, but most public spaces are free conditionally, because any action allowed in these spaces falls within the law of the locality in which the space is located. This definition lacks the symbolic significance associated with public spaces as peaceful arenas of self-expression and enjoyment (fairs, sports), as well as spaces of contestation (quarrels, fights, riots, demonstrations) and appropriation by "illegitimate" users (homeless people, illegal hawkers, or criminals). For an in-depth discussion of the term, see Ceccato (2016).

Public transport operators—Private or public companies that run and maintain public transport services. Västtrafik is an example of a public transport authority in the Västra Götaland Region of Sweden.

Public transport stakeholders/actors—Public transport stakeholders include government bodies, transport operators, employees, businesses, and environmental groups. All of these influence, or are affected by, transportation policies, operations, and infrastructure. Their collaboration ensures efficient, accessible, and sustainable public transport services.

Private transport operators—These include ride-sharing services or private bus companies that may supplement public transport; in Sweden, an example of a private transport operator is MTR or Nobina (bus services).

Safety—In this book, safety refers to the absence of crime or the fear of crime and other triggers that make an individual feel unsafe. It is used interchangeably with security. Research on transitions and sustainability (e.g., Mapar et al. 2017) adopts "safety" as a synonym for "security" due to the interdisciplinary nature of the term "safety", recognising that these dimensions often overlap.

Safety from the perspective of city users—This is a multifaceted concept that blends physical, social, psychological, and cultural dimensions. It involves the objective presence of crime or hazards and how individuals feel and perceive their environment (even when the risk is not present). In this book, safety from the perspective of city users means considering the diverse needs and perceptions of different groups, which can sometimes be conflicting.

Safety perception—In transit, 'safety perception' is used here as a general concept of 'declared perceived safety' for individuals who answered questions about victimisation in public places. Thus, perceived safety can be at a high level when an individual feels safe or at a low level when an individual feels fearful or unsafe. Although the focus is on crime victimisation, safety perception can reflect individual factors, other local threats, and global concerns (real or perceived). Examples are fear of contamination or fear of terror attacks.

The whole-journey approach—In the context of transit safety, the whole-journey approach refers to planning strategies that focus on the individual's entire trip experience, from the moment the person begins a trip until they reach their destination. Rather than isolating individual safety measures at specific points in the journey (e.g., vehicle safety, station security), the whole-journey approach includes the path to and from public transportation. See, for instance, Natajaran et al. (2017).

Travellers/passangers/riders—This includes all individuals who use public transportation, more specifically, rail bound transportation, in this book.

References

Ceccato V (2016) Public space and the situational conditions of crime and fear. Int Crim Justice Rev 26(2):69–79
Ferraro KF (1995) Fear of crime: Interpreting victimization risk. SUNY press.
Gabriel U, Greve W (2003) The psychology of fear of crime. Conceptual and methodological perspectives. Br J Criminol 43(3):600–614.
Mapar M, Jafari MJ, Mansouri N, Arjmandi R, Azizinejad R, Ramos TB (2017) Sustainability indicators for municipalities of megacities: Integrating health, safety and environmental performance. Ecol Indicat 83:271–291. https://doi.org/10.1016/j.ecolind.2017.08.012
Natarajan M, Schmuhl M, Sudula S, Mandala M (2017) Sexual victimization of college students in public transport environments: a whole journey approach. Crime Prevent Community Saf 19:168–182
Németh J (2012) Controlling the commons: How public is public space?. Urban Aff Rev 48(6):811–835.
UNHSP, United Nations Human Settlements Programme (2007) Enhancing urban safety and security: Global report on human settlements. Earthscan.

V. Ceccato et al., *Transit Safety for a Sustainable City*, Sustainable Development Goals Series,
https://doi.org/10.1007/978-3-031-97427-4

Index

V. Ceccato et al., *Transit Safety for a Sustainable City*, Sustainable Development Goals Series,
https://doi.org/10.1007/978-3-031-97427-4

MIX
Papier aus verantwortungsvollen Quellen
Paper from responsible sources
FSC® C105338

If you have any concerns about our products,
you can contact us on
ProductSafety@springernature.com

In case Publisher is established outside the EU,
the EU authorized representative is:
Springer Nature Customer Service Center GmbH
Europaplatz 3, 69115 Heidelberg, Germany

Printed by Libri Plureos GmbH
in Hamburg, Germany